W9-DEW-806

# The Psychology of Science and the Origins of the Scientific Mind

# The Psychology of Science and the Origins of the Scientific Mind

Gregory J. Feist

Yale University Press

New Haven and London

Copyright © 2006 by Gregory J. Feist. All rights reserved. This book may not be reproduced, in whole or in part, including illustrations, in any form (beyond that copying permitted by Sections 107 and 108 of the U.S. Copyright Law and except by reviewers for the public press), without written permission from the publishers.

Set in Adobe Garamond type by The Composing Room of Michigan, Inc.
Printed in the United States of America.

Library of Congress Cataloging-in-Publication Data

Feist, Gregory J.
The psychology of science and the origins of the scientific mind / Gregory J. Feist.
p. cm.
Includes bibliographical references (p. ) and index.
ISBN-13: 978-0-300-11074-6 (alk. paper)
ISBN-10: 0-300-11074-X (alk. paper)
1. Classification of sciences. 2. Science—Psychological aspects. 3. Science and psychology. I. Title.
Q177.F45 2006
501′.2—dc22

2005018887

A catalogue record for this book is available from the British Library.

The paper in this book meets the guidelines for permanence and durability of the Committee on Production Guidelines for Book Longevity of the Council on Library Resources.

10 9 8 7 6 5 4 3 2 1

For my parents, Mary Jo and Jess Feist,
and my wife, Erika Rosenberg

Science is like sex. Sometimes something useful comes out,
but that is not why we are doing it.
—*Richard Feynman*

# Contents

# Preface

Scientific thinking is a hallmark intellectual achievement of the human species. Science involves myriad cognitive and intellectual processes, including abstract and symbolic thought; reasoning and logic; pattern recognition; planning; problem solving; creativity; hypothesis testing; mathematical, analytical, and spatial reasoning; intuitive hunches; chance associations; and the art of coherent and cogent verbal expression and persuasion, to mention but a few of its qualities. Science is first and foremost a cognitive activity of the highest order.

Scientists also think and behave in social contexts; have particular talents and aptitudes; grow up in specific households with particular family structures and influences; have unique personalities that make scientific thought and behavior more rather than less likely; and are motivated by curiosity, intrinsic pleasure of discovery, and the triumph of figuring out how things work. That is, scientific behavior, interest, talent, and achievement stem from basic topics of focus in the field of psychology. Psychological principles are at work with all scientific thought and behavior. Simply put, there is a psychology behind science.

The chief objective of this book is to justify the need for a fully developed discipline of the psychology of science and to lay the foundations for such a field. To this end, I have two related yet distinct ambitions. One is to organize and codify the nascent discipline of the psychology of science and thereby demonstrate the field's potential for joining the ranks of the major science studies disciplines (history, philosophy, and sociology). The second is to examine the evolutionary and historical origins of the scientific mind. If we wish to understand something as complex as scientific thinking and behavior, a basic understanding of how the human mind evolved is in order. The book is divided according to these two goals, with part 1 focusing on the development of scientific interest and talent within certain groups of individuals, and part 2 on the development of science within our species.

The guiding assumption behind the psychology of science is that a complete understanding of scientific thought and behavior requires a psychological perspective. As one prominent psychologist of science, Dean Keith Simonton, wrote in *Scientific Genius:* "Without the addition of a psychological dimension, I believe, it is impossible to appreciate fully the essence of the scientific imagination. And without this appreciation, the origins of science, the emergence of new ideas about natural phenomena, must escape our grasp. Psychology is mandatory if we wish to comprehend the scientific genius as the generator of science." This is what the psychology of science is all about: to understand scientific thought and behavior we must apply the best theoretical and empirical tools available to psychologists. And what psychology has to offer the studies of science is indeed unique. For instance, only psychologists of science bring the experimental method (that is, random assignment of participants to conditions and manipulation of an independent variable) to the study of scientific thought and behavior. Also, in contrast to the history and philosophy of science and in common with the sociology of science, psychology tests hypotheses by means of statistical analysis of data.

In addition to the experimental technique and hypothesis testing, psychology can borrow from historians and examine case studies and apply principles of behavior gleaned from the laboratory to the analysis of great figures in science. Consider the case history of one of the best-known and most influential scientists of all time, Charles Darwin. In *The Descent of Man* he wrote: "I have no great quickness of apprehension or wit . . . my power to follow a long and purely abstract train of thought is very limited . . . [but] I am superior to the common run of men in noticing things which easily escape attention, and in observing them carefully." Darwin's own self-evaluation of his strengths and weaknesses gives a glimpse into his own self-concept—clearly a psychological concept. Moreover, ability with abstract thought, attention, and focus on details are very much psychological in nature; cognitive psychologists among others have much to say about these aptitudes. What precisely is the association between Darwin's life and personality and his science? In this book I propose that we can fruitfully apply the methods and theories of modern psychology to shed light on these sorts of questions.

To a psychologist of science it is obvious that scientific thought and behavior are the outcomes of a person's cognitive style and aptitudes; affective, motivational, and developmental histories and proclivities; and their unique and stable personality traits and social influences. These topics, after all, are the bread and butter of current psychological inquiry and psychological science. And given the importance and uniqueness of scientific thinking and behavior over the course of history, one would think that a large number of psychologists would have long ago systematically applied their theories and empirical methods to understanding science. Surprisingly, until the late 1970s there was little accumulated knowledge concerning topics in the psychology of science. As Michael Mahoney wrote in a 1979 article in *Social Studies of Science,* "In terms of behavior patterns, affect, and even some intellectual matters, we know more about alcoholics, Christians, and criminals than we do about the psychology of the scientist."

Twenty-five years later, however, this paucity of psychological research on the nature of scientific interest, thinking, creativity, and achievement no longer holds. This book summarizes, organizes, and critiques the vast literature on the psychological processes of science and scientists by offering one of the first comprehensive views of a nascent discipline. One major thesis throughout the book is that numerous studies exist that inform questions of the psychology of science, but until now they have not been contextualized as such.

If the psychology of science has been late in developing, the same does not hold for the three major studies of science, namely, history, philosophy, and sociology. For instance, the history of science began to emerge around the 1840s, the philosophy of science around 1900, and the sociology of science around 1930. These "studies of science" (or "metasciences") devote systematic attention to such questions as what distinguishes scientific from nonscientific knowledge, what is the historical context to great scientific discoveries (for example, the theory of evolution or quantum mechanics), and what are the sociological and political forces behind becoming a have or a have-not in science. By understanding how and when other major studies of science emerged and became viable independent disciplines, psychologists of science will be better positioned to facilitate their own field's development and independence. In chapter 1 I consider these issues.

In chapters 2 through 6 I review the empirical literature in the psychology of science by summarizing and organizing it along the lines of the major subdisciplines of psychology, namely, biology-neuroscience, development, cognition, personality, and social. In addition to reviewing and organizing the literature, I also argue that scientific thought and behavior deserve more attention from psychologists and that the psychology of science deserves more attention from philosophers, historians, and sociologists of science. These chapters show how the psychology of science has grown beyond the amorphous field it was just twenty-five years ago. I also propose some parameters for the psychology of science and trust that in doing so I might inspire researchers at the core and periphery of the field to codify their interests and to stimulate the field's emergence as a major player in science studies.

In the last chapter of part 1 (chapter 7), I explore the applications that an informed and well-developed psychology of science might stimulate, as well as what needs to be accomplished before we have journals, societies, and conferences on the psychology of science. Specifically, it behooves the gatekeepers of science (scientists, teachers, mentors, editors, grant administrators) to be well informed of the empirical research that demonstrates which specific psychological qualities (neuroscientific, cognitive, developmental, personality, and social) are the most reliable and robust predictors of real-world creative achievement in science, mathematics, and technology. In other words, if we are to recognize, recruit, and retain the best young scientific talent available to the science professions, we must understand the psychology behind scientific talent, how to identify it early on, and ultimately how to encourage those with high-level skills and talent to enter the math and science workforce. Accurate, reliable, and valid psychological measures can only aid this process.

In addition to exploring the evolutionary and historic origins of scientific thinking, I address in part 2 of the book the complex interplay between scientific, pseudoscientific, and antiscientific thinking in modern life. More specifically, in the second half of the book I ask the questions "Why do humans—and no other species—do science?" and "How did we go from *Australopithecus* (non-homo hominid species) to early *Homo* (for example, *habilis, erectus,* and *neanderthalensis*) to living in a world of high-energy subatomic particle physics, sequencing the entire human genome, being able to send space craft out of our solar system, and having machines that can outplay any human in the world in chess?" In examining these questions I was taken much further and deeper than I expected into many areas beyond psychology—archeology, neuroscience, genetics, anthropology, history, philosophy, and sociology, to name but the most obvious ones. One lesson I have learned from this foray into the evolutionary origins of scientific thinking is that formal science—science as we know it—is but one specific expression of scientific thinking. Elements of scientific thinking have their origin in our distant preverbal ancestors, with most of these elements taking implicit rather than explicit form. As I argue in chapter 8, a basic grasp of principles of evolution in general and human cognitive evolution in particular allows one to explore and even provide answers to the fascinating and otherwise unanswerable question of how scientific thinking came to be in our species.

Of course, an evolutionary perspective takes us on a journey that is not specific to science and scientific thinking, but rather on a journey that explores the modern human mind in general. Symbolic, abstract thought, language, literature, art, music, and other pinnacles of human cognitive and aesthetic capacities are also unique expressions of the modern human mind. In chapter 9, therefore, I delve into the prehistoric and historical developments that made modern scientific thinking possible—in all of its forms and variations—as well as what distinguishes science from other higher-order cognitive capabilities. Science and scientific thinking consist of developing and testing mental models of how the world works, be they of the physical, biological, or social worlds. The essence of these mental models is coordinating theories (models) with the evidence (data). Specifically, it is a process of observing events, recognizing patterns, testing hypotheses, and making causal connections between the observed events. Early in the development of our genus (*Homo*) and now early in the development of modern individuals (that is, *Homo sapiens sapiens*), these processes were and are mostly implicit—outside conscious awareness. With both phylogenetic and ontogenetic development, however, they gradually become more and more explicit, part of conscious awareness, and ultimately we developed the capacity to be aware of our awareness; that is, to think metacognitively. Science as we now know it is a metacognitive act, one that combines logic and reason with empirical observation. The outcome of such reasoning is the complex melding of innate skepticism with openness to go wherever the evidence takes us. In chapter 9 I explore in more depth both the phylogenetic and historic origins as well as the trademark characteristics of scientific thinking.

There are other important questions related to a psychological understanding of scientific thinking. One is, "How do we distinguish it from pseudoscientific thinking?" Some individuals in the modern world claim to be doing science and even have co-opted the name "science." Upon further examination, these methods and practices are little more than ideology couched in scientific-sounding language. Why might such "pseudoscientific reasoning" hold such strong appeal for a large section of the population? Again, a psychological perspective allows us to address questions such as these, and I do so in chapter 10.

In chapter 10, I also examine the psychological foundations for the anti-science movement in and out of academia. In particular, I explore the attraction for some scholars and lay people to knock science off its "privileged" pedestal and argue that science is little more than stories or fictions of how the world works that are afforded hegemonic control over other forms of knowledge. This control, they argue, comes from the status and power scientists are awarded in modern society. Scientists, these scholars continue, can make no more valid claims for understanding how the world actually works than stories by children, artists, writers, musicians, and philosophers. Science is socially constructed—like all knowledge—and therefore devoid of any inherent meaning and validity. Deconstruction is an act of meaning the reader not the author/scientist places on the scientific text.

Needless to say, many scientists as well as others in the studies of science and even some in the humanities take issue with these claims and counter that scientific knowledge is of a special kind, not inherently but rather because the methods on which its knowledge is based are socially shared, open, reproducible, systematic, and empirical. The scientific method is neither capricious nor a mere social construction. Scholars like Paul Gross and Norman Levitt in *Higher Superstitions,* for instance, defend science, reason, and rationality against claims of meaningless, absurdity, and extreme constructivism. I examine this debate not so much as to offer a solution to it, but rather to again demonstrate how psychological theory and empirical evidence from psychology can better inform such debates. I end the chapter and the book with an analysis of the current state of the psychology of science and make recommendations for what must be done if the discipline is to become the full-fledged discipline it is capable of becoming and, from my vantage point, should and must become.

*The Psychology of Science* is my attempt to uncover some of the mysteries of the scientific mind and how it came to be, both within individuals and within our species. If I have done my job, then you—the reader—will come away convinced that psychological research and theory add a crucial and even necessary perspective to our understanding of the scientific mind, and that other studies of science can no longer turn a deaf ear to what psychologists of science have learned. Psychologists of science now know too much about the nature of scientific thinking, the developmental origins of theory construction, scientific personality, scientific motivation, scientific interests, and scientific creativity and achievement for these insights not to be integrated and synthesized in one place. Science is a fascinating accomplishment of the human mind, and so, too, is the psychology of science.

# Acknowledgments

As with all books, this one has many origins and people whose influence has been crucial; hence, it is almost arbitrary where I begin in acknowledging them. The most obvious points of origin are my dissertation research on the psychology of science, conducted between 1989 and 1991 at University of California at Berkeley, and the review article on the topic that Michael Gorman and I published in 1998. Just prior to my dissertation research, I was fortunate to be part of a seminar given by Daniel Kahneman and Phillip Tetlock on decision making, which was a direct spark for a problem that was to become my dissertation, namely, the cognitive (integrative) complexity of truly creative and revolutionary scientists. Moreover, I was fortunate to have Gerald Mendelsohn serve as the chair of my dissertation, with Phil Tetlock and the historian of science Roger Hahn also on the committee. It was a few years and a few publications after my dissertation research—and meeting and collaborating with Michael Gorman—before I was able to integrate the entire field of the psychology of science in a review paper published in *Review of General Psychology.* The current book is an expanded, elaborated, and updated version of that paper. In particular, part 1 of *Psychology of Science* builds directly on the paper I published with Michael Gorman. Michael's encouragements and insights were critical in helping me form my ideas, especially in the chapter on cognitive psychology of science. Indeed, in him I found someone who believes as I do that the psychology of science is a rich but underappreciated and underdeveloped discipline. Two books appeared as I was working on my dissertation that demonstrated to me how the psychology of science was a viable, if infantile, field. These books were Dean Simonton's *Scientific Genius: A Psychology of Science* published in 1988 and the edited volume by the so-called Memphis Group, *Psychology of Science: Contributions to Metascience* published in 1989. Simonton and also Will Shadish have personally been important in my evolving interest in the psychology of science.

A more distant yet no less important point of origin was my senior year of high school, spent as an exchange student in Bielefeld, Germany. By that time, I had already developed quite an interest in philosophy, but my guest father (Dankwart Vogel) was a physicist with a disposition for the philosophy of science and he had an extensive library. On his shelves I found, among others, books by Popper, Kuhn, and Lakatos, the philosophers of science. As a young intellectual, reading those works and having conversations with Dankwart about them proved both enlightening and inspiring. There began in earnest my passion for the nature of knowledge and of scientific knowledge in particular. Even before that year in Germany, the topic of creativity had captured my fascination: why are some people so creative and able to see things and solve problems in ways that others couldn't? Indeed, the question of creativity took hold of me at age sixteen and has not let go to this day.

I also want to make known my gratitude to Lee Kirkpatrick and an anonymous reviewer for having provided helpful and sharpening feedback to earlier versions of the book. The book is better and clearer because of them. In addition, my editor at Yale University Press, Jean E. Thomson Black, has provided just the right balance between hands-on and hands-off guidance that all great editors seem to posses intuitively. In a project as big and at times as overwhelming as this one has been, her promptness in responding to queries and her ideas for how to better organize and present the material has been most appreciated. Eliza Childs, as copyeditor, has also been most helpful in tightening and clarifying the language throughout the book.

Portions of this book were published previously in modified form. Portions of part 1 were published by Feist and Gorman as "Psychology of Science: Review and Integration of a Nascent Discipline" in *Review of General Psychology* (volume 2). Parts of chapter 8 were published as "The Evolved Fluid Specificity of Human Creative Talent" in Sternberg and colleagues, eds., *Creativity: From Potential to Realization,* and as "Domain-Specific Creativity in the Physical Sciences" in Kaufman and Baer, eds., *Creativity across Domains: Faces of the Muse.* In addition, I want also to acknowledge and thank Susannah Paletz for allowing me to make use of some of the data from her unpublished senior honors thesis (Wesleyan University, Middletown, Connecticut).

I can acknowledge but never repay the debt I owe to my parents, Jess and Mary Jo Feist. Their love and support has been felt at every stage of my journey. My two wonderfully lively sons, Jerry and Evan, have been more inspiring than I could ever imagine, even if Jerry is a bit disappointed that my book is not as many pages (and hence not as good!) as any of J. K. Rowling's Harry Potter books. Lastly, I gratefully acknowledge and cannot begin to repay the love, support, editing skills, and encouragement of my talented wife, Erika Rosenberg.

# Part One **Psychology of Science**

# Chapter 1 Psychology of Science and the Studies of Science

Science and scientific thinking, as prototypes of human thought and understanding, have long fascinated scholars and thinkers in philosophy, history, and more recently, sociology. Indeed, philosophy of science, history of science, and sociology of science are well-developed disciplines. By contrast, psychology of science is an infant that has much to learn from the other, more mature metasciences. My intent in this chapter is to examine the developmental paths taken by the three major players in science studies—history, philosophy, and sociology—as a means for understanding what may be necessary for psychology of science to establish itself as a viable discipline. If psychology of science is to learn from these other more codified studies of science and develop its own identity, then it must knowingly proceed through similar stages.

As a precursor to discussing the stages of development that other studies of science have taken, I first must be clear on what the psychology of science is. Although the heart of this book is an elaborate answer to that question, for now suffice it to say that the psychology of science applies the empirical methods and theoretical perspectives of psychology to scientifically study scientific thought and behavior (hence, it is a "metascience"). At its core, psychology of science is the empirical study of the biological, developmental, cognitive, personality, and social influences of scientific thought and behavior. Scientific thought and behavior are not limited to scientists per se but also encompass thought processes of children, adolescents, and adults who are simulating scientific problem solving and developing mental models of how the world works. Just as science can be either implicit or explicit, so too can be the psychology of science. In fact, I view much of the work discussed in this book as implicit psychology of science: the psychologists doing it would not label it "psychology of science" or think of themselves as "psychologists of science." One of my goals, therefore, is to convince these researchers that they are in fact doing psychology of science.

## PSYCHOLOGY AND THE PSYCHOLOGY OF SCIENCE

Psychology in general is and has been the model for the psychology of science. That is, all the major questions and perspectives for an informed psychology of science derive directly from the parent discipline and its subdivisions. Psychology is a field that currently has five or six major perspectives: biological-neuroscience, developmental, cognitive-perceptual, personality, social, and clinical-mental health. Biological-neuroscience psychology explores the link between brain, mind, and behavior; cognition examines how we perceive, think, remember, speak, and solve problems; developmental psychology explores how humans change and grow from birth to death; personality psychology investigates how dispositions influence one's unique responses to the environment and the consistency of these dispositions over the lifespan; and social psychology explores how individuals are perceived and influenced by the real or imagined presence of others.

In table 1.1, I have listed some examples of major questions addressed by each of psychology's subdisciplines and whether each might also be a topic for psychologists of science. These questions are rather general and meant only to give a taste of the kinds of questions each subdiscipline addresses. For instance, to the degree that biological-neuroscience uncovers the neural mechanisms and anatomical architecture of sensory, perceptual, and cognitive processes involved in abstract, spatial, and quantitative reasoning, it sheds light on the neural and anatomical basis of scientific thought. Because cognitive psychology is concerned with perception, concept formation, learning, memory, problem solving, and creativity, it has the most obvious possible connection with a psychology of science. The only subdiscipline I do not take up in the book is clinical psychology, simply because there is little to no empirical work directly related to scientific thought and behavior. The one fascinating clinical topic that has garnered some empirical attention and could justify including a clinical subdivision in the psychology of science in the future would be the extent to which particular mental disturbances (for example, autism, manic-depression, or obsessive-compulsive disorder) help or hinder interest or creative achievement in science. For example, as I discuss in the chapters on development and evolution, Simon Baron-Cohen and his colleagues have found a connection between high functioning autism (Asperger's Syndrome) and scientific interest and talent.

Not only does psychology suggest general questions and topics, but it also offers the psychology of science guidance in research methodology. For instance, one method that psychologists of science bring to the study of science that no other metascientific field does is the experimental method. The two essential ingredients for the experimental method are random assignment of participants to conditions and manipulation of the main variable in question (holding all else constant). Cognitive and social psychologists in particular make use of the experimental method because cognitive and social factors are relatively easy to manipulate.

Just as psychology is the model for the psychology of science, the latter can also be a model for the former. Over the last fifty years, all major subdisciplines in psychology have become more and more isolated from each other as training becomes increasingly specialized and narrow in focus. As some psychologists have long argued, if the field of psychology is to mature and advance scientifically, its disparate parts (for example, neuroscience, developmental, cognitive, personality, and social) must become whole and integrated again.[1] Science advances when distinct topics become theoretically and empirically integrated under simplifying theoretical frameworks. Psychology of science will encourage collaboration among psychologists from various sub-areas, helping the field achieve coherence rather than continued fragmentation. In this way, psychology of science might act as a template for psychology as a whole by integrating under one discipline all of the major fractions/factions within the field. It would be no small feat and of no small import if the psychology of science could become a model for the parent discipline on how to combine resources and study science from a unified perspective.

Table 1.1 Subdisciplines of Psychology and the Psychology of Science

| Subdiscipline | Sample Questions | Possible Topic for Psychologists of Science? |
| --- | --- | --- |
| Biological-neuroscience | Which brain regions are most active in sensing, remembering, speaking, reading? | Yes |
| | Is the human brain structured (biased) to solve problems in specific domains (such as with people, objects, animals, plants, number) or in a generalized, domain-free manner? | Yes |
| | How much of the variability in intelligence is heritable? | Yes |
| | Are neural connections and brain structures shaped by environmental input? | Yes |
| Developmental | Does development occur in discrete stages or continuously? | No |
| | What are the major domains of development (e.g., sensory, cognitive, emotional, etc.)? | No |
| | How and when do infants and children build implicit models of how the world works? | Yes |
| | Are cognitive models domain specific or general? | Yes |
| | When do specific emotions first emerge and how do they change with age? | No |
| | Do infants come into the world predisposed to solve particular problems and to be interested in particular category of things? | Yes |
| | Do talent and precocity tend to be expressed in specific domains? | Yes |
| Cognitive | How does the brain affect what we see, hear, feel, taste, and smell? | Yes |
| | How is the brain changed by what we see, hear, feel, taste, and smell? | Yes |
| | What is learning? How does it occur? | Yes |
| | Do humans learn things equally easily or are some topics more readily learned than others? | Yes |

| | | |
|---|---|---|
| | Are cognitive heuristics rational and adaptive or irrational and maladaptive? | Yes |
| | Why are some things/events more readily remembered than others? | Yes |
| | What role does awareness of our awareness (metacognition) play in the development of knowledge? | Yes |
| | Is intelligence one general thing or many specific things? | Yes |
| | What cognitive processes and talents are most often involved in coming up with novel and adaptive solutions to problems? | Yes |
| Personality | When do unique personality traits first emerge? | No |
| | What are the basic dimensions of personality (how is personality structured)? | No |
| | Is personality stable or capable of fundamental change? | Yes |
| | To what extent to traits lower behavioral thresholds? | Yes |
| Social | What is the relation between attitudes and behavior? | No |
| | What are the conditions whereby individuals are most influenced by others? | Yes |
| | What role do perceptions and attitudes play in how we treat others? | Yes |
| | Are person or situational forces more important in explaining behavior? | Yes |
| Clinical-mental health | What are the major categories of mental illness (thought, mood, personality)? | No |
| | To what extent do biological, environmental, and personality factors interact to result in disturbances of the mind? | Yes |
| | Are some forms of therapy more effective than others at treating particular kinds of illnesses? | No |
| | Is creative achievement helped or hindered by certain mental disturbances (e.g., autism, manic depression, obsessive-compulsive disorder)? | Yes |

## LESSONS LEARNED FROM OTHER STUDIES OF SCIENCE

As the least developed study of science, psychology has much to learn from the more established metascientific disciplines of history, philosophy, and sociology of science. The most important lesson comes from knowing the general stages that any scientific discipline goes through in its path toward maturity. Guiding the discussion of the development of each study of science, I make use of but modify Nicholas Mullins's stage model of theory or network development.[2] Mullins argued for four potentially overlapping stages of development in theories and/or scientific networks in sociology. I propose only three stages and apply them not just to one field (sociology), but to all of the metasciences (history, philosophy, sociology, and psychology). In addition, I simplify the components of each stage and focus only on each stage's intellectual leaders, social-organizational leaders, research-training centers, and intellectual successes.

In stage 1, *Isolation,* scholars work on the same problem in isolation, with the founding intellectual figures setting the stage. There is no social organization in terms of training centers, conferences, or societies. Late in stage 1 and early in stage 2, a core group of scholars may be working in the field, but doing so implicitly rather than explicitly, not yet labeling themselves as members of the field.

In stage 2, *Identification* is reached, as the intellectual success of the founding figures provides explicit theoretical and conceptual parameters for the field that attracts a wider range of students and other scientists who start to explicitly identify themselves with the field. Semi-regular meetings are organized and the first training-research centers may form. Such training centers are usually highly centralized around an intellectual leader, whose students have begun to have a major impact on the field. A leading journal becomes necessary as the outlet for the increased level of productivity of the field.

In stage 3, *Institutionalization,* the field becomes well established and institutionalized. Meetings become annual conferences because societies have now formed with their own social structure and hierarchy. Often multiple societies, some of them international, become necessary. Training centers proliferate and become less centralized, and at least one journal is now required for the expanding productivity of the field. Indeed, splinter movements, with different foci or agendas, may form and either break away or stay on the edge of the central field.

**Brief History of Metasciences**

*Philosophy of Science.* Although philosophy of knowledge (that is, epistemology) was a central theme in ancient Greek philosophy, the field of the philosophy of science is a much more recent development. Its origins are seen in three trends: classification of the sciences, methodology, and the philosophy of nature. The intellectual leaders, in the sense of writing the first books on the topic, were William Whewell (1794–1866) in England and Auguste Comte (1798–1857) in France—both of whom wrote in the 1840s. Whewell actually wrote two books on the philosophy of science and coined the terms "scientist" and "physicist" in the process.[3] He took a modified Kantian view that there are laws of nature independent of our understanding and that by our inductive intuitions, rather than raw empiricism, we can come to understand the laws of nature (see table 1.2).

John Stuart Mill (1806–73) developed his own positivist position in reaction to Whewell's inductivist position. Indeed, the two major proponents of *positivism* were Comte and J. S. Mill. Positivism holds that nature has no ultimate purpose and there is no "essence" to be discovered a priori. All scientific knowledge must be based in observable and positive facts. Positivists, at their core, are refuting the purely reflective method of acquiring knowledge, believing that only what comes through the senses is valid, scientific knowledge. Comte, in particular, put a historical spin to the positivist argument and claimed that the history of ideas passes through three phases—theological, metaphysical, and positivist (scientific)—with positivism being the penultimate stage of knowledge. In so doing, Comte was taking a classic empiricist stance by arguing that human nature was modifiable and capable of progress. During the second half of the nineteenth century, the publication of books in the philosophy of science went from a trickle to a fast drip, with some major works, including William Jevons, Ernst Mach, and Karl Pearson.[4]

At the turn of the century scholars began to organize more formally and establish the philosophy of science as an independent field of study (stage 2, Identification). For instance, the first congresses on the philosophy of science were held in Paris in 1900 (as sections of the First International Congress of Philosophy), and the first manifestations of what later became the Vienna Circle began in 1907.

The Vienna Circle (formed officially in 1922) played a big role in establishing the parameters of philosophy of science, gave it an empiricist-positivist orien-

Table 1.2 The Three Stages of Field Development for Philosophy of Science

| Stage of Development | Decades | Major Representative Event |
|---|---|---|
| Isolation: articles, books | 1840s–1890s | • Publication of William Whewell's *The Philosophy of the Inductive Sciences,* 1840, and *On the Philosophy of Discovery,* 1856<br>• Auguste Comte publishes *Discours sur l'ensemble du positivisme* (*A General View of Positivism*), 1848<br>• William Jevons publishes *The Principles of Science,* 1874<br>• Ernst Mach publishes *Die Mechanik in ihrer Entwicklung Historisch-Kritisch Dargestellt* (*The Science of Mechanics),* 1883<br>• Karl Pearson publishes *Grammar of Science,* 1892 |
| Identification: conferences, centers | 1900s–1950s | • First International Congress of Philosophy, with a section devoted to philosophy of science, 1900<br>• Henri Poincaré publishes *La science et l'hypothèse,* 1908<br>• Vienna Circle (Neurath, Mach, Schlick, Feigl, Gödel, Carnap, et al.) and logical positivism, 1910s to 1930s<br>• Hans Reichenbach publishes *Philosophie der Raum-Zeit-Lehrer,* 1928<br>• Karl Popper publishes *Logik der Forschung* (*Logic of Scientific Discovery*), 1935<br>• Philosophy of Science Group (later becomes British Society for the Philosophy of Science) has first congress, 1935<br>• First International Congress for the Unity of Science, 1935<br>• First International Congress of the Philosophy of Science, 1949<br>• Minnesota Center for the Philosophy of Science founded by Herbert Feigl, 1953 |
| Institutionalization: journals, societies, degrees | 1930s–1950s | • *Philosophy of Science* (journal) and Philosophy of Science Association, 1934<br>• *British Journal for Philosophy of Science,* early 1950s<br>• British Society for Philosophy of Science |

tation, and established the first modern answers to basic questions in the philosophy of science. The members of the Vienna Circle provided logical positivism its clearest and most cogent voice. Their fundamental argument was that empirical statements and their verifiability take priority over all other forms of knowledge, especially metaphysical and ethical. Can an idea be empirically verified? If not, it is meaningless—a conclusion they came to concerning ethics, morality, and metaphysics. The primary advances by the logical positivists over the positivists were the law of verification and the addition of mathematical and logical methods of analysis. Hence the "logical" descriptor to the positivist name. The major figures in forming the Vienna Circle were Max Neurath, Moritz Schlick, Ernst Mach, Herbert Feigl, Kurt Gödel, and Rudolf Carnap, all of whom were influenced by Auguste Comte's positivism as well as Bertrand Russell's logic and mathematical precision. Logical positivism went on to shape the entire field until the 1960s, and indeed all of modern philosophy of science owes its origin to logical positivism (if nothing else as a critical jumping-off point).[5] By the early 1930s, the founding members of the Vienna Circle began to disperse, landing in positions throughout Europe and the United States. In no small part because of the dispersal of these members of the Vienna Circle, the third stage in philosophy of science's development became established internationally in the decades between the two world wars, a movement that was codified with the first dedicated journal (*Philosophy of Science*) and the founding of the Philosophy of Science Association in 1934.

Karl Popper (1902–94) was a mathematics- and physics-oriented philosopher who early in life was influenced by the rational and scientific attitude of the Vienna Circle (especially Carnap and Schlick) as well as its intellectual father, Bertrand Russell.[6] He published a critique of logical positivism that set the stage for philosophy of science for the next forty years. That book was originally published in German in 1935 under the title *Logik der Forschung* (*Logic of Research*) and was not published in English until 1959 under the somewhat different title *Logic of Scientific Discovery.* Whether Popper's book is seen as a death knell for logical positivism (as he claimed) or as a critical variant of the position (as many others have claimed) is still a matter of debate.

Popper tackled head on one of the major questions in the philosophy of science: what makes one form of knowledge "scientific" and another "nonscientific" or "pseudoscientific"? To this demarcation question Popper did in fact provide a different answer than the logical positivists. Instead of verifiability, Popper argued it was falsifiability that separated science from nonscience. If a theory makes clear, unambiguous assumptions and predictions that can be put to both logical and empirical test, and if a negative result contradicts the theory, then the theory is falsifiable and therefore scientific. If a theory does not do these things and explains away (post hoc) both positive and negative results then it is not falsifiable and not scientific.

In the 1930s, when Popper was developing his ideas, Einstein's theory of relativity, Freud's and Adler's psychoanalytic theories, and Marx's theory of dialectical materialism were the predominant theories of the day. Each claimed to be scientific, but Popper felt that there were real differences between Einstein's and each of the social science theories, which propelled him toward the solution of falsifiability. It was especially after the experimental corroboration in 1919 of Einstein's theoretical and risky predication that gravity should bend light that Popper realized that real scientific theories prohibit rather than allow. They make risky and specific predictions, and if they are not supported empirically, their validity is inherently undermined and challenged. Theories by Freud, Adler, and Marx made no such claims and indeed contradictory findings could each be subsumed (and often were) post hoc under their respective theories. They could explain everything; scientific theories, on the other hand, must forbid some events. In Popper's mind, therefore, they were pseudoscience rather than science.

Moreover, Popper was to foreshadow a position that was to directly contradict the positivist position, namely, constructivism. Popper in his autobiography wrote: "I was reading Kant's *Critique* again and again. I soon decided that his central idea was that *scientific theories are man-made, and that we try to impose them upon the world.* 'Our intellect does not derive its laws from nature, but imposes its laws upon nature.' . . . Our theories, beginning with primitive myths and evolving into the theories of science, are indeed man-made, as Kant said."[7] The man-made element is a fundamental tenet of constructivism. The other fundamental tenet of constructivism is the inherently and unavoidably theoretical nature of all observation, a view that Popper also clearly espoused. The inextricably theoretical nature of observation is closely aligned with a principle that was antithetical to logical positivism, namely, that science must have metaphysical (unobservable) statements and assumptions. Popper's Kantian view had begun in *Logic of Scientific Discovery* but reached its clearest expression in *Conjectures and Refutations,* in which he expanded his anti-inductivist view that science works deductively, from theory down. One of Popper's main arguments, therefore, was that pure observation (that is, without some preconceived theory) is impossible. Theory guides our every observation. The essence of the scientific process, therefore, is a development from conjectures and hypotheses being put to logical and empirical test and either being temporarily corroborated or refuted.

The major figure most readily identified with constructivism, however, is Thomas Kuhn (1922–96). In many ways Popper and Kuhn were diametrically opposed, mainly in their attitudes toward social science perspectives of science.

But there was one important way in which they were quite compatible: they both developed a neo-Kantian view that the mind cannot observe and perceive without theory and expectation being an active part of the process. Kuhn argued in *The Structure of Scientific Revolutions* that scientific theories, too, must always include unobservables (metaphysical constructs) and are man-made.[8] But he then came to a different conclusion from this premise: any two competing theories may be incommensurable, especially during revolutions, and empirical observations cannot mediate the conflict. Moreover, new theories, revolutions even, develop only after gaps or shortcomings (anomolies) gradually become clear in the older theories. Here Kuhn made his well-known distinction between "normal science" and "revolutionary science." The former consists of working within a paradigm, whereas the latter creates a new paradigm. Most working scientists, needless to say, work within normal science. If theoretical assumptions widely diverge between competing theories, then empirical testing cannot be the arbiter in determining the superiority of one theory over the other. They are talking different languages, making different assumptions, and have different criteria—in short, different theories are incommensurable. Any given scientist's resistance to a new theory or paradigm stems more from nonscientific reasons. Competing theories gradually win or lose converts but less so for empirical than for aesthetic, social, or psychological reasons. With such a position, Kuhn moves away from Popper's focus on logic and rationality and exposes a major difference between the two major thinkers of twentieth-century philosophy of science.

I must, however, make clear that Kuhn and, especially, Popper each stop quite a bit short of the sort of cultural or social constructivism that was to develop in sociology in the 1970s: a social constructivism that argues that science is *nothing but* a social construction and therefore cannot be distinguished from and does not have any intellectual superiority to all other forms of knowledge, whether historical, literary, ethical, or political. The social constructivism position was furthered in the 1970s and 1980s more by sociologists than philosophers (Latour, Knorr-Cetina, Collins, and so on), and therefore I save the discussion of social constructivism for later in this chapter as well as later in the book (chapter 9).

In addition to Kuhn, another major response to logical positivism in the twentieth-century philosophy of science has been one that opposes, at times rather forcefully, the constructivist position, namely, scientific realism. The core argument of realism is that scientific theories are as real as we can approximate and the models should be accepted as being real. To use Kantian language, scientific theories can approximate the nomenal world (reality). One assumption of realism is based on the observation that if science did not approximate reality on some level, it simply would have taken the path of other pseudoscientific or nonscientific forms of knowledge (alchemy, astrology, phrenology, mythology, and so on) and not have been successful. As a modern realist, Richard Boyd, has put it: "If scientific theories weren't (approximately) true, it would be miraculous that they yield such accurate observational predictions." So there is a pragmatic criterion to realism. Science works and is the only form of knowledge to accumulate, and therefore it must at least approximate reality.[9]

What is the least man-made element of science is its descriptive component rather than its theoretical component. When science describes physical phenomena and structures—be they genes, cells, proteins, chemicals, plants, animals, or stars—it is not simply constructing a text with only local validity (as social constructivists would have us believe). Otherwise, there would be no meaningful distinction between hallucination and perception. Therefore, as the realists claim, "reality" exists independently of our thoughts, perceptions, and theories. For example, the structure and functioning of the neuron is real to the extent that all observers looking under an electron microscope could observe the structures and after many studies could determine some of the functions of the different structures. This is not simply a construction with only local validity. In this sense, positivists are correct: sensory experience and empirical results do play a deciding role in science. Different observers of very different backgrounds would corroborate these structures and processes in the laboratory.

A final movement in the modern philosophy of science has been dubbed "natural epistemology," and one of its most basic assumptions is that philosophy of science must be based more on what scientists actually do rather than what they ideally should do (as Popper's logical analysis would have us believe). Two major representatives of this perspective are Aharon Kantorovich and David Hull. Natural epistemology, and in particular the writings of Kantorovich and Hull, demonstrate the rapprochement that exists between many current philosophers of science and psychology in general and psychology of science in particular. The field has come a long way since Popper's "antipsychologism" of the mid-1930s. There are many ways in which Kantorovich's position in *Scientific Discovery* merges philosophical and psychological perspectives: first, instead of simply focusing on the product of science (knowledge), he examines the process of science (discovery); second, by focusing on discovery, he acknowledges the nonrational elements to the scientific process; third, he examines not prescriptive and normative forms of knowledge (how scientists *should* think), but descriptive ones (how scientists *actually* think). Finally, he takes his major inspiration and metaphor of knowledge creation from Darwin's theory of natural selection and develops a theory around evolutionary epistemology. Similarly, Hull contends that the scientific enterprise mirrors Darwinian competitive and inclusive fitness strategies seen in nature. That is, scientific ideas are generated and get selected through the same principles that occur in biological evolution. In addition, movements and camps in science stand in competition with those holding opposing views and vie for limited resources (funding, students, jobs), whereas those of similar views form collaborations and act to minimize their "inclusive fitness."[10]

Even more important, philosophers such as Hull have begun to blur the distinction between philosophy and science by arguing for applying the primary tools of the scientist—hypothesis testing and empirical evidence—to the study of science. As I elaborate on later in the chapter, the historian Frank Sulloway and psychologist Dean Simonton both have argued forcefully that historians must begin to test their ideas if they are to avoid the relativist conclusion of the constructivists that nature and reality play no role in the creation of knowledge—it is all a matter of politics and public relations strategies. Hull has made the same argument for philosophers. "Throughout most of its history, philosophy has been defined in such a way that evidence cannot possibly bear on it. Increasingly, however, those of us who are philosophers by training have begun to interpret philosophy in such a way that evidence does bear on the views that we express. Testing strictly scientific hypothesis is very difficult. Testing claims about science is even more difficult, but test them we must."[11]

From this review, it should be clear that since the 1920s (and especially since the 1950s), the philosophy of science has been a very visible force within philosophy, some might say an overly visible force. For at least eighty years now philosophy of science has been fully established intellectually, socially, and institutionally. Like its parent discipline, the philosophy of science began as an analytic field that examined the logic, limits, and structure of scientific knowledge. The questions of induction versus deduction, science versus nonscience, the importance of sensory data, whether scientific knowledge is real or socially constructed, and how scientific knowledge actually rather than ideally comes about have been the field's major contributions. As we will see later in this chapter, the major shortcomings of a philosophical analysis, as was true with the history of science, has been its resistance to test its own assumptions empirically—despite the best efforts of philosophers like Hull.

*History of Science.* History is the discipline of documenting and contextualizing the major events and trends that occur over time. The history of science documents, describes, and explains the developments of science, from its origins to its most contemporary forms of expression. It is not hard to understand that of all the disciplines that examine science, history would be the oldest. In fact, as George Sarton documents, the earliest known treatises on the history of science go back to ancient Greece and Syria and have been parts of official national academies of science since the 1700s.[12] For instance, in 1758 Jean E. Montucla published a two-volume work entitled *Histoire des Mathématiques,* and in 1834 Baden Powell published *Historical View of the Progress of the Physical and Mathematical Sciences from Earliest Ages to the Present Time.* Shortly thereafter (1837), William Whewell's major work, *The History of the Inductive Sciences,* was published. The works of Powell and Whewell, in particular, put the history of science on solid ground, but it remained an intellectual rather than social enterprise. Powell, Whewell, and others worked in isolation with no social or organizational structure, and therefore their work is indicative of the Isolation stage (see table 1.3).

The second stage of development (Identification) consists primarily of conferences, centers, and departments forming, and for the history of science these developments started in earnest right around the turn of the century and up through World War I. The number of scholars and works on the history of science had steadily grown by the end of the nineteenth century, such that the first congress on the topic was held in 1900 as a section of International Congress of Philosophy in Paris. The first stand-alone international congress for the history of science took place in 1929, also in Paris. The History of Science Society (HSS) formed in 1924 in Boston and began holding its annual conferences that year.

Identification occurs when the persuasiveness and charisma of intellectual or organizational leaders begin to attract students and scholars, who organize the field. For the history of science, if the intellectual leaders were Powell and Whewell, the organizational (and intellectual) leader was clearly George Sarton. He organized the first journal (*Isis*) as a regular publication outlet, organized many of the national and international congresses, and formed the leading department in the field at Harvard. By 1952 there were six universities that granted the PhD in the history of science: London, Harvard, Cornell, Columbia, and Wisconsin.[13] It would be misleading, however, to paint a picture of the organization of the field as being mainly an American phenomenon, for many of the congresses and publications were in Europe rather than the United States.

The third stage of a field's development is epitomized by the formation of so-

Table 1.3 The Three Stages of Field Development for History of Science

| Stage of Development | Decades | Major Representative Event |
|---|---|---|
| Isolation: articles, books | To 1890s | • Some histories of science originated as early as ancient Greece and Syria, and many scientific societies in Europe had their own history from the 1700s on[a]<br>• Jean Etienne Montucla publishes *Historie des mathématiques,* 1758<br>• Baden Powell publishes *Historical View of the Progress of the Physical and Mathematical Sciences from Earliest Ages to the Present Time,* 1834<br>• Publication of William Whewell's *The History of the Inductive Sciences,* 1837<br>• Royal Academy of Bavaria begins publication of its *Geschichte der Wissenschaften in Deutschland,* 1864<br>• William Youmans publishes first history of American science, *Pioneers of Science in America,* 1896 |
| Identification: conferences, centers | 1900–1930 | • First congresses of the history of science, as a subsection of the International Congress of Philosophy, 1900 and 1904<br>• First International Congress for the History of Science, 1929<br>• History of Science Society (U.S.) starts holding annual conferences, 1924 |
| Institutionalization: journals, societies, degrees | 1910–1930s | • Institutes, museums, and libraries of science and the history of science open all over the world, especially in the 1920s and 1930s[a]<br>• *Isis* (journal) first published, 1912, by George Sarton<br>• History of Science Society forms, 1924 (3,000 members in 2003)<br>• Académie Internationale d'Historie des Sciences forms, 1928<br>• Society for the History of Natural History, 1936<br>• British Society for the History of Science (BSHS) forms, mid-1940s<br>• Currently dozens of different societies and more than 100 journals devoted to the history of science, technology, medicine, or mathematics[b] |

[a]See Sarton 1952b.

[b]See Sarton 1952b for extensive documentation of history of science up to 1950; for more current information, see Web sites in text or links on the BSHS Web site.

cieties and the publication of regular research outlets, namely, journals and periodicals. Scholars identify with the field and solidify their professional identities by joining societies of like-minded scholars. In short, the field is institutionalized. By these standards, the third phase for the history of science also began in earnest in the first decades of the twentieth century, culminating in 1912 in the first international journal (*Isis*) devoted to the history of science. Moreover, the need for organization grew steadily, and by the turn of the century most European countries and one Asian country (Japan) had formed societies for the history of science.

There is no doubt that of all the fields that study science, the history of science is the most developed and institutionalized in terms of journals and societies. As Sarton's *A Guide to the History of Science* made clear, by the late 1940s there were already dozens of journals and many organizations throughout Europe and the United States devoted to the history of science; the history of a particular science; or the history of medicine, mathematics, or technology. Currently there are at least sixteen societies in the history of science (including history of technology and medicine) and at least fifty-seven different journals currently being published on the topics of the history of science, technology, medicine, or mathematics.[14]

The major intellectual contributions that the history of science has made to the study of science are its documentation and analysis of both individual scientists and the trends and themes that cut across time. By focusing on the life and times of individual scientists as well as on the scientific contributions of cultures, historians of science can paint a rich and complex picture of how science emerged in a particular time and place. The history of science contributes most to our understanding of the specifics of time and place. Historians of science, like all historians, document and describe what happened, where it happened, and often make causal claims about why it happened.

But it is precisely these explanatory systems and claims of causality that require the most systematic testing to determine their validity and generalizability. And yet very few historians do such testing.[15] Like many fields of social science where experimental manipulation is impossible, historians could test their ideas by doing correlational analyses to determine the direction and strength of relations between two or more variables. More important, however, this would require a different way of thinking about history, and many historians actively resist the idea that history should be scientific (that is, that their ideas should be subject to hypothesis testing or that evidence is relevant). To give a flavor of such a tension, the historian and psychologist of science Frank Sulloway has reported a review he received to a grant proposal to National Science Foundation (NSF) on a study on aging and creativity in science. The NSF program director summarized the thrust of the panel's criticism of Sulloway's proposal: "Many panelists thought that applying a heavy-duty statistical analysis to history is naive, inappropriate, and even peculiar. Is it really the case that generalizations in history should be tested with statistics, rather than be tested through a detailed examination of the sources? Some panelists noted that it seemed as if the Principal Investigator was going back to nineteenth-century beliefs that history is a science which could uncover laws. Panelists were opposed to such a narrow view of history."[16]

*Sociology of Science.* To a large extent, the sociology of science developed out of the philosophy and history of science and the sociology of knowledge. Perhaps the first major intellectual contribution to the field (stage 1) would have to be William Ogburn's *Social Change* from 1922. Ogburn (1886–1959) was more generally interested in social change, but scientific and technological change played a key role: technological invention is the precursor of cultural change, which in turn is followed by social disorganization, which in turn is followed by social and cultural adjustment. One major idea introduced by Ogburn was that of "cultural lag," namely, that human behavior lags behind scientific and technological innovation. He further argued that analysis of social trends would require large-scale statistical databases (see table 1.4).

But it was not until the mid-1930s that the founding intellectual works really began to appear: in 1935 the undisputed intellectual father of the sociology of science, Robert K. Merton, finished his PhD dissertation entitled "Science, Technology, and Society in Seventeenth-Century England." Two years later he published a chapter in *Sociology and Social Research* entitled "Civilization and Culture," in which he laid bare the need to include knowledge (especially scientific) as a focus of sociological investigation. Moreover, he argued that "civilization" (that is, theoretical knowledge) was more accumulative than "culture" (values and norms), but progress in the former was not linear.

At the same time (1937), one of Merton's mentors, Pitirim Sorokin (1889–1968) published *Social and Cultural Dynamics.* This work was a massive analysis of the development and evolution of culture and civilization in general, only part of which was devoted to the development of scientific knowledge. One main idea put forth by Sorokin was that of locating scientific knowledge less in the minds of individuals and more in "cultural mentalities" and that the history of knowledge sees dynamic back and forth between periods. Lastly, another major work that intellectually put the sociology of science on the map was John

Table 1.4 The Three Stages of Field Development for Sociology of Science

| Stage of Development | Decades | Major Representative Event |
|---|---|---|
| Isolation: articles, books | 1920s–1950s | • William Ogburn publishes *Social Change,* 1922<br>• R. K. Merton's PhD thesis "Science, Technology, and Society in Seventeenth Century England," 1935<br>• Karl Mannheim establishes the field of sociology of knowledge with publication of *Ideology and Utopia,* 1936<br>• Pitirim Sorokin publishes *Social and Cultural Dynamics,* 1936<br>• Merton publishes "Civilization and Culture" in *Sociology and Social Research,* 1936, and "The Sociology of Knowledge" in *Isis,* 1937<br>• John Bernal publishes *The Social Function of Science,* 1939<br>• Merton lays out the paradigm for the field in his chapter "Sociology of Knowledge," 1945, in *Twentieth Century Sociology*<br>• Bernard Barber publishes *Science and the Social Order,* 1952 |
| Identification: conferences, centers | 1960s–1970s | • Division for Sociology of Science and Technology in the International Sociological Association forms, 1966<br>• National Association for Science, Technology, and Society (NASTS) forms, 1988 |
| Institutionalization: journals, societies, degrees | 1960s–1970s | • Three of Merton's students at Columbia University (Stephen and Jonathon Cole and Harriet Zuckerman) receive their PhDs between 1965 and 1969; each becomes a major figure in sociology of science<br>• *Science Studies* (journal) is founded in 1971 by Roy MacLeod and David Edge, later is renamed *Social Studies of Science*<br>• Society for the Social Studies of Science (4S) forms in 1975 and currently has about 1,000 members |

Bernal's (1901–71) *The Social Function of Science* in 1939, which argued for a Marxist integration of science, philosophy, and society.

The first wave of intellectual formation in the sociology of science came to a close at the end of the 1930s, reflecting the decline in liberal- (even left-) oriented political views. With the exception of a review chapter by Merton in 1945 and a book by Bernard Barber in 1952, there were few major works on the sociology of science during the 1940s and 1950s.[17] Indeed, in the introduction to Barber's book, Merton remarked, "Among the several thousand American sociologists, not even a dozen report [sociology of science] as their field of primary interest." Although Merton focused more on broader sociological issues during the 1950s, by the early 1970s a compilation of his major papers on the sociology was published that truly established the parameters for the field, *The Sociology of Science: Theoretical and Empirical Investigations.* To this day it is still the definitive work outlining the early stages of development of the sociology of science.

By the mid to late 1960s, things began to change dramatically for the sociology of science. The political zeitgeist clearly was not only sympathetic toward, but even demanding of a sociological perspective. Social and cultural forces were *the* focus of many scholars, not just sociologists and social psychologists. It is safe to say that the second stage (Identification) of the sociology of science began in earnest by the mid-1960s. It was then that societies and centers began to form that explicitly investigated sociology of science. For instance, in 1966 the Division for Sociology of Science and Technology in the International Sociological Association was officially established. By the mid-1970s the major society for the sociology of science, the Society for the Social Studies of Science (4S) formed; it currently has about one thousand members and holds annual conferences in the United States and in Europe. In terms of first degrees that were granted in the sociology of science, three of Merton's students received their PhDs in sociology of science at Columbia University and went on to be major figures in codifying the field: Harriet Zuckerman in 1965, Stephen Cole in 1967, and Stephen's brother Jonathan Cole in 1969.

The sociology of science is somewhat unusual in that stage 2 and stage 3 (Institutionalization) occurred more or less simultaneously, both happening during the 1960s and 1970s. Moreover, once the field became more established, it also incurred a bit of a name change and merged with the more general science studies field labeled "Science, Technology and Society" (STS). In 1988 another society formed that is devoted to sociology of science concerns, namely, the National Association for Science, Technology, and Society (NASTS). More formally, as of 2003 there are at least thirty departments in STS in North America (with four of these thirty being more in philosophy and history than sociology), eight in Europe, and a handful in Asia and Australia.[18] In 1971 the major journal for the field, *Science Studies,* was founded by Roy MacLeod and David Edge. With this development, coupled with departments, societies, and degrees, the institutionalization of the sociology of science was complete.

One of the field's earliest and most consistent contributions concerns the analysis of "multiple discoveries," namely, the historical cases where more than one person independently hits upon a particular scientific or mathematical insight at the same time. For the sociologist such a phenomenon is of utmost importance because it suggests the individual is rather unimportant; what matters most is the social-cultural context. Merton, for instance, writing in 1952, was already quoting work from 1885 by Babcock and Pierce, who wrote of the "synchronism of inventions," which "shows that the individual man is of less importance in invention than his environment."[19] Viewing social and cultural factors as primary and individual and psychological factors as secondary or even tertiary is a relatively common attitude among sociologists—more of which will be discussed later in the chapter.

Three Mertonians, Zuckerman and the Cole brothers, have made perhaps the single biggest contributions to the field. Jonathan and Stephen Cole wrote one of the definitive books for the field entitled *Social Stratification in Science* in 1973, in which they argued for institutional and social forces being the pre-eminent factors behind the reward system (jobs, promotions, prizes, and awards) in science. As they write in their preface: "This book examines several aspects of a single basic question: is the stratification of individuals in science based upon the quality of scientific performance, or does discrimination obtain in the processes of status attainment?" Their answer is both matter, but science (physics at least) is mostly meritocratic, with quality being the prime predictor of status. However, pedigree, training, mentorships, and prestige of institution do also matter. For instance, following the lead of their mentor Merton, Cole and Cole analyzed "cumulative advantage" (the Matthew Effect), in which reward and recognition early in one's career snowballs and has a cumulative effect throughout one's career. As the saying goes, and it is not restricted to economics, "the rich get richer, and the poor get poorer." The Coles' overall conclusion is a complex one: science is more meritocratic than most any other institution, but at the same time it is not the ideally rational system it might claim to be. Gender, race, age, religion, and institutional affiliation do affect, often in a cumulative way, the reception and impact awarded to any particular scientific idea or discovery. It would be naive to think otherwise. This being said, Cole and Cole conclude "that the single most important variable in influencing the distribution of rewards is the quality of one's work as it is perceived by colleagues." It is when two works are roughly equal in quality that extrascientific traits matter most.[20]

Harriet Zuckerman first published *Scientific Elite* in 1977 in which she analyzed "the Nobel laureates in the United States, the Nobel Prize as an institution, and the stratification system in science."[21] Zuckerman's conclusion is not quite as meritocratic as the Coles', arguing that extrascientific (that is, demographics, mentoring, and pedigree) factors matter quite a bit in the reward system of science. For instance, concerning the religious origins of American laureates, 72 percent are Protestant, 27 percent are Jewish, and 1 percent are Catholic, and these figures compare to base-rates in the population of 67 percent, 3 percent, and 25 percent, respectively. So the overrepresentation of scientists from Jewish backgrounds and underrepresentation of scientists from Catholic backgrounds is quite obvious in the scientific elite. One somewhat surprising finding of Zuckerman, however, is the fact that productivity of scientists after winning the Nobel Prize almost always declines as the nonscientific demands on their time increase dramatically.[22]

*Psychology of Science.* Recall that stage 1 (Isolation) involves major intellectual works being published before the institutionalization of the field. For psychology of science, this period could be dated as having started in the 1930s and lasting up until the 1980s. Although Francis Galton wrote the first book that may be considered to explore questions of a psychology of science (*English Men of Science,* 1874), it was Stevens who coined the phrase in the 1930s. These works, however, did little to jumpstart the field and it was not until the 1950s that more systematic works on the psychology of science began to appear (see table 1.5). In particular, Anne Roe's classic work, along with that of Raymond Cattell, foreshadowed the burst of research on psychological attributes of scientists that occurred in the early 1960s. In general, studies in the 1960s placed a heavy emphasis on creativity in science. In fact, it would be more appropriate to view these works as dealing with the psychology of scientific creativity rather than the psychology of science, but nevertheless they demonstrated quite clearly the importance of psychological factors (developmental, personality, motivational, and social) behind scientific achievement and creativity. In addition, Maslow published a book in 1966 with the title *The Psychology of Science,* in which he argued for expanding the scope of traditional mechanistic, reductionistic views of science to include a broader, more humanistic, and psychological conceptualization of science. Finally, a precursor to the entire discipline of cognitive psychology of science can be seen in Herbert Simon's 1966 chapter on scientific discovery and the psychology of problem solving.[23]

But the 1950s and 1960s were a false start for the psychology of science. During the 1970s there was a decline in research on the psychology of science, and few major works were produced on the topic. One exception was a conceptual article by Singer, who pointed out that although a new "science of science" was

Table 1.5 The Three Stages of Field Development for Psychology of Science

| Stage of Development | Decades | Major Representative Event |
|---|---|---|
| Isolation: articles, books | To 1980s | • Francis Galton publishes *English Men of Science,* 1874<br>• S. S. Stevens publishes first attempts at identifying psychology as a metascience, 1936, 1939<br>• Abraham Maslow publishes the first book entitled *Psychology of Science,* 1966<br>• Rudolf Fisch publishes first bibliography on the psychology of science, with more than 300 references, 1977<br>• Michael Mahoney publishes "Psychology of the Scientist" in *Social Studies of Science,* 1979<br>• Dean K. Simonton publishes *Scientific Genius: A Psychology of Science,* 1988<br>• "Memphis Group" (Gholson, Shadish, Neimeyer, and Houts) edits *Psychology of Science,* with chapters by numerous psychologists reviewing literatures in many different topics<br>• Review of field is published by Gregory Feist and Michael Gorman, 1998 |
| Identification: conferences, centers | 1980s– | • Conference, "The Psychology of Science," Memphis State University, 1986<br>• Gregory Feist receives one of the first PhDs granted on the topic at University of California, Berkeley, 1991<br>• Conferences on "The Cognitive Basis of Science" held at Rutgers University (1999) and University of Sheffield (2000); a book by same name is published in 2002 |
| Institutionalization: journals, societies, degrees | Not yet | n/a |

a nascent discipline as far back as the 1930s, "some 30 years have passed, and we do not as yet have a developed, self-conscious discipline of a science of science. We are now, however, in a better position to anticipate its arrival." Another exception from the 1970s was the first major review of the field by Rudolf Fisch. Toward the end of the decade Fisch echoed Singer's concern and opened his review by pointing out the disparate and unsystematic nature of investigations into the psychological attributes of scientists. He concluded his review pessimistically: "Having now reviewed the field, it is lamentably clear that basic concepts are diffuse and contradictory, and rarely become common to several investigations. For this and other reasons, results cannot really be compared, and little scholarly cumulation has resulted." In another review of the literature, just two years later, Mahoney reached similarly pessimistic conclusions about the state of the field.[24]

Since the early to mid-1980s there has been a steady surge in works devoted to the psychological underpinnings of science.[25] One such work was a small volume by Sonja Grover who was inspired by Mahoney, T. Kuhn, and Feyerabend. Her main thesis is that science is a function of subjective and nonrational (intuitive, imaginative, and creative) processes more than rational ones, and therefore a psychology of the scientist is required if we are to understand the scientific process, including the justification and empirical testing stages. There were conferences in 1986 at Memphis State University (now University of Memphis) and a panel discussion (with Feist, Simonton, Shadish, Fuller, and Gorman) at American Psychological Association's annual conference in 1992. These were followed by two conferences on "the cognitive basis of science" in 1999 (Rutgers, N.J,) and 2000 (Sheffield, U.K.) and a panel on the psychology of science at the International Congress of Psychology in Stockholm in 2000. Together, the events that began in the mid to late 1980s led Shadish, Fuller, and Gorman in 1994 to proclaim that the "psychology of science has finally arrived." The edited book that resulted from the conference by the "Memphis Group" (Gholson, Shadish, Neimeyer, and Houts) was an important development for the psychology of science: "Substantively, psychological contributions to science studies are increasing in frequency and quality. Sociologically, psychologists are beginning to identify themselves as interested in the topic. But much work needs to be done if the psychology of science is to achieve its potential. In the present book, I plan to further this agenda—to examine the history of and justification for a psychology of science, to outline its possible content and methods, to document some of its accomplishments and its potential, and most of all, to intrigue and encourage fellow psychologists to bring their expertise to bear on the study of science."[26]

For instance, in their opening chapter, Shadish, Houts, Gholson, and Neimeyer proposed a systematic set of problems for the psychology of science. The two dimensions, "domains of psychology" and "dimensions of scientific work," were each composed of multiple categories and when combined created at least ninety distinct problem areas for a psychology of science. For example, the domains of creativity, cognition, personality, motivation, and social psychology could be combined with dimensions of scientific work, such as career choice, problem selection, question generation, and obtaining funding. Moreover, the chapter by Houts outlined many different empirical and theoretical questions from the history, philosophy, and sociology of science that can and should be addressed by psychologists. One goal of the Memphis Group was to expand psychology of science beyond its three main topics of personality, creativity, and cognition and apply it more generally to most any problem within psychology.[27]

Dean Keith Simonton expanded his chapter for the Memphis conference into a book entitled *Scientific Genius: A Psychology of Science.* Influenced by Darwin and D. T. Campbell, Simonton furthered his theory that creativity is a Darwinian chance configuration process, with highly creative people being able to produce a high number of ideas, some of which "hit" and some of which "miss."[28] As a graduate student, I was quite inspired by the books by Simonton and the Memphis Group, and the current book can be seen as a direct outgrowth of that inspiration.

Even more indicative of the growth of the psychology of science than the increase in works explicitly devoted to the topic has been the explosion of works implicitly devoted to the psychology of scientific thought and behavior. Leading the way have been the developmental psychologists. For instance, the developmental psychologist Deanna Kuhn is a perfect example of a psychologist of science who may not explicitly identify with the field. She has put forth one of the most systematic and impressive research programs on the development of scientific reasoning, addressing questions like what is the essence of scientific reasoning and when and how it develops. Similar implicit psychology of science comes from the work of such developmental psychologists as Alison Gopnik and Elizabeth Spelke, who investigate how conceptual knowledge of the physical and social world develops in infants. As I demonstrate in chapter 3, although these developmental programs of research are very much implicit scientific thinking, they nonetheless are very much part of the psychology of science.

One could legitimately ask, "Were the 1980s another false start?" given that some of the excitement of the late 1980s and early 1990s has dissipated. Although Feist, Simonton, and cognitive psychologists (Tweney, Klahr, Gorman, K. Dunbar, and so on) have continued to publish regularly on the topic, others have actively or passively left the field. My own evaluation of where the field stands is in the transition between stages 1 (Isolation) and 2 (Identification). More psychologists than ever before are doing work on the psychology of science, broadly and narrowly defined, but quite a few of these scholars are only implicitly identified with the psychology of science. Since the 1950s intellectual works have been appearing, and since the late 1980s there have been some early signs of identification, with more and more psychologists identifying with the field. There are some researchers who have a core group of students (for example, Tweney, Gorman, Klahr, D. Kuhn, and so on) consistently investigating cognitive and developmental processes in the formation of scientific knowledge. Full development of stage 2 (Identification), however, requires centers or departments dedicated to the psychology of science, as well as graduate degrees and regular conferences. None of these yet exist with enough regularity to claim the psychology of science is fully into the Identification stage of its development. One goal of this book is to spur on some of these developments toward explicit identification of the field.

### The Reaction of Other Metasciences to the Psychology of Science

My objective in this chapter is to examine why the psychology of science has been relatively slow in getting established and how it might learn from the process by which other studies of science developed. History, philosophy, and sociology of science, however, have also played an active role at times in slowing the development of the psychology of science because some scholars in these other areas have not always been accepting of a psychological perspective.

Historians of science traditionally have not been much concerned with the general psychological traits and motives involved in the history of science. Ever since Whewell, many historians have documented and described the major events and trends in the history of science, and some, of course, have documented the lives of particular scientists, delving into psychological forces. But on the whole historians have not moved beyond narrative histories of individuals' lives; they have generally avoided "psychologizing" about individual scientists or even samples of scientists. Yet psychological forces—concept development, motivation, ambition, creativity, imagination, and so on—are implicit in much of this history of science. Historians view these aspects at a cultural rather than an individual level. Moreover, as discussed above with Hull's comments about philosophers, Sulloway has argued that historians often actively resist a fundamental tool in the psychologist's arsenal, namely, hypothesis testing, believing instead that their analyses are narrative explanations that neither require nor can make use of testing and evidence.[29]

My response to such conflicting perspectives is to point out that studies of science are not in a zero-sum game with one another; one must not win at the other's expense. These are complementary, not contradictory, perspectives. Indeed, there have been historians who have begun to look at history of science by integrating psychological methods and ideas. There are, in addition to Sulloway, Arthur I. Miller, who argues for the power of visual imagery in the history of scientific creativity, and Gerald Holton, who argues for the role of psychological forces in scientific discovery (in contrast to scientific justification and knowledge per se). In addition, the mutuality of psychological and historical perspectives is evidenced by a number of psychologists who have developed the field of "psychohistory" and have conducted detailed psychological analyses of major figures in the history of science.[30]

Historically, philosophers of science have probably been more actively disdainful than historians of the psychological perspective in the study of science. This disdain is perhaps nowhere seen more clearly than in Popper's arguments against "psychologism." Here we get to Popper's real concerns of introducing psychological factors into the analysis of science. "The initial stage, the act of conceiving or inventing a theory, seems to me neither to call for logical analysis nor to be susceptible of it. The question of how it happens that a new idea occurs to a man—whether it is a musical theme, a dramatic conflict or a scientific theory—may be of great interest to empirical psychology; but it is irrelevant to the logical analysis of scientific knowledge."[31]

Taken at face value, Popper is making an important distinction, one first made by Hans Reichenbach in 1928. On one hand, Popper is arguing that *science as a product* must be evaluated completely independently of the psychological, social, and historical context. Personalities, historical contexts, and sociological influences are irrelevant to the evaluation of a scientific idea. On the other hand, he implies that *science as a process* is precisely the topic of psychological, social, and historical analysis. Of course the stage of inspiration and creativity is not amenable to logical analysis, for the process itself is intuitive and implicit. Popper was perfectly correct to point this out, but the distinction and boundary between process and product is not quite as clean as the philosophers would have us believe.

By making this distinction between product and process, it would appear that Popper was amenable toward a psychological perspective if it focused on process rather than product. Later writings, however, made quite clear that Popper had no sympathy for a psychological (or historical or sociological) perspective under any circumstance. In 1970, for instance, T. Kuhn wrote a critique of Popper's work, entitled "Logic of Discovery or Psychology of Research?" in which he critiqued some of Popper's most basic assumptions. The sparks really started to fly when the issue of logic versus psycho-logic arose, the essence of Kuhn's chapter. For Popper, logical analysis is absolutely paramount and any evaluation of scientific theory must be limited to either logical or empirical criteria. Kuhn, on the other hand, argued that Popper has provided not a logic of knowledge but rather an ideology of knowledge and that there are forms of scientific knowledge to which logical analysis does not apply. "Already it should be clear that the explanation must, in the final analysis, be psychological or sociological. It must, that is, be a description of a value system, an ideology, together with an analysis of the institutions through which that system is transmitted and enforced. Knowing what scientists value, we may hope to understand what problems they will undertake and what choices they will make in particular circumstances of conflict. I doubt that there is another sort of answer to be found."[32]

In a response to Kuhn's chapter, Popper stated with surprising frankness his disdain and even abhorrence of psychological, historical, and sociological analyses of science:

> In fact, compared with physics, sociology and psychology are riddled with fashions, and with uncontrolled dogmas. The suggestion that we can find anything here like "objective, pure description" is clearly mistaken. Besides, how can the regress to these often spurious sciences help us in this particular difficulty? Is it not sociological (or psychological, or historical) *science* to which you want to appeal in order to decide what amounts to the question "What is *science?*" or "What is, in fact, normal in science?" For clearly you do not want to appeal to the sociological (or psychological or historical) lunatic fringe? And whom do you want to consult: the "normal" sociologist (or psychologist, or historian) or the "extraordinary" one?
>
> This is why I regard the idea of turning to sociology or psychology as surprising. I regard it as disappointing because it shows that all I have said before against sociologistic and psychologistic tendencies and ways, especially in history, was in vain.
>
> No, this is not the way, as mere logic can show; and thus the answer to Kuhn's question "Logic of Discovery or Psychology of Research?" is that while the Logic of Discovery has little to learn from the Psychology of Research, the latter has much to learn from the former.[33]

Such unbridled defensiveness and dogma are truly surprising, if not shocking, coming from such a sharp and luminary mind as Popper's. In one fell swoop he dismisses three major fields of investigation as "lunatic fringe" and as having nothing to offer logic and philosophy, ignoring all the while his own distinction between the process and product of science. "Logic, logic, and nothing but logic!" Popper seems to be shouting. By his own admission, he was very much at home in logic and biased against emotional and subjective experience: "The whole experience, and especially the [question of whether Marxism could ever be scientific] produced in me a life-long revulsion of feeling."[34] This aversion to the social science perspective lasted to the end of his life. Speaking at the University of California at Berkeley in 1989 and when asked about Kuhn's criticism of his work, Popper retorted "sociologists are low on the totem pole of science and try to boost themselves up by studying real scientists."

This conflict between Popper and Kuhn can be resolved relatively easily: not only does Popper focus on product and Kuhn on process, but they are debating different aspects of scientific behavior. Popper is more concerned with the normative and prescriptive question of how science *should* be carried out, and Kuhn is more concerned with the descriptive question of how science *is* carried out. More recently philosophers of science have indeed begun to be more descriptive and naturalistic as well.[35]

Another source of conflict between historians and philosophers of science on the one side and the psychologists of science on the other has been attitudes toward testing one's own theories of science, that is, toward evidence and hypothesis testing.[36] As Hull has pointed out, philosophers are not trained in experimental and statistical methods and do not see their work as one that requires evidence or involves formal or statistical testing. That is not even what philosophers are supposed to do; their analysis is and always has been logical and conceptual, not empirical. Psychologists, in contrast, are trained in experimental and statistical methodology and view evidence and hypothesis testing as both necessary and self-evident. So an inherent tension between these perspectives was bound to arise.

More surprising, if for no other reason than that they both are social sciences and empirically based, has been the hostility of sociologists toward a psychological perspective. Two major figures, the Cole brothers, wrote the following in their book *Social Stratification in Science:* "Perhaps the most important contribution of the sociology of science has been its challenge to the psychologistic view of scientific development. This view holds that science moves forward as a result of the idiosyncratic creativity of isolated geniuses. The sociological perspective sees scientific ideas as a creation of individuals within a community."[37] Here, in an early founding work on the sociology of science, two leading thinkers see the thrust of sociological perspective as a direct challenge and contradiction to the "psychologistic view of scientific development." The opposing perspectives on the individual and the group or community are presented as conflicts between the psychological and sociological perspectives on science rather than two legitimate and worthy perspectives on the same phenomenon.

Starting perhaps with the phenomenon of multiple discoveries, the movement to denigrate the individual and psychological factors reached a crescendo with social constructivism beginning in the 1970s and continues to this day. The defining tenet of social constructivism is that science is *nothing but* socially and group-constructed knowledge and therefore the individual is the wrong level of analysis. In fact, social constructivists go so far as to argue that the adjective "social" is meaningless because no scientific behavior is anything other than social. Although the denigration of the individual's importance is seen most clearly in social constructivists, to some extent it is inherent in the sociological perspective in general.[38]

From the perspective of psychology, and in particular a psychology of science, the denigration of the individual is misguided. That *people* become interested in science and make creative contributions may be the position of sociologists and historians of science; but that particular *individuals* do so is the position of psychologists of science. To understand why one person does and another person does not become a scientist requires a psychological perspective. Personality, temperament, motivation, cognitive ability, intelligence—all characteristics of the individual—do matter in determining who will be interested in science and, of those, who will be more likely to make creative contributions. There is no doubt that it is often the creativity of the individual that drives most innovations. The group, as Einstein often vigorously pointed out, cannot be given priority over the individual. Pretending that individuals are simply interchangeable pawns and to be understood only as players at the mercy of larger social and institutional forces with no individual differences to speak of is naive at best and dangerous at worst. Of course, individuals exist only in social groups, but to argue that an understanding of the psychological factors behind the individual is irrelevant is simply narrow and disciplinocentric (that is, the inherent tendency to believe that one's discipline is the best way of analyzing a phenomenon).

Just as is true of Popper's antipsychologism, the sociological and psychological views are not zero-sum games. They are not mutually exclusive but rather different ways of looking at and contributing to an understanding of the same phenomenon. More important, although disciplinocentrism is still quite evident, it is also clear that since the 1980s the distinctions between history, philosophy, sociology, and psychology of science have become increasingly blurred as each borrows from and is influenced by the other. As these fields develop, they become inherently more interdisciplinary. Philosophers like Kantorovich and Hull, historians like Miller and Sulloway, and sociologists like Fuller have all taken psychological perspectives in their own analysis of scientific thought and behavior. Also, psychologists like Simonton have combined philosophical, historical, and sociological perspectives with their psychological training. One conclusion is clear: we need all studies of science to fully understand the nature of science.

## THE CONTRIBUTIONS OF A PSYCHOLOGY OF SCIENCE

Psychologists have developed sophisticated theoretical models and empirical strategies for studying behavior and thought, and scientific behavior and thought are simply part of a more general human psychology. From the brain structures involved with recognition of and reasoning about physical, mathematical, biological, and social phenomena, to the development of mathematical and scientific reasoning, to confirmation bias and creative problem solving; from personality traits that predict scientific interest and achievement, to the influence of groups and laboratories—all of these topics are relatively well-developed lines of psychological research and form the core topics of a psychology of science. More important, scientific thought and behavior provide insight into a very core dimension of human nature: observation, pattern recognition, expectation and theory construction, hypothesis testing, and causal thinking, not to mention intuition, imagination, and creativity. Scientific interest, thought, and behavior provide an excellent testing ground for our most basic principles of human behavior. Psychology simply cannot afford to ignore one of the most important human activities, one that has transformed the very world we live in. Studying the scientist will force psychological theories into an important new domain, leading to changes in psychological concepts. In short, because scientific thought and behavior are among the most recent and complex of human capacities, understanding it will move forward our understanding of what humans are capable of at their best.

The other metasciences—philosophy, history, sociology (and more recently, anthropology)—are trying to supply their own answers to psychological questions concerning conceptual change, theory choice, and motives and personal styles of scientists. The two oldest studies of science, history and philosophy, are not empirically based. Psychology, along with sociology, is inherently empirical and therefore will not simply propose ideas about scientific thought or knowledge or historical development but will test its models of scientific thought and behavior against empirical evidence. Psychology, however, is the only empirical study of science that also makes uses of the experimental method (that is, manipulates variables and randomly assigns participants to conditions). Moreover, it is the only metascience to combine empiricism with an examination of the individual, and in doing so it includes the biological, neuroscientific, sensory, developmental, cognitive, personality, and social factors behind scientific thought and behavior.

In addition to the pure, theoretical, and methodological reasons why the psychology of science can contribute to the science studies field, there are applied ones as well. Teachers and parents stand to gain a much better understanding of the conditions that foster scientific interest and talent once psychologists advance our understanding of the psychological (developmental, personality, cognitive, social, and neuroscientific) underpinnings of scientific thought and behavior. A more fully developed psychological understanding of science is already leading to improvements in pedagogy, both for those who will become scientists and for those who need to understand science in order to be informed citizens (that is, politicians in charge of budgets). Along similar applied lines, by understanding the actual psychological processes behind science, and in particular the best science, perhaps a psychology of science can have a loud and clear voice about selection criteria for potential graduate students and faculty (see chapter 7). Psychological assessment and testing can aid in the identification of young people who might be our best scientists of the next generation, but we need empirical evidence for the predictive validity of such measures. A well-developed psychology of science can provide such evidence.

It is of utmost importance to understand the history of science, the philosophy of science, and the sociology of science. Historical, philosophical, and sociological perspectives are required, and each can bring its own unique contributions to the study of science. Scholars in those disciplines have been working systematically toward an understanding of science for forty to one hundred years. Psychology has learned too much about scientific thought and behavior for it not to develop its own voice and inform these other studies of science.

## INTEGRATING THE PSYCHOLOGY OF SCIENCE AND ORIGINS OF THE SCIENTIFIC MIND

One could argue that the two major themes addressed in this book—the psychology of science and the origins of the scientific mind—are separate and distinct enough to warrant separate books. Although there is some truth to this view, I see real benefits to bringing them together in one book, not least of which is that when considered together they will provide a wider perspective for how scientific thinking came to be at the species level, which in turn informs us of how it develops in individuals. The inverse is also true: knowledge of how scientific thinking develops in individuals provides clues for how it may have developed in our species. Indeed, it is not difficult to see how the book's themes are connected if one thinks about the related concepts of recapitulation and levels of analysis.

Recapitulation is the idea that ontogeny and phylogeny are related processes. Developmental processes that occur at the species level are occasionally reflected at the individual level and vice versa (recapitulation). This principle holds true for the development of scientific thinking. If we want to fully understand how and why scientific thinking is possible in individuals, then we must study its development at the species level.

Moreover, human behavior can and should be examined on different levels of analysis. In the 1930s, an anthropologist (Clyde Kluckholm) and a psychologist (Henry Murray) wrote that everyone is (1) like all other people, (2) like some other people, and (3) like no one else.[39] These three levels correspond to the species, the group, and the individual, respectively. The psychology of science as presented in part 1 deals primarily with the individual and group levels of analysis, whereas the origins of the scientific mind as presented in part 2 focus on the broader species level perspective ("like all other people").

In figure 1.1, I present a general conceptual diagram summarizing the major evolutionary-historical (external) and psychological (internal) factors that lie at the foundation of scientific interest, talent, and achievement.[40] Evolutionary theory has proved a useful theoretical foundation insofar as the human brain is both a product of evolutionary forces and is responsible for all human thought, including science. Evolutionary forces have shaped the human brain to be biased toward certain categories of information processing and knowledge, for example, people, objects, plants, animals, number, and language (see chapter 8). These categories of knowledge in turn have become domains of science (social, physical, biological-natural history, and math). Simply put, the domains

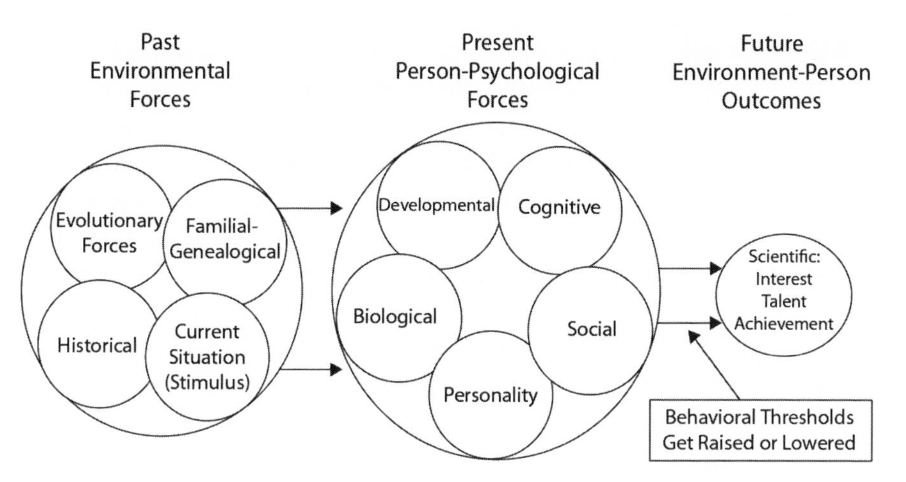


Figure 1.1. Summary of the situational and personal forces behind scientific interest, talent, and achievement.

of science that exist today are not arbitrary but rather fall along evolved domains of knowledge.

The major environmental forces presented in figure 1.1 consist of distant (evolutionary and historical) as well as proximal (familial-genealogical and current situation) influences. To develop a full and complete understanding of individuals, groups, and the species, one needs to have an appreciation of the evolutionary influences that have produced the unique traits and abilities of our species; the historical and cultural events of our more recent cultural evolution; the specific family genealogy of the individual; and finally, the current situation the person finds him or herself in at any given moment. For instance, the three major stages of human cognitive evolution discussed in chapter 8 go a long way in explaining how we went from ancestral hominid thinking to modern human thinking capable of symbolic abstraction and science. In addition, each of the four prehistoric and historic phases of science discussed in chapter 9 shed light on the origins and development of, first, implicit "folk" science and, then, explicit and modern scientific thinking.

External environmental forces causally act on the individual and all of his or her biological-genetic, developmental, cognitive, personality, and social processes. For instance, evolutionary and genealogical factors directly contribute to one's genotype, which make up, among other things, each person's central nervous system in all its neurochemical and neuroanatomical uniqueness. The variability in biological structures contributes to temperamental differences in infancy and early childhood, which in turn become the foundation for individual differences in intelligence and adult personality.[41] Some creative people develop talents in particular domains, be they social-psychological, physical-spatial, numeric-quantitative, or biological-natural history. Being thing- or people-oriented starts different people down different paths of science, namely, physical or social. Children are inherently incipient scientists and construct implicit domain-specific and domain-general theories of their social, physical, biological, and quantitative worlds. Self-image, personality, and demographic forces (gender, birth order, religious background, and immigrant status, for example) influence scientific interest as well. The earlier a child shows an interest in and talent for science, the more likely that person is to have a creative and productive career. Scientific creativity and productive achievement change over the life course, with peaks generally occurring twenty years or so into one's career. One's ability to use metaphor and analogy, to separate and coordinate theory and evidence, to systematically test hypotheses, to think complexly, to solve problems intuitively and by working forward each facilitate scientific reasoning and creative problem solving. Moreover, people with a certain cluster of traits, such as intelligence, openness, introversion, confidence, and independence, have lower thresholds for developing interest in and talent for science. Lastly, social and group forces (parents, teachers, mentors) lower thresholds for scientific interest and talent as well and depending on their structure, make interest and talent either more or less likely.

# Chapter 2 Biological Psychology of Science

The reasoning behind this chapter on the biological psychology of science stems from two basic observations. First, I start with the most fundamental, if not obvious, of observations: the brain controls almost every single behavior we exhibit—from breathing to writing a symphony (the only exception is reflexes). If we want to understand any aspect of human behavior, we must start with the brain. Second, contrary to what some think is true, our genotype is not our destiny but rather our starting point. That is, what happens between the formation of the genotype and the expression of phenotype is experience, and genotype and experience together influence phenotypic outcomes. Indeed, it is now quite clear from research on genetics (more specifically epigenetics) and neurogenesis (the development of neural structures) that the environment and experience can and do change gene function without changing gene sequence. Just to provide one example: one identical twin might develop Alzheimer's disease and the other might not, although they both have completely identical genotypes. Something environmental must explain the difference. Moreover, neuroscience is now demonstrating how experience plays an absolutely vital role in the formation of neurons as well as the number and density of neural structures. The best way to put this general argument is that our genes and brain structures are simultaneously "hardwired" and flexible or "plastic." Variability in our genotype does lead to variability in brain processes and structures, but so, too, does variability in our environment. The more we understand of genetics and neuroscience, the more we come to see the power of the environment in molding and shaping our biological systems, most prominent of which are genes, neurons, and brain structures. The flexibility of humans to learn virtually whatever our environment throws at us may be related to such genetic and neural malleability and may be, when all is said and done, the signature trait of the human species.[1]

The study of scientists per se is not required for a biological psychology of science, for much of scientific thinking is in fact implicit and stems from basic neuroscientific processes. The "implicit biological psychology of science" is indeed a rich and exploding domain of investigation due mostly to its connection with two major revolutions in science over the last ten to fifteen years, namely, genetics and neuroscience. To be clear: the genetic and neuroscientific foundations of human thought are not specific to scientific thought. But specific domains of mind (especially the social, physical, natural-historical-biological, and mathematical domains) are the origins of the systematic and formal physical, biological, and social sciences that exist today (see part 2). To the extent that current genetic and neuroscientific research sheds light on these cognitive facilities, they shed light on foundations of higher human cognitive functioning, including scientific reasoning, problem solving, and creativity.

## BEHAVIORAL GENETICS AND INTELLIGENCE

Behavioral genetics is the best method for acquiring knowledge of the genetic basis of intelligence and behavior. Because genes build proteins and proteins make up each structure of the body, including the nervous system, and given temperamental differences between people in how they respond to every kind of stimulation, an understanding of some of the fundamental principles of genetics is necessary before attempting to elucidate why different people have different skills and talents, including scientific.[2]

### Principles of Genetics

The "central dogma of molecular biology" (as coined by Francis Crick) is that genes code for proteins and that is all. More specifically, every string of three base-pairs (triplets or codons) in the gene codes for a particular amino acid, and these chains of amino acids are particular proteins. The corollary to this is that the journey from gene to behavior is never direct. Genes build proteins, some of which form enzymes, some of which form neurotransmitters, some of which form neurons. The pathway from gene to amino acid to protein to neurotransmitter to neuron to brain region to behavior is long and complex. Genes do not cause behavior—they simply build proteins. Similarly, what happens to us during development influences how genes get expressed. The fact that the path from genotype (gene) to phenotype (trait) is not immutable is captured by the new subfield of genetics called "*epigenetics*," that is, how events after conception alter gene expression without any change in gene structure or DNA structure.[3] The field of epigenetics has been bolstered by the recent findings that heritable gene function can be changed by such environmental events as maternal diet or exposure to carcinogens. Genes are not our destiny; they are just the starting point.

Another principle, and perhaps the most commonly misunderstood principle in popular culture about genetics, is that individual differences between people are more a result of genes having different forms (alleles) than they are of someone having a particular gene or not. In terms of existence of genes, every human is about 99.9 percent genetically the same as every other.[4] It is simply not true that someone has "the gene" for something and others do not. Rather it is the form of genes that varies between individuals, a phenomenon known as "polymorphism." It may simply be that one version of a gene leads to one expressed trait (red petal in a flower), whereas another version of that gene results in a different form of that trait (white petal). What often differs from one allele to another, and hence one individual to another, is the number of repeating sequences of a particular base-pair motif, otherwise known as "short-tandem repeats" (STR). For instance, one person may have twenty-six repeats of the ATGT base-pair sequence in a particular location of the gene and someone else may have thirty-five repeats of that sequence in that location. It is variability in the number of repeats in an STR that leads to different amino acids being built and hence different forms of the protein being made. These differences in proteins ultimately lead to differences in the phenotypic trait, whether physical or psychological.

A related principle is that very few phenotypic traits are the result of one gene. Single gene transmission is what classic Mendelian inheritance is all about, but as it turns out, most traits are more complex and the result of many genes. Only traits that have a few distinct categories, such as hair or eye color, are the result of single genes. Most traits, and certainly all psychological traits, are the result of many dozens or hundreds of genes, a phenomenon known as "polygenic transmission." The implication here is that any one gene will account for only a very small amount of the variability in any given trait. In essence, any trait that is expressed in a normal bell-shaped distribution—with a few people having a little or a lot of the trait and most having an average amount of it (like weight or extroversion)—is a result of many genes. These traits are known as "quantitative traits." The genes that are responsible for a particular quantitative trait are not usually found on one chromosome but rather randomly on many different ones. These gene locations are known as "quantitative trait loci" (QTL). New techniques have been devised that compare statistically the association between QTLs and particular traits. In simple terms, this technique is a helpful first stab at uncovering genetic influence on behavior because it narrows the locations on chromosomes where target genes may be. These QTLs, then, provide a sense of where the "candidate genes" for a particular trait may be. QTLs, however, do not tell us which specific genes are associated with which specific traits. Once the locations are narrowed down, more direct molecular genetic techniques can be applied to uncover the actual genes involved.

The connection between genes and brain size and structure is long but direct. Recently neuroscientists have reported that about 65 percent of the variability in brain volume is genetically influenced.[5] The brain is made up of trillions of nerve cells, both white (myelin) and gray matter (neurons), which cluster into numerous brain structures. How brain structures differentiate from the neural tube is primarily due to genes, and by all recent accounts, neural development is an incredibly dynamic and complex process. Moreover, there are hundreds of neurochemicals active in transmitting signals between neurons. Neurons and neurochemicals are of course made up of proteins, which are a direct result of genetic coding. Individual differences in structure and efficiency of these neural structures and neurochemicals must by necessity be at least partly a function of differences in genes and more specifically the form of the genes. Which genes are involved in building neural structures and neurochemicals is only now beginning to be discovered; the vast majority of these functions await future research. To the extent that intelligence is a result of brain structure, efficiency of neural transmission (mental speed), and the ability of these structures to process, retain, and integrate sensory input, then intelligence must be in part under genetic influence.

### Intelligence

Although somewhat controversial, the topic of genetics and intelligence is important to examine under the rubric of the biological psychology of science because intelligence is a necessary if not sufficient ingredient in scientific interest, ability, and achievement. Intelligence is a notoriously difficult concept to define, but a panel of experts in psychology recently defined it this way: "Individuals differ from one another in their ability to understand complex ideas, to adapt effectively to the environment, to learn from experience, to engage in various forms of reasoning, to overcome obstacles by taking thought." Similarly, a survey of 661 experts in psychology and education on the topic of intelligence reported the following components were consensually agreed upon by at least 50 percent of those surveyed: abstract thinking and reasoning, problem-solving ability, capacity to acquire knowledge, memory, adaptation to one's environment, mental speed, linguistic competence, mathematical competence, general knowledge, and creativity.[6]

There are three major techniques for assessing genetic influence on intelligence and behavior: the first pinpoints the quantitative trait loci associated with traits, the second maps brain structures related to intelligence, and the third estimates how heritable a trait is via twin studies. Regarding the genetics of intelligence in general, one of the only programs of research using the QTL method has been carried out by Michael Chorney and his colleagues. In 1998 they reported differences between high and low IQ groups on the frequency with which they had one allelic form of the growth factor gene IGF2R. This difference could account for but 2 percent of the variability in general intelligence, which reminds us that there are many, perhaps hundreds, of other genes responsible for this trait and that the IGF2R marker is but one of these. At least one attempt at replicating this finding failed.[7] Nevertheless, in this research we see the principles of quantitative traits and polygenesis.

A second technique, using mapping of brain structures to examine genetic influence, has confirmed the significant role that genetics plays in brain structure development and function. Paul Thompson and colleagues reported greater brain structure similarity in people of increasingly similar genetic affinity. Specifically, comparing identical to fraternal twins, Thompson and colleagues found nearly the same amount of gray matter (neurons) in identical but not fraternal twins in language and frontal cortices. These regions of the brain are closely associated with general intelligence. Moreover, Jeremy Gray and Thompson reported two other interesting findings: first, a modest correlation between intelligence and amount of gray matter (total brain volume), especially in the frontal lobes; and second, the amount of variance in intelligence that can be explained by genetic factors (that is, heritable) actually increases with age.[8]

A third and more established method for examining the role that genetics plays in intelligence has been twin studies, which do not look for specific genes or gene locations but rather at the amount of variability in a behavior that may be due to genes. The basic logic of twin studies is to parse the relative contributions of genetics and environment by taking advantage of the fact that twins are a natural experiment. Identical twins are 100 percent genetically alike and fraternal twins are 50 percent genetically alike, and yet both sets of twins usually grow up in much the same environment. If genes play a large role in a trait, we should see a higher correlation between identical twins on a given trait than between fraternal twins. By comparing similarity on a trait between identical and fraternal twins, researchers can get an estimate of how much of the variability in that trait is attributable to genetic factors. By this measure, the general conclusion from twin research on intelligence has been that about 70 to 80 percent of the variability in intelligence is due to genetics.[9] When one takes into account that two of the three subtests for IQ are science related (quantitative and spatial reasoning), one does not have to make a very large leap to conclude that a similar amount of the variability in scientific interest and aptitude is under genetic influence. This assumption has yet to be tested.

There is a phenomenon within genetics and intelligence worth noting: extreme precocity and genius—for instance, in math or music—most often do not appear to run in families. The question, therefore, that arises is, if extreme ability is to some extent genetically determined, how is it possible to have genius spring from nongenius families? As Bouchard and Nancy Segal argued in 1990, such innate genius may demonstrate the principle of "emergenesis," which is the idea that genetic influence is interactive and multiplicative rather than merely additive. In other words, some genetic traits are so complex and are made up of "a configuration—rather than by a simple sum—of polymorphic genes" that even though they are genetically influenced they are not likely to run in families. There is some evidence that creativity and genius are such complex traits. Similarly, researchers other than geneticists have argued for multiplicative models in explaining talent and genius that are quite consistent with the idea of emergenesis. In contrast to more traditional additive models, the multiplicative models argue that in order for talent to be manifested, any number of critical components must be present. Any one missing component exerts veto power over all other components (that is, it is a 0 in the equation) and no talent will be expressed. Such multiplicative models predict with a high degree of accuracy the exceptional and rare nature (that is, highly positively skewed distribution) of observed talent.[10]

## NEURAL DEVELOPMENT AND BRAIN PLASTICITY

For years and years scientists thought that the brain was unique in an important way: only it and the spinal cord have permanent cells, cells that generally do not die or regenerate. How else would we learn or remember if the cells involved in those processes kept dying? We now know that the brain is much more malleable and open to environmental influence; much of the wiring (neural connections) in the brain requires experience to be formed. Experience literally shapes neural connections and hence the brain. In other words, the environmental is not completely separable from the biological.

The study of neural growth (that is, "neurogenesis") has led to major new insights into how the brain grows and changes over time.[11] We now know, for instance, that new nerve cells can in fact grow in adults, and such growth is especially likely in the hippocampus, the brain structure most active in learning and memory. The underlying mechanism of such growth is the presence of neural stem cells. Researchers have documented that stem cells act initially as precursors to brain cells, divide, and then make a month-long migration to deeper regions in the hippocampus (and sometimes the olfactory bulbs), and about 50 percent survive to become fully functioning neurons. Various growth factors are involved throughout this process from stem cell birth to fully functioning adult nerve cell. As the neuroscientist Terrence Deacon has made clear, the process of neural growth is not prewired by genetic factors, nor are clusters of neurons terribly specific—they can end up doing many different things. Much neural growth is general and nonspecific with neurons growing at first in somewhat indiscriminate directions, then being guided by many different mechanisms, such as growth factors, cell adhesion, and synchronicity of neural firing. The nonspecificity and flexibility of neurons has been most powerfully demonstrated by "xenografting" experiments, in which cells from certain regions of the brains of fetal pigs are transplanted into other regions of the brains of adult rats, where they function almost as if nothing had happened! Moreover, early in development cells from one region of a brain (for example, frontal cortical-motor) can be moved to another region of the same brain (for example, occipital-visual) and can adapt completely to take on the new function.[12]

All of which leads to a second important principle of neuroscience: neurogenesis is Darwinian in nature. In other words, like all living things neurons compete for survival, and generally the strongest and best adapted to their environment survive while those least adapted die. Up through and following birth there is a tremendous explosion in the number of neurons in the brain. For example, in humans, neural volume peaks around age two to three. Ironically, up to a point, as we learn more and more we lose neurons. The brain becomes more efficient as it learns. Neural pruning, in other words, is closely associated with "learning." At the neuronal level, learning involves strengthening synaptic connections, and this occurs when certain neurons consistently fire together (synchronicity). Psychologist Donald Hebb first described this principle in the 1940s with a phenomenon that has been described with the phrase "cells that fire together, wire together." It is not too difficult to see how clusters of synaptic connections ("cell assemblies") that "win out" form the foundation for "learned associations" (although how precisely these neural clusters and "associations" become "ideas" and "thoughts" is still anybody's guess).[13]

What also makes this Darwinian nature of neurological growth so profound is its implication for how malleable and *plastic* the brain can be. The process of overproduction and then selective pruning lays the foundation for neural plasticity; by overproducing and selectively pruning neurons, the brain is molded by its specific environment. It is during this period—up through the peak in neural growth—that the brain is most "plastic," that is, when particular events and experiences can change density, organization, and amount of neurons in particular brain regions. It is also important to point out that although plasticity slows down after the period following birth, it does not stop completely over the life span.[14] Research on neural plasticity had its origins in the 1960s in Mark Rosenzweig's lab at the University of California, Berkeley. Rosenzweig and his colleagues conducted a series of rather simple studies in which they put one strain of genetically identical rats in one of two kinds of environments, namely, "enriched" or "impoverished." "Enriched" environment simply meant the rats were in cages with ten or so other rats and given exercise wheels and other objects of play. Rats in the "impoverished" conditions were housed alone in a small cage with nothing but food and water. Rats that received enriched environments for the first thirty days of their lives had a greater density of dendrites ("fingerlike" connections emanating from cell bodies that form synapses with other nerve cells) and greater overall cortical weights (about 5 percent). In short, environmental stimulation resulted in heavier brains with more neurons. Other researchers have since replicated and extended the Rosenzweig group's findings.[15]

Not only the environment but also the length of time the brain takes to mature is critical in its final number of neurons. The longer the brain has to form synaptic connections, the greater the final number of neurons, not just linearly, but exponentially. To bring home this point, neuropsychologists Barbara Finlay and Richard Darlington remarked that having an additional seventeen doublings of precursor cells can yield more than 130,000 more neurons. Humans have the distinction of needing the longest time for brain development, a phenomenon some have dubbed "elongated immaturity." Our brains grow more both in absolute and relative amount after birth than those of any other species of animal. Indeed, we are born with a relatively low percentage of our adult brain size (23 percent), whereas the average for great apes (orangutan, chimp, and gorilla) is right at 50 percent. There is an advantage to being defenseless and immature for such long periods of time—our brains can continue to grow and be shaped by their environment. In humans, evolution has indeed created a species whose brain development allows for tremendous flexibility and adaptation to whatever environment they find themselves in.[16]

Taken as a whole, I contend that these phenomena (pruned neurogenesis, Darwinian neural growth, plasticity, and elongated immaturity) go a long way toward explaining why the human brain is capable of symbolic, abstract, and metacognitive thought. Once we add language, written thought, some basic mathematical development, and a culture that sees the need to test ideas systematically and explain things naturally, we have the essential ingredients of science. The human brain does indeed build theories to explain its experiences. To be sure, this is simply what brains of all species do: organize and interpret sensory experience. The human brain, however, has a uniquely developed frontal lobe that in large part is what allows the unique human cognitive capacities of integrated, abstract, symbolic, and metacognitive thought. In the end, science, in the modern sense of the word, is simply an extension—albeit not inevitable—of the most rudimentary and basic capacities and functions of the brain, namely, organizing sensory experience.

## NEUROSCIENCE AND THE ARCHITECTURE OF THE HUMAN BRAIN

In addition to the general processes of neurogenesis, plasticity, and elongated maturity, there are many specific brain regions that are most active in the domain-specific tasks involved in folk psychology, physics, biology, and math. Moreover, recent evidence from neuroscience is showing us how the frontal lobes and the two major hemispheres are somewhat specialized and perform distinct functions involved in the different folk scientific domains of thought.

### Brain Specificity

*Folk Psychology.* The evidence of brain specificity in the folk psychological domain comes from many different sources, but two of these—face recognition and theory of mind—have garnered the most attention. Because particular human faces and their memories hold a special place in our hearts, they have become the object of many poems and songs. Scientists are now trying to determine where in our brains these perceptions and memories of human faces are occurring. As it turns out, the findings of these studies are consistently showing that the right temporal lobe directly above the right ear (fusiform gyrus) and the frontal lobes are where we process familiar faces; the fusiform gyrus is where we process unfamiliar faces; and the amygdala is where we process emotion (especially fear) in facial expression. People with damage to the fusiform gyrus often are unable to recognize faces, whether familiar or unfamiliar, a condition known as prosopagnosia. Systematic brain imaging technology has provided further evidence of this region's activity in face recognition. As is true of many brain developmental processes, early experiences shape the neural growth of these structures, "whereby exposure to faces during a sensitive period of development likely leads to perceptual and cortical specialization."[17]

Theory of mind (TOM) is a capacity to recognize the internal states of other people and to attribute their behavior to these states. It also involves being able to use and detect deception, self-awareness, and perspective taking. Imaging research on theory of mind has demonstrated the four brain regions most involved in the acquisition and maintenance of social knowledge, namely, the orbitofrontal cortex (OFC) and the anterior cingulate of the prefrontal cortex (PFC), the amygdala, and the anterior temporal lobes. The evidence for anterior cingulate cortex and the OFC will be reviewed in the section on frontal lobes (below). Evidence for the amygdala suggests that it becomes especially active when people are asked to judge the mental states of others by looking into their eyes as well as when people view increasingly fearful expressions on other people's faces.[18] People with damage to their amygdala can sometimes distinguish different emotional categories (for example, sad and happy), but they appear to be incapable of using this information in making social judgments. In other words, the amygdala is involved in interpreting the emotional meaning of different facial expressions.

Two regions of the temporal lobes, the anterior temporal (AT) region and the superior temporal sulci (STS), both closely interconnected with the OFC, have also been implicated in theory of mind functioning by numerous brain imaging studies. Along with the OFC, the AT and STS regions of the temporal lobes are related to social-emotional behaviors and their dysfunction, such as theory of mind, face recognition, self-awareness, impulse control, irritability, and antisocial and violent behavior.[19]

*Folk Physics.* Folk physics consists of multiple abilities, such as object recognition, understanding the causal relationships between objects, mechanics, spatial ability, tool use, and navigational ability. Of course, recognition of objects is simply part of the general visual system, and vision is a very complex brain activity that involves many brain regions, but vision is mostly located in six specific areas (that is, V1 to V6) of the occipital lobe. The general layout of the visual cortex is as follows: V1 is the primary visual cortex involved in general scanning, V2 in stereo vision, V3 in perception of depth and distance, V4 in color perception, V5 in motion, and V6 in objective (rather than relative) position of objects. For such physical tasks as mechanics, spatial ability, and navigational ability, V3, V5, and V6 seem to be especially important. Spatial and navigational ability also involve subcortical structures in the middle of the brain, namely, the hippocampus and the parahippocampus, as well as parts of the parietal cortex and basal ganglia. Researchers, for instance, have reported that the parahippocampal cortex or parahippocampal place area (PPA) is involved in perceiving photos of indoor or outdoor scenes but not of faces or objects in general.[20]

*Folk Biology.* Knowledge in the folk biological domain concerns the animate-inanimate distinction, landscape evaluation and preference, and plant and animal taxonomies and behavior. There is something intuitive and automatic about the distinction between the animate and inanimate. After all infants make such a distinction, but neuroscientists have yet to examine more systematically the brain regions that are involved. In one of the few studies of its kind, Rosaleen McCarthy and Elizabeth Warrington report that a sixty-three-year-old man with left temporal lobe damage had grave impairment in his knowledge for living things but not for inanimate objects.[21]

Folk taxonomy is also a general categorization function, but categorizing the world into plant and animal is not enough for successful hunting and gathering. To be efficient and successful food gatherers we need to make many further distinctions within each of these categories, and developing detailed and accurate knowledge of these local groupings is exactly what folk biology is all about. The main brain region involved with general categorization is the frontal lobes, but the processing of stimuli first as plant and/or animal involves integrating information from the visual, auditory, tactile, and olfactory sensory cortices. The integrative and complex nature of processing such stimulation is one reason why research on specific brain regions for folk biology is not as advanced as it is for the other folk domains.

*Folk Math.* Folk math involves ability with seriation, number, and calculation. The general conclusion concerning number-math processing and the brain is that the parietal lobes are the primary brain region involved, in particular the angular gyrus of the left parietal lobe. The neuropsychologist Brian Butterworth describes clinical evidence for this, involving an Italian woman who suffered a stroke to her left parietal lobe area and thereafter could not recognize or reason about any number greater than four. Even quantities from one to four gave her trouble: she could not look at three objects and immediately tell you there were three; she had to count them one by one. Another man who suffered a left parietal lobe stroke had systematic problems with subtraction. Other researchers have reported brain-damaged patients who can add and subtract and multiply simple numbers but cannot multiply multiple digit numbers. For instance, one patient was multiplying $45 \times 8$ (on paper, with the 8 below the 45, but between the ones and tens column) and came up with 1213. This person clearly added 8 to both 4 and 5 rather than multiplying ($8 + 5 = 13$ and $8 + 4 = 12$). Neuroimaging research has confirmed the importance of the left parietal lobe, in particular the angular gyrus region, in mathematical reasoning and has also added the left prefrontal lobe for complex calculations and the bilateral temporal lobe for solving math problems, at least in men if not for women.[22]

### Frontal Lobes

Capacities that make us most human—planning, abstract thought, executive control, attention, control of impulse, complex theory of mind, creative combination of ideas, self-awareness, and consciousness—all require a high degree of frontal lobe activity. Such executive and integrative functions are at the foundation of our ability to make creative mental associations as well as engage in metaphorical and symbolical thought, or what Steven Mithen calls "cognitive fluidity." Supporting this view is the finding that the front-most region of the frontal lobes (the prefrontal cortex) "is arguably the best connected of all cortical structures." As Donald Stuss and others have pointed out, the brain region above and behind the eyes are what make our species unique, being the seat of the "higher reaches" of human nature, namely, consciousness, creativity, personality, and morality.[23]

Recently "biologists have begun to find new evidence for the idea that developmental processes—the complex mechanisms by which an individual organism grows to its full size and form—can provide a window into the evolution of a species' anatomy." Frontal lobe development is one area that provides such evidence, for this part of the brain is the last to develop ontogenetically (not reaching full maturity until adolescence) as well as phyologenetically (not reaching their current size in humans until the emergence of modern humans roughly 150,000 years ago).[24]

The forward-most region of the frontal lobe, the prefrontal cortex, has three subregions, the dorsolateral (that is, the back side), the orbitofrontal (lowest section above the eye sockets), and the paralimbic (innermost region around the limbic system) regions. The dorsolateral area of the prefrontal cortex is involved in higher-order thought (maybe even consciousness) and executive functioning, while also integrating specific information from other brain regions. The paralimbic system includes the anterior cingulate cortex, a region that numerous studies have shown to be active in theory of mind. Lastly, the orbitofrontal region is most intimately involved in social knowledge and skills, such as self-awareness, empathy, perspective taking, and deception detection. For instance, patients with frontal lesions often are unable to pick up on facial cues of emotion, being oblivious to rather obvious signs of distress or anger or simple cases of deceit. Additionally, the fact that TOM (false-belief) tasks can be performed by four-year-olds but not three-year-olds is consistent with a possible growth spurt in frontal lobes around age four. In other words, the anterior cingulate and OFC may well be a major focal point for our "theory of mind."[25]

It may have first become clear with the case of Phineas Gage in the 1840s that the frontal lobes have something to do with social skills and impulse control, but this certainly was not the last piece of evidence. Hundreds if not thousands of other case studies, as well as systematic laboratory investigations, have been reported in the clinical literature, especially in the realm of the attention deficit disorders (ADD and ADHD), schizophrenia, and antisocial and sociopathic cases. The major behavioral pathologies of the OFC are the failure to concentrate, cognitive disinhibition, and lack of impulse control. Other intriguing, if somewhat disturbing, social implications of frontal damage are seen in extremely violent and antisocial actions. In neuropsychological evaluations of violent criminals, frontal lobe damage early in life is almost always observed in conjunction with abusive environments and psychiatric illness. Volumetric assessments of prefrontal gray matter show significant reductions in men with antisocial personality disorder compared with normal controls. These findings suggest not only plasticity of the brain, in that abuse leads to less neural volume and neural connectivity in the brain, but they also suggest that a deficit in one's ability to empathize or take another person's perspective, coupled with extreme impulse control deficits, contributes most directly to violent and antisocial behavior.[26] These abilities and disabilities are most localized in the frontal lobes.

As maladaptive as some expressions can be, there are, to be sure, some adaptive expressions of disinhibition, most important of which is creative thought. Going back to the early 1950s, most major theories of creativity have included looseness and remoteness of association, disinhibited thinking, and fluency of ideas.[27] Although none of these theories has been explicit in locating creative cognitive functioning in the frontal lobes of the brain, each is quite consistent with what neuroscience has learned about frontal lobe functioning over the last ten years and its central role in executive planning, inhibition of both thought and behavior, abstract reasoning, attention, and memory. A few researchers have examined directly the connection between frontal lobe activity and creativity. For instance, Tiffany Chow and Jeffrey Cummings reviewed neuropsychological evidence that demonstrates loss of creative thought, impaired set shifting, and an increase in stimulus-bound behavior as a result of dorsolateral and anterior cingulate lesions in the frontal lobes. Bruce Miller and his colleagues reported a case of woman with frontal-temporal dementia whose paintings went from amateurish to quite sophisticated and more creative. One explanation offered by the authors was that certain forms of visual creativity may be somewhat inhibited by the language regions in the left frontal-temporal area and these inhibitions were removed with the dementia. Although coming from a cognitive rather than neuroscience perspective, a recent study reported decreased latent inhibition (that is, disinhibition) in highly creative and intelligent students relative to less creative students. Others have demonstrated that latent inhibition is primarily frontal lobe activity. Some research has examined frontal lobe activity in science directly and found that frontal lobe development may be a precondition for scientific reasoning.[28]

### Hemispheric Laterality

One of the oddest facts about the brain is that its two hemispheres carry out somewhat distinct functions, at least in certain domains. Why the human brain would evolve to have two hemispheres that are somewhat distinct in function may at first glance be a puzzle, but in the context of many natural phenomena such laterality is not that rare or unexpected. Our bodies in general and faces in particular are subtly but clearly asymmetrical. As it turns out, in addition to the well-known association between the left hemisphere (LH) and language, there are some other more recently uncovered differences between the hemispheres: anatomical (with the frontal lobes of the RH being larger than the LH and the reverse being true of the occipital lobes), neurochemical (the LH has more dopamine, the RH has more norepinephrine), and cognitive. The RH is more active when processing novel, diffuse, heuristic, and global (early-stage) information, whereas the LH is more active when processing routinized, analytic, and focused (late-stage) information.[29]

The distinction between analytic-focused processing and heuristic-global processing is quite relevant for reasoning in general and for scientific reasoning in particular. Beginning with Sigmund Freud, many theorists and researchers have argued for the importance of dual systems of processing information or two modes of thought along the analytic-heuristic continuum. Applied to scientific reasoning, researchers such as Paul Klaczynski and Deanna Kuhn have argued that people who are able to reason scientifically (that is, metacognitively and distinguish theory from evidence) are better able to use focused, analytical reasoning strategies compared to people who are not able to reason scientifically. Science, therefore, may involve well-developed LH functioning, especially of the frontal lobes. Albert Katz has been one of the only researchers to directly investigate laterality and scientific thought. Using a pen-and-paper test of hemispheric dominance, he reported more LH dominance in scientists and mathematicians relative to architects. Additionally, research on subitizing (sudden enumeration of smalls sets of objects) shows a left hemisphere advantage (and an RH advantage for the more complex task of counting). Recall, too, the neuroscientific evidence implicating the left parietal lobe in mathematical reasoning. This confirms the logical-mathematical and more sharply focused nature of the LH. But the evidence on this is not always consistent. For instance, based on a high incidence of left-handedness in the mathematically precocious and in particular the precocious males, and the greater bilateral or diffuse cognitive functioning of left-handed individuals, Camilla Benbow concluded that bilateral and/or a strong right hemispheric functioning may be involved in extreme mathematical ability. Other researchers have also found the right hemisphere is active in the abstraction of numerical relations.[30]

Scientific problem solving, however, involves not just mathematical-logical thought, but in the early stages of creative and intuitive insight, it also involves the more diffuse, visual, nonverbal forms of thinking.[31] The early stages of insight, problem solving, and hypothesis generation require the diffuse and coarse approach of the RH, whereas the later verification stages and hypothesis testing require the sharper more analytic LH. Science involves both creative and intuitive insight as well as logical-mathematical analytic thought. The greatest scientists have abilities in both of these domains. In our discussion of creativity we saw that integration of loose and yet adaptive ideas is commonplace. Such integration of the sharp and focused with the fuzzy and diffuse is a hallmark of creative people, including creative scientists.

In sum, principles and discoveries from behavioral genetics (genetics and intelligence) and cognitive neuroscience (neurogenesis, brain plasticity, frontal lobe functioning, and hemispheric laterality) have begun to shed light on the structure and function of the human brain in general and even somewhat on scientific and mathematical thought in specific. Indeed, a few principles of the genetics of intelligence are now in place: about 80 percent of the variance in intelligence is genetic; brain structures are more similar in people who are more genetically alike; and genius-level ability appears to be the result of multiplicative rather than additive genetic influences. It is also increasingly clear from neuroscientific evidence that specific regions of the brain are predominantly active in thought, reasoning, and problem solving within folk scientific domains of mind, namely, psychology, physics, biology, and math. Although brain-imaging technology has revolutionized our understanding of the brain regions responsible for different kinds of perceptual and cognitive activities, it is important to underscore that there is usually no one-to-one correspondence between anatomy and function because many integrated brain regions are active with each specific task.

Again, I am not arguing that these brain structures evolved to do math and science, or that those structures that solve math and science problems are unique to math and science. The brain's basic function is to organize and make sense of sensory experience. To the extent that specific parts of the brain make sense of different domains of sensory experience and solve problems of intuitive and folk psychology, physics, biology, and math, the neuroanatomical foundation for these sciences is clear. It took much else, however, for these intuitive and implicit forms of thought to become explicit and formal math and science.

# Chapter 3 Developmental Psychology of Science

Developmental psychology is very well suited for studying scientific behavior and thought because scientific reasoning, interest, and achievement do develop; they do not just come out of nowhere. Developmental psychologists are good at studying not only change, but also the kind of change (linear or nonlinear, positive or negative), the mechanisms of change, and the structure of the change. Scientific thought and behavior in individuals do change and evolve with age. Developmental psychology in general and developmental psychology of science in particular has quite a story to tell concerning how people reason about science and math, how they construct theories about the world, how they test hypotheses, how they evaluate evidence, and how all of these change over an individual's life course.

The field does indeed have much to say about the development of scientific interest, thought, and behavior, even if much of it is often more implicit than explicit. One reason why many scholars—even developmental psychologists of science—do not recognize the value of the developmental psychology of science is that they fail to acknowledge very implicit cognitive and behavioral *processes* as components of science. Recall the definition of science and scientific thinking put forward in the preface: the process of developing and testing mental models of how the world works, be it the physical, biological, or social world. The essence of these mental models is coordinating theories (models) with the evidence (data). Specifically, this is a process of observing events, recognizing patterns, testing hypotheses, and making causal connections between the observed events. Infants, children, adolescents, and adults all do this but with varying levels of explicitness and metacognition. Once the implicit processes are incorporated under the purview of science, the whole field opens up and becomes a very rich and vital source of knowledge concerning the developmental origins of scientific behavior.

Because infants and children (and even adolescents) are not yet scientists, all developmental psychology of science through late adolescence is concerned with implicit science and therefore is implicit psychology of science. At a very general level, developmental psychologists since Jean Piaget, the most famous of all developmental psychologists, have been implicit psychologists of science because they have been interested in the extent to which children act and think like scientists, an approach that has become known as the "child as scientist metaphor." There is, in addition to the "implicit" developmental psychology of science, the more "explicit" developmental psychology of science that we see when developmental psychologists start to focus on adolescents and adults and address the questions of how and why certain individuals become interested in science and ultimately become scientists.

## IMPLICIT DEVELOPMENTAL PSYCHOLOGY OF SCIENCE

### Domain Specificity: Children as Incipient Scientists

How do completely helpless human infants become such sophisticated perceivers and knowers of their social, physical, biological, and linguistic worlds in such short order? By three or four months of age, for instance, the human infant is more interested in faces than in almost any other object and understands that physical objects cohere and cannot pass through each other, that some things are animated and some things are not, and that changes in focal intonation mean different things. The answer comes from having a mind that evolved over millions of years to solve these problems and therefore comes into the world predisposed with perceptual and cognitive constraints that stem from first principles of perception and theory construction. These predispositions facilitate quick and efficient learning and understanding of the social, physical, and biological worlds. Humans, in other words, starting in infancy, are surprisingly sophisticated folk psychologists, physicists, biologists, and mathematicians—even if many "commonsense" theories turn out to be limited or wrong.

*Child as Folk Psychologist.* The developmental criterion by which a purported domain of mind must be evaluated is whether the capacity in question develops relatively automatically and spontaneously in children. In other words, for a domain to exist infants and children must perform certain domain-specific tasks spontaneously, with little or no training. Capacities that develop automatically and spontaneously suggest a nervous system with a built-in tendency (that is, evolved) to do such things.

Spontaneous and automatic ability has been demonstrated in the social-psychological domain. The first cluster of folk psychological skills revolves around our ability to discriminate and prefer certain facial configurations and facial expressions over others. Even when nearsighted, newborns less than two months old distinguish between people (faces) and things and prefer the former to the latter. The human face seems to be intuitively and automatically of interest to us, some would even say it is imprinted.[1] For discrimination, researchers habituate the infant to a happy or sad face and then show them that same face. The infant is not interested. But show a different emotional expression (for example, sad if habituated to happy), and the newborn will become interested again, that is, will look longer at the new face. At as early as four months and quite consistently after six months, infants can discriminate certain facial expressions and can perceive changes in emotion.

Newborns are also somewhat physiologically disturbed when the face does not have the right configuration, such as eyes on the bottom and mouth on top. Moreover, newborns show preferences for some expressions over others, and even more remarkably, they prefer attractive to less attractive faces. They are obviously not being told or reading in a magazine that they should look longer at the attractive face, and yet this is what they do quite consistently. Moreover, what is attractive is not nearly as arbitrary and culturally specific as many people may think. The standards of human attractiveness are quite consistent across cultures, arguing against the maxim that "beauty is in the eye of the beholder." A rather universal finding is that faces with average configurations, for example, eye size and spacing, nose size, and so on, are deemed consistently the most attractive.[2]

In addition to the built-in fascination and preferences we have for the human face and its configurations and expressions, we develop other social-psychological abilities early in life, although later than the newborn facial discrimination and preference skills. Two of these are "perspective taking" and "false belief." The ability to have a cognitive representation of other people's perspective, thoughts, or feelings appears to develop around the age of four.[3] For example, at around two years of age, children explain and think about other people's behavior only in terms of perception and desire. That is, three-year-olds always explain why someone else does something in nonrepresentational terms of want or desire (for example, someone else either wanted X or did not want X) and never in terms of what the other person may believe or not (representational). Four-year-olds, on the other hand, begin to explain why someone did something in terms of their beliefs (for example, someone else did X because *they thought* they wanted it). Such behavior implies an ability to imagine oneself from another person's perspective.

Representational belief is also seen in the false-belief task. Joseph Perner and colleagues conducted a classic study of the development of false belief in humans.[4] A child is shown a box of candies and asked what is inside. Unsurprisingly, the child says "candies." Next, the experimenter takes out the candies and places a pencil inside the box. The child sees this and is told that a playmate, Bobby, is going to come into the room. The experimenter then asks the child the critical question: what do you think Bobby is going to say is inside the box? In order to answer this question correctly (candy), the child has to know that Bobby can hold a belief different from her or his own and different from what she/he knows to be true. If the child is a three-year-old, she will consistently answer incorrectly and say "pencils," unable to disentangle her own belief from Bobby's. If she is four years old, however, she will consistently answer "candy." She is able to hold a belief of what someone else will believe even though she knows it to be wrong! That is a rather complex cognitive task and as far as we know, only humans can do such a thing—and we can do it at four years of age!

Another dramatic source of evidence for a nativist view of a theory of mind is the fact that these social-psychological skills and preferences are so lacking in autistic children. Autism is first and foremost a disorder of social functioning and language, most often accompanied by repetitive behavior and limited skills of imagination.[5] Delayed and disordered language and cognitive functioning, along with repetitive behavior and little or no joint attention, are the hallmarks of autism. Autistic children seem to be limited to behavioral forms of perspective taking and joint attention, that is, they tend to try to change another person's behavior not his or her thoughts.

*Child as Folk Physicist.* Why do magic tricks entertain us so? Human bodies float or get sawed in half. Solid objects pass through other solid objects and back again. Birds and rabbits appear from thin air or from a hat much smaller than their body. None of these things of course can or does happen, and we know this even as we see them apparently being done. But we love to think they are happening before our very eyes. They entertain us so because they play with and appear to violate the most fundamental principles of our intuitive understanding of physics. What is equally amazing is that children, sometimes very young children, especially love these tricks because they too have the same intuitive understanding of the physical world, what's possible and what's impossible.[6]

Knowledge of the physical world (objects and their permanence, movement, causation, and spatial arrangements) also develops automatically, spontaneously, and early in infancy. Regarding the permanence of objects, Piaget was of course one of the first to systematically investigate the development of children's understanding of the physical world.[7] He argued that in the first year the infant understands the world only through operating (manipulating) objects and experiencing them through the senses. Knowledge is strictly and directly sensory. If an object is not being sensed, it does not exist ("out of sight, out of mind"). There is no cognitive representation independent of sensation, and this was the basis for Piaget's well-known concept of object permanence. Because infants are tied to the sensory, they will not grab for an object that has just disappeared from their view. According to Piaget, it is not until around the age of nine to eleven months that internal representations of the external world start to develop, that is, object permanence.

Piaget's assessment of cognitive ability during the first year of life was primarily motoric—whether the child continued to reach for an object just hidden from sight. But psychologists and neuroscientists had known for years about the principles of "center-out" and "head-to-toe" development of motor skills. That is, motoric command at the center of the body matures before the extremities and the head before the arms, which mature before the legs and feet. It was not until the 1970s, however, that developmental psychologists started to take advantage of such knowledge and developed assessment techniques that involved movement of the eyes and mouth rather than the arms and hands as a measure of interest and/or surprise. One of the major figures who has been using eye-preference technique and has come to challenge some of Piaget's basic arguments is Elizabeth Spelke formerly of MIT and now at Harvard. She argues that principles of cohesion, contact, and continuity guide and constrain perception and understanding of physical objects in early infancy. Four-and-a-half-month-old infants already understand how objects are distinct from one another and how they move about in space based on principles of cohesion, contact, and continuity. Single objects are recognized if their parts are connected, if their surfaces are in contact, and if the object moves continuously over one path and one path only in space and time. Spelke, also one of the pioneers in the now ubiquitous preferential eye-gaze technique, showed that infants look much longer at "impossible events," such as two objects apparently taking up the same space. Furthermore, she demonstrated that these principles override and develop before the Gestalt principles of perception, such as similarity, closure, and simplicity.[8]

One of Spelke's students, Rene Baillargéon, challenged Piaget's claim that the onset of object permanence occurred around nine to twelve months of age, arguing for around four months.[9] She did this by first habituating four- and five-month-old infants to a screen that moved in a 180-degree fashion, from flat pointing toward the child to flat pointing away from the child. Then, in clear view of the infants, a box was placed in the path of the screen (at the point where the screen reaches 112 degrees of its arc; it should also be pointed out, that the infants were positioned in such a way that once the screen reached 90 degrees, the box was completely occluded from view by the screen). Then half of the infants saw a perfectly normal event: the screen stopped when it got to the now unseen box. But the other half (who had not seen that an adult surreptitiously removed the box) saw something that was "impossible," namely, the screen continued to move all the way past the box at the 112-degree mark until it was perfectly horizontal. If Piaget were correct, and four-month-olds possess no object permanence, they should not be at all surprised by this "impossible" event. After all, the box was out of sight and therefore out of mind. But they were surprised (by the eye-gaze attentional method), which suggests they do have some degree of understanding that objects still exist after their direct perception. That is, they have some kind of mental representation of the object.

*Child as Folk Biologist.* Besides the physical world, another part of the world that our mind seems automatically attuned to and develops knowledge of is the biological world—geography, landscapes, flora, and fauna. We intuitively understand that some things are alive and others are not, and early on in life we develop an implicit understanding for what it means to be alive and how to categorize the natural world into useful nonarbitrary distinctions. The most basic distinction is between animate and inanimate. One simply has to look at children's literature and cartoons to know that animals play a major role in their representation and understanding of the world.

By the age of three, children implicitly, if not explicitly, understand at least two principles of what makes something alive: possession of innards (even if they have misunderstandings of their actual function) and self-propelled motion. These two things are the sine quo non of all living things, and children know this at a very young age.[10] Moreover, Frank Keil has argued for four or five other distinctive properties of living things, such as reproduction, growth, and interconnection of working parts. He and his colleagues have conducted a series of experiments with children of different ages and shown that children have distinct theories concerning growth, inheritance, reproduction, disease, and contagion. For instance, children as young as four years of age judge that certain psychological states, such as fear of disease, are not contagious. In addition, some research in the early 1980s showed that children as young as three do not attribute animal properties—such as breathing, moving, thinking—to inanimate objects, only to animate ones. Also three-year-olds cannot consistently describe the brain as being in the head, but by age four they know that the brain is what makes thoughts, dreams, and knowing the alphabet. Four-year-olds, however, still do not think the brain is necessary for telling a story, making a face, or picking up a glass. In short, children understand the brain is for mental events, but they do not intuitively understand its role in sensory or motor activity, an understanding that usually develops only in adolescence or later.

In the mid-1980s the developmental psychologist Susan Carey published one of the first books on the development of children's understanding of animate-inanimate and in particular the concept of "animal."[11] She found that for children, "animal" is distinct from "plant" primarily because they attribute human properties, such as breathing, eating, and moving, more to animals than to plants. In addition, Carey investigated knowledge of animal properties in four-, five-, and seven-year-olds, as well as in adults. She asked them whether unfamiliar (at least to the children) objects had animalistic properties, such as breathing, sleeping, getting hurt, having a heart, or eating. The animals were humans, aardvarks, dodos, stinkoos (stinkbugs), and hammerheads (and for older children and adults, annelids [worms] were added to the category); plants were orchids and baobabs; and inanimate objects were harvesters, garlic presses, volcanoes, and clouds. One overall finding was that young children underattributed animal properties to nonhuman animals. For example, four- and five-year-olds say that aardvarks do not breathe or even eat. Young children were also less likely to make the implicit distinction between vertebrates and invertebrates by attributing, for instance, more mental activity to insects than sharks. With these underrepresentations, young children are much more likely to attribute animalistic qualities to animals than nonanimals, even in objects completely new to them. Four-year-olds, for instance, attributed animalistic properties (eating, breathing, sleeping) to animals between 58 percent of the time (stinkoo) and 100 percent of the time (humans) but only 0 to 3 percent of the time for inanimate objects like harvesters or garlic presses. Children under the age of seven categorize animals along prototypic class lines (for example, mammals and reptiles) and exclude such things as people and insects. Preschool children, therefore do not use the word "animal" and "alive" interchangeably. In short, young children are rather clear on the distinction between animate and inanimate.

More recently Rochel Gelman has argued that there are first principles specific to each domain that constrain early perception and understanding, for instance in number and biology (animate-inanimate).[12] The principles and constraints facilitate fast acquisition of knowledge and are automatic, implicit, and most often nonverbal. They answer the question of how very young children can focus on relevant aspects in one domain and ignore the irrelevant ones. When shown objects that none of them had ever seen before—a displaying lizard, an echidna, a mythical vessel, an insect-eyed figurine, or an old-fashioned bicycle—and asked whether each object could move up and down a hill by itself or not, three- and four-year-old children invariably got things right: animals they had never seen before and that may not look like any animal they had seen (for example, echidna) could move up a hill by themselves, whereas inanimate objects, even if it had feet (for example, the figurine) could not. The principle that animals have self-propelled motion and physical objects do not seems well ingrained, and even young children have the ability to extract it from perceptions new and unfamiliar to them.

*Child as Folk Mathematician.* The development of math in any higher sense of the word is neither automatic nor spontaneous in infants and children. But the concept of number and seriation is. As with most topics in the development of cognitive ability, Piaget was the first to systematically tackle the problem of how the concepts of number and math develop. Based on a simple task that has been replicated thousands of times, his now classic finding was that before the age of five, children lack the ability to conserve number. The task involves placing two series of, say, five objects in front of the child. The first time through the objects are paired and their one-to-one correspondence is quite obvious. Ask the child whether one row has more or if they are the same, and the child quickly says they are the same. Next, spread one row of objects out a little and ask the same thing, and the four-year-old almost invariably gets it wrong: the spread-out series has more. Even this classic finding, however, has been modified because experimenters neglected to consider the absolute amount in each series. Once they used a small number of objects, less than five, the child's capacity for numerosity became clear. For instance, when only three or four objects are used, children under age five do conserve number quite consistently.[13]

Moreover, conservation is a relatively complex ability and clearly not the only one relevant to numeric ability. Recall the argument of first principles by Rochel Gelman and her colleagues in the previous section. Innate first principles constrain and focus attention within domains, and number is one of these domains. Evidence from two phenomena other than conservation supports the first principles argument: discrimination of quantities and "subitizing." For discrimination, researchers have used visual preference (eye-gaze) methods and have shown that immediately after birth (one- to six-day-old) newborns become interested in changes to small numbers but not to changes in larger numbers.[14] For instance, if two things become three a newborn will show interest, but if six things become seven she will not. The most compelling implication about this finding is that the researchers controlled for line length and density and were able to show that number only affected neonates' interest.

Another demonstration of ability rather than disability with number in very young children is seen in "subitizing" or the ability to know very quickly (it literally means "arrive suddenly") how many objects are in a limited series. One way this ability is tested is by showing a series of two or three or four or five dots for a period of time below counting threshold, say 200 milliseconds (ms), and asking how many dots there are. Recent research has shown that the subitizing range increases with age, with infants being able to accurately subitize one to three objects presented at the 200 millisecond range; three- to five-year-olds, one to four objects; and adults, one to five objects. There is some debate, as there is with object permanence, about whether subitizing is a low-level perceptual process or a higher-level cognitive heuristic.[15]

As amazing as it is that newborns and infants can subitize quantities and detect changes in quantity, an even more fascinating finding is they may be able to reason about number. Karen Wynn has been the primary researcher responsible for such work, and again the method used is eye-gaze, based on the assumption that gaze represents interest and on the evidence that infants look longer at events and objects that violate their expectations.[16] If infants can reason as to how many objects should be there, they will look longer at a series that is inconsistent with their expectation. After habituation trials, infants as young as five months do in fact look longer at series that violate their expectations.

The general design of the studies is simple: put a doll figure in a box stage in view of the infant, then have a screen come up that blocks the view of the doll but has a side opening that allows the infant to see a human hand adding a second doll. One of two things will happen next: either two dolls are revealed when the screen drops or only one doll is revealed. Infants look longer (that is, show more interest) when the event violates their expectations. Wynn argues that this is, in a very real sense, implicit arithmetic and is the developmental origin of formal adding and subtracting that comes a few years later. Moreover, she and others in the field argue that these capacities are a naturally selected (evolved) mental "number mechanism" found throughout the animal kingdom because they have provided adaptive benefits.[17] First, they allow animals to calculate and keep track of food returns for different locations, and second, they allow animals to keep track of other animals over space and time.

If these basic principles of subitizing, counting, and numerosity have evolved, one would be hard pressed to argue that formal mathematical ability (algebra, geometry, calculus, and so on) is an evolved ability—it is much too recent historically and archeologically (being only a few thousand years old at best), is not universal among the cultures, and does not develop automatically without cultural training and transmission. This poses the difficult evolutionary question of whether math is an evolved adaptation or not. It also poses the difficult developmental question of how individuals get from the intuitive and implicit forms of a sense for and reasoning about number to formal mathematics. Obviously, many people find this transition quite difficult and by looking at the percentage of "math phobic" adults, one could say many do not make the transition very well. As Rochel Gelman and Kimberly Brenneman argued, innate first principles indeed can hinder as well as facilitate learning.[18]

The answer to these problems lies in the distinction between an "adaptation" and a "co-opted adaptation" or "by-product" of an adaptation.[19] Numerosity is an evolved adaptation, but formal mathematical reasoning is co-opted. First principles in math and other domains may in fact have evolved and be adaptations, but they are just that, first principles, and hence the beginning, not the end, of what we know. Although first principles may be the foundation on which formal mathematical reasoning is built, they may collide with formal mathematical principles, and this is where the difficulty that so many people have with math arises. For instance, the concepts of zero, fractions, and irrational and imaginary numbers do not stem directly from our number sense and indeed contradict it. Much more has to happen after the first principles have guided our attention. As Stanislas Dehaene put it: "Mathematics consists in the formalization and progressive refinement of our fundamental intuitions. As humans, we are born with multiple intuitions concerning numbers, sets, continuous quantities, iteration, logic, and the geometry of space. Mathematicians struggle to reformalize these intuitions and turn them into logically coherent systems of axioms." In other words, mathematical first principles are part of human nature, whereas higher mathematical reasoning is more culturally determined. To demonstrate how this process might work, Steven Pinker borrowing from Saunders MacLane argued that most every intuitive mathematical skill has an analogue in formal mathematics: counting → arithmetic and number theory; measuring → real numbers, calculus, analysis; shaping → geometry, topology; estimating → probability, measure theory, statistics; moving → mechanics, calculus, dynamics; calculating → algebra, numerical analysis; proving → logic; puzzling → reasoning ability, spatial reasoning ability.[20]

### Domain Generality: Development of Scientific Reasoning

In addition to the specific domains of implicit scientific thought in psychology, physics, biology, and math, there are the more domain-general reasoning skills that deal with science as a process of acquiring and revising knowledge, such as theory revision and conceptual change; coordinating theory and evidence; hypothesis testing; and using heuristics, metaphor, and analogy to solve problems. As is true for the domain-specific research with children, these psychological processes have been studied quite extensively, yet most often implicitly without scientists themselves being the object of investigation. Developmental psychologists are in a unique position to unpack two crucial questions about scientific reasoning in general: what are the mental components necessary for scientific thought and how do these develop and change from early childhood to adulthood?

The starting point for our discussion on general knowledge and the development of scientific reasoning in children is a brief summary of Piaget's seminal work on the topic. When it comes to the development of scientific reasoning, Piaget was clearly a pioneer. He offered a milder form of recapitulation, namely, that the child's development of scientific thought recapitulated the history of science. Children, in this view, begin with a kind of Aristotelian view of how the world operates, and—if they reach the highest level of formal operations—end up with a Newtonian or perhaps even an Einsteinian view, one they have internalized, not merely memorized. Piaget not only examined how children reason about number and physical entities, but he also proposed a general model of cognitive development that had as its final stage abstract reasoning and the ability to systematically test hypotheses, to make inferences, and to hold variables constant, that is, to think scientifically. According to Piaget, the components of scientific reasoning were abstract and inferential reasoning that allows for systematic testing of hypotheses by holding variables constant, and these abilities do not really develop until the formal operations stage of adolescence. He made an enormous methodological contribution; he inspired researchers to question children closely and to pose problems for them that would reveal not only what they knew, but how they knew it and how their knowledge could be changed by further experimentation. For instance, conservation of liquid was assessed by pouring liquids from one shaped beaker to another; conservation of number by placing two rows of coins down and then spreading one farther apart. But because he had only relatively crude means of empirical investigation (mainly verbal, motoric, or physical) at his disposal, many current developmental psychologists believe that Piaget underestimated what and when children know and understand certain things.[21]

It is safe to say that most subsequent research and theory on the development of scientific reasoning are extensions, elaborations, or refutations of Piaget. Research on the development of scientific reasoning since Piaget has begun to converge on two general conclusions about what constitutes scientific reasoning and how it is similar and yet different from other forms of reasoning. First, fully developed scientific reasoning involves consciously and explicitly distinguishing theory from evidence. Second, it involves recognizing that other people's or one's own ideas may be wrong (skepticism and exclusionary thinking) and therefore subject to empirical testing. Both of these capacities in turn suggest a more general ability to step back and explicitly think about one's thinking, that is, to think metacognitively.[22]

The most influential person in the field of the development of scientific reasoning since Piaget—Deanna Kuhn—has been instrumental in exploring both the question of what scientific reasoning is as well as how it changes over the life span. Kuhn and her colleagues have been at the forefront of arguing that the coordination of theory and evidence are the essence of scientific reasoning. "Accordingly, the development in scientific thinking believed to occur across the childhood and adolescent years might be characterized as the achievement of increasing cognitive control over the coordination of theory and evidence. This achievement, note, is metacognitive in nature because it entails mental operations on entities that are themselves mental operations."[23] Evidence, to be clear, consists of empirical observations distinct from theory that bear on the theory's veracity (more about Kuhn's theory and research in the next section).

But what, then, distinguishes those who are capable of reasoning scientifically from those who are not? As mentioned above, in order to separate theory from evidence, one must also be able to reflect on the theory as an object of one's thinking (metacognition). To coordinate and change theory to fit new and especially disconfirming evidence one must be able to stand back from one's ideas and see them as things that can and should be tested, or to quote a recent car bumper sticker, "Don't believe everything you think!" Many people are not able to question their own thinking, and therefore scientific reasoning is difficult to achieve. If one were able to do this, one could more readily dismiss a line of thinking as wrong when evidence suggests no relation between two events.

This pushes the question back: who develops metacognitive thinking and why? One characteristic that seems to make metacognitive thinking more likely is general intelligence. Highly intelligent and gifted students tend to have higher metacognitive skills than less intelligent and less gifted students. Metacognition, however, can indeed be learned and is not the privileged domain of only the most gifted. A significant body of literature now exists demonstrating the effectiveness of teaching metacognitive skills in helping students to better understand mathematical and scientific concepts and therefore to think more scientifically.[24]

*Similarities between Child and Adult Reasoning.* The answer to what is scientific reasoning influences the second major question: how does such reasoning change and are children categorically the same or different from adults in their reasoning. Over the last twenty years, many in the field have begun to use the "intuitive scientist" metaphor. In this view, children and nonscientist adults construct cognitive models, evaluate evidence, and modify their conceptualizations of how the world works in a similar but less developed manner to scientists.[25] For example, Alison Gopnik and her colleagues agree with the use of the metaphor but argue that it goes in the other direction: scientists are big children. In their words: "We think there are very strong similarities between some particular types of early learning—learning about objects and about the mind, in particular—and scientific theory change. In fact, we think they are not just similar but identical . . . We think that children and scientists actually use some of the same machinery. Scientists are big children. Scientists are such successful learners because they use cognitive abilities that evolution designed for the use of children." They go on to argue that the chief difference between theories of children and of scientists is in the kinds of things the theories are concerned with. Children develop theories about things that are immediately perceived, common, and midsized, whereas scientists develop theories about all sorts of things from subatomic particles to hidden diseases to galaxies. Because scientific objects are not so sensory-bound, scientific theory formation and revision are more culturally based and culturally specific compared to folk theory formation. That is, people all over the world have not formed these sorts of scientific theories because the objects on which they are based are not immediately sensed without instrumentation.

Developmental psychologist Barbara Koslowski also argues that the literature on scientific reasoning has underestimated people's ability to think scientifically.[26] Many researchers have focused too narrowly on domain-general, atheoretical (theory-independent) covariation and hypothesis testing and ignored theory-dependent mechanisms and hypothesis discovery and revision. When researchers focus on hypothesis discovery and revision, people do better at scientific thinking than some have argued. Emphasis on theory-based knowledge is tied to understanding of mechanism, and all scientific knowledge in this sense must be theory based. Because people sometimes have an intuitive understanding of cause and effect mechanisms they are slow to give up their theories with the first contradictory evidence, and up to a point, this persistence can pay off. Scientists also do not abandon theories with the first disconfirming evidence; otherwise, no theory would stay around. Indeed, Ryan Tweney has shown that some of the great figures in the history of science (such as Michael Faraday) have used the most effective strategy possible: being biased toward confirmation in the early stages and then switching to a disconfirming strategy after the theory has survived the first few attempts at confirmation.

Some scholars have argued that scientific reasoning also involves pattern recognition and cause-and-effect reasoning.[27] If this is so, then once again there is a fundamental similarity between children and adults because seeing patterns and cause-and-effect relations in the world is not unique to scientists. As Robin Dunbar has pointed out, children, as well as many other animals and mammals, share this capacity for cause-effect thinking because that is what brains do—they make associations and connections between events. Children have the ability at a very young age to think via hypothesis testing and to see causal connections between events (two critical elements of science). In this sense children are doing science, but it is "cookbook science" rather than "explanatory science."

Finally, William Brewer and his colleagues argued that children's explanations do not fundamentally differ from those of scientists.[28] Explanations are conceptual frameworks for specific phenomena, they integrate various facets of the world, and they show how the event or behavior follows from the framework. Children do all of these things every day in many domains of life, and their explanations of the world work reasonably well. Brewer and colleagues, however, are careful to point out that there are also differences between children and scientists, and that these mainly involve the socialization (that is, explicit education) of learning to do science. Indeed, more recently Brewer has moved to the position that emphasizes the differences more than similarities between the reasoning of children, adults, and scientists.

*Dissimilarities between Child and Adult Reasoning.* There are other developmental psychologists, however, who argue that children and adults are categorically different in their reasoning. Deanna Kuhn and David Klahr, for instance, argue that although there is some validity to the similarity between children and adults, it is quite misleading to argue that children and scientists use the same cognitive strategies and processes.[29] Overall, there are at least three reasons for arguing that children do not think like scientists.

First, although children develop theories about how the world works, Kuhn argues that these theories are very implicit, nonconscious, and sensory-bound, and children are not very good at distinguishing theory from evidence. Initial theoretical thinking is observational and sensory-bound, such as "plants are living things." Later theoretical thinking becomes more explanatory, such as "this plant died because it had inadequate sunlight and nutrients from the soil."[30] In other words, in childhood one is not capable of having one's beliefs be the object of one's thought. This leads to the confusion and confounding of theory and evidence.

According to Kuhn, when confronted with disconfirming evidence, children often unknowingly distort it to try to fit it into their theories of how the world should work, or selectively make use of the evidence, or unconsciously adjust theory to fit with the evidence (of course, scientists occasionally have been known to do such things as well).[31] Similarly, when asked how they know something, nonscientific adults often say that this is the way it should be or the way that makes most sense, that is, they repeat their theoretical assumptions and fail to distinguish theory from evidence. They are not able to provide the kind of evidence that would falsify their theory. The truth and self-evidence of their theoretical assumptions are taken for granted. Adult scientists—as both Piaget and D. Kuhn have argued—are somewhat unique in the regularity with which they clearly demarcate their theoretical ideas from the empirical evidence. This is not to suggest that only adult scientists can separate theory and evidence, but they do so most consistently.

An example of the research on which these conclusions are based comes from Kuhn's investigation into the connection between scientific and informal (everyday) reasoning in adults.[32] She asked 160 people (teenagers and people in their twenties, forties, and sixties) their theories on three topics, namely, what causes prisoners to return to a life of crime, what causes children to fail in school, and what causes unemployment. Participants were then asked for evidence, alternative theories, counterarguments, and rebuttals. Only 40 percent of the people could give actual evidence (that is, differentiated from theory and something that bears on the theory's correctness). Sixty percent gave nonevidence or pseudoevidence. Pseudoevidence cannot tell us whether the theory is correct because it can never conflict with the theory. For instance, a man in his twenties whose theory for school failure was poor nutrition said (to the question "what would show that?"), "[they would get poor grades because] they are lacking something in their body." This is nonevidence because the outcome presumes the cause. College participants were much more likely to provide evidence than noncollege participants (approximately 55 percent versus 20 percent). About 60 percent of the participants were able to offer alternative theories, and 50 percent could provide counterarguments. But many actively resist the idea of counterarguments. As one said, "If I knew the evidence that I'm wrong, I wouldn't say what I am saying." Others get even more stubborn and say things like "they'll never prove me wrong." Only 20 percent could offer rebuttals, which are essential because they integrate argument and counterargument (original and alternative theories).

These conclusions are in line with the work on searching in two spaces (hypothesis and experiment) by David Klahr and his colleagues, in that both Kuhn and Klahr conclude that children fail to distinguish theory from evidence in ways that some adolescents and adults are capable of doing. For instance, Klahr and colleagues performed a fairly sophisticated experiment in order to test whether there are developmental differences in scientific problem-solving heuristics. They tested four groups that varied in age and in scientific/technological skill: third graders, sixth graders, community college students with little technical training, and college students with technical training. Results rather clearly showed that under some circumstances children can perform cognitive processes that are similar to those of adults, but under other circumstances they cannot. When the actual hypotheses are plausible or when the experimental alternatives are few in number, children perform similarly to adults. But when the actual hypotheses are implausible or the alternatives are not few in number, adults' performance is categorically superior to children's. Children were not able to consider two alternative hypotheses when the actual hypothesis was implausible, and they stuck with their original plausible but incorrect hypothesis. Adults, in contrast, were able to search for solutions in two spaces simultaneously, namely, hypothesis and experiment, even when the hypothesis was implausible. More recently, Klahr has begun to argue that the differences between children and adults are greatest at the more domain-specific level of scientific reasoning. At the more domain-general levels of reasoning, basic similarities do exist between child and adult reasoning.[33]

A second reason for the dissimilarity between children's and scientists' thinking is that to reason scientifically one must be capable of recognizing the importance of disconfirming evidence and of realizing that one's thinking may be wrong and in need of revision. Children simply are not really capable of this (nor are many adolescents and adults). One must be able to infer correctly that a relationship exists when one outcome consistently follows one event and also to rule it out when the events do not consistently follow each other. In scientific reasoning, disconfirming evidence ultimately forces a change in theory (as Thomas Kuhn, no relation to Deanna, argued with his notion of "anomalies" and "paradigm shifts"). Inhelder and Piaget recognized this in 1958, calling it "the false causal inference." Scientific thinking requires avoiding false causal inferences ("superstitions"), and correlation—as we teach our students in introductory psychology and statistics courses—does not imply causation. Covariation is a necessary but not sufficient condition for causality, and sound scientific reasoning acknowledges this. Children simply have a difficult time realizing that covariation is not enough for causation, as do many adolescents and adult. Scientists, however, do not. Because our minds are so good at seeing cause and effect, there is no doubt that we sometimes see causal connections when there are none (aka superstitions). It is the ability to realize that we may be wrong and may have to revise our theory that is the essence of scientific thinking. Intellectual honesty is a humbling and sometimes painful prerequisite for thinking like a scientist. Although not automatic and intuitive, exclusion thinking can develop in most people by late adolescence with practice and training.[34]

Finally, the third and related reason for the dissociation between children's and scientists' thinking is the extent to which, when confronted with evidence that contradicts their beliefs, people often unknowingly ignore or distort the evidence, or unwittingly adjust the theory to fit the evidence. Adults and even scientists, of course, are not immune to such distortions and denials, but they are less likely to resort to these tactics than children are. Paul Klaczynski and his colleagues, for instance, have shown empirically that adolescents are not very scientific in their thinking because of the degree of defensiveness and distortion of evidence seen when they are reasoning about evidence that disconfirms their previously held beliefs. They seldom adjust (accommodate) their thinking, but instead they distort, ignore the evidence, and defend their original conceptualization through ego-protective and complex and detailed analysis of why the evidence, rather than their thinking, is wrong. Klaczynski argues there are two classes of explanation for these distortions: motivational reasons (that is, to protect one's self-esteem) and cognitive reasons (that is, cognitive efficiency and to protect one's theoretical system). These distortions, however, do seem to decrease from late adolescence to early adulthood, at least on personally engaging tasks.[35]

The developmental changes in cognitive ability is almost certainly tied in with brain maturation, in particular maturation of the frontal lobes. Recent advances in cognitive neuroscience have begun to clarify the brain mechanisms involved in complex and analytic reasoning, including components of scientific reasoning. In particular, cognitive neuroscientists have argued for the critical role of executive function (and therefore the frontal lobes) in complex and analytic reasoning. Neuroimaging research has generally supported this notion.[36] Executive functions consist of such distinct yet related skills as working memory, inhibition of prepotent responses (impulses), shifting between cognitive sets, planning, and organizing and achieving future goals. Most every one of these cognitive skills could and probably does play a role in scientific reasoning, and yet few have empirically examined the connection between executive function and scientific reasoning. Joaquin Fuster comes close to explicitly stating the connection between frontal maturation and scientific reasoning when he argued that coincident with frontal lobe maturation around ages eleven to fifteen "the child begins to utilize logical reasoning for the construction of hypotheses and for the testing of alternative solutions." Yong-Ju Kwon and Anton Lawson have been among the only researchers to propose that such executive functioning might map onto the last stage of Piagetian reasoning, formal operations (and its related functions of scientific reasoning and hypothesis testing).

A few others have linked executive-frontal functioning to mathematical reasoning and demonstrated that children with the highest executive functioning scores tend to score highest on math tests.

Cognitive development most certainly involves the ability to make ideas more and more explicit and under conscious control. The developmental psychologist Annette Karmiloff-Smith put forth perhaps one of the clearest descriptions of the development from implicit to explicit thought with her notion of "representational redescription." First, the child just knows something implicitly and then with development, skill, and experience becomes capable of "redescribing" his or her previously implicit representations explicitly (that is, consciously and verbally). "[Representational redescription] involves a cyclical process by which information already present in the organism's independently functioning, special-purpose representations, is made progressively available, via redescriptive processes, to other parts of the cognitive system. In other words, representational redescription is a process by which implicit information in the mind subsequently becomes explicit knowledge to the mind, first within a domain and then sometimes across domains."[37] With age a person is increasingly able to represent ideas consciously and redescribe them explicitly.

Indeed, I would argue that Karmiloff-Smith's theory of cognitive development can be fruitfully applied both to the cognitive development between and within the human lineage (see chapter 8) and to the cognitive development in the history of science (see chapter 9).[38] In other words, I propose a mild form of recapitulation: the development of the species is reflected in the development of the individual. In both, thinking becomes less and less immediate and sensory-bound and more and more consciously represented, explicit, and metacognitive. We see this as we move from ancient to modern human thought, as we move from the preverbal, to verbal, to applied, to pure forms of science, and as we move from infant to child to adolescent to adult thought.

## EXPLICIT DEVELOPMENTAL PSYCHOLOGY OF SCIENCE

Although one might be able, as I have tried, to make the argument that all people are capable of implicit science (construct theories, find patterns, test hypotheses), one obviously cannot make that claim for explicit science. We do not all become scientists and in fact only about 2 percent of the population does.[39] That begs the question of who becomes a scientist and why. The literature tends to point to at least two major psychological influences on scientific interest (self-image and personality) and three familial influences (birth order, religious background, and immigrant status).

### Psychological Factors in the Development of Scientific Interest

*Self-Image.* Self-perceived ability is indeed crucial to the development of interest in and motivation in a given career. If one cannot imagine oneself in a career, then there is little chance one will attempt to pursue that career. Occupational interest research has demonstrated that congruency between talent, performance, self-perception, and drive is the best predictor of career interest. The principles of cognitive consistency, self-efficacy, and stereotype threat can fruitfully be applied to models that explain interest and attrition in science. Can one easily envision and imagine oneself as a "scientist"? For some the label fits well with their projected self-image and for others it does not. Moreover, especially in the field of science and math, gender would appear to moderate the relation between self-image and scientist, given the historical association between gender and science. Similarly, math skills have been seen as part of the male domain, and if one is female, there is less congruency between self-image and a career in math or science.[40]

*Personality.* In a quantitative review (meta-analysis) of twenty-six published studies comparing the personality scores of scientists and nonscientists, Feist reported that being open to new and alternative ideas and/or experiences (openness), being somewhat introverted (extroversion), and being organized and self-disciplined (conscientiousness) are important dispositions for the development of scientific interest. Moreover, in a study of more than six hundred college students, Feist and colleagues recently found that openness and conscientiousness most strongly covaried with scientific interest.[41]

### Familial Factors in the Development of Scientific Interest

*Birth Order and Theory Acceptance.* Research has demonstrated that birth order is related to both interest in science and one's predilection toward accepting or rejecting revolutionary and novel theories. The research on birth order has a long history, beginning with Francis Galton in the 1870s. One of the first questions addressed in terms of science was simply whether scientists were more likely to be one birth position or another compared to nonscientists. Numerous studies consistently reported that compared to nonscientists, scientists are disproportionately firstborn.[42] But birth-order research has been replete with inconsistent and even contradictory findings. As Frank Sulloway has pointed out, however, much of these inconsistencies have been methodological.

Additionally, in his 1996 book *Born to Rebel,* Sulloway makes a persuasive case that birth order is a fundamental influence on an individual's disposition to accept or reject authority, be it familial, educational, political, social, or scientific. The fundamental finding, and one that he puts in the context of evolutionary theory of sibling rivalry and competition for resources, is that firstborns are disposed toward accepting the power structure they are born into because they are the oldest and strongest and identify most with the authority of their parents. Due to their temporary only-born status, they once garnered all the parental resources of attention and care, and when siblings come, they are thrust into positions of responsibility and power. Later-born children, in contrast, are inherently disposed toward questioning and challenging the innate power structure of the family, given their built-in inferior status within the family. What makes Sulloway's argument persuasive is the extensive historical documentation and systematic testing of the basic hypothesis that firstborn individuals are more likely to accept intellectually and politically conservative theories and revolutions, whereas later-born individuals are more likely to support liberal theories and revolutions. From the perspective of the psychology of science, most relevant is Sulloway's detailed analysis of revolutionary theory acceptance in the history of science, focusing mostly on Charles Darwin's theory of evolution by natural selection but also on Copernicus's heliocentric theory and dozens of other radical, technical, or conservative theories in the history of science. To give but a few examples: laterborns were 4.6 times more likely to accept Darwin's theory of natural selection in the sixteen years after it was first published than firstborns. They were also almost ten times as likely to accept evolutionary theory prior to Darwin and more than five times as likely to accept Copernicus's sun-centered theory. On the other hand, firstborns were more likely than laterborns to endorse and support conservative scientific theories, such as vitalism or eugenics. Perhaps even more telling is the fact that creative revolutionary thinkers themselves, at least in the ideological revolutions of Copernicus and Darwin, are more likely to be laterborns than firstborns. No such effect held for technical revolutions (for example, Newton, Einstein, Quantum theory), or conservative theories (for example, vitalism, idealist taxonomy, or eugenics).

*Religious Background and Science.* Another way family may influence scientific development is through its dominant religious orientation. Many researchers have reported that a disproportionate number of eminent and creative scientists come from Protestant or Jewish families compared to those with

Catholic backgrounds.[43] For example, whereas only 2 to 3 percent of the American population comes from Jewish backgrounds, the percentage of eminent and elite scientists from Jewish backgrounds ranges from 9 to 38 percent. In fact, among the most creative and elite groups of scientists most estimates suggest that 20 to 30 percent come from Jewish families.[44] Some have speculated that the Jewish culture and history of critically evaluating ideas and not basing belief on hierarchical dogma is more conducive to scientific reasoning and hypothesis testing.

Religious background, however, does not tease apart variability due to religious orientation, culture, race, or even genetic influence. Moreover, we must make clear that these data refer to the religious faith of one's family background and upbringing, not one's current behavior. Scientists in general, and eminent scientists in particular, are conspicuous in their rejection of organized religion.[45] The few studies that have asked scientists about their current religious practices have reported an almost complete absence of current religious faith.

*Recent Immigrant Status.* A small but consistent body of evidence suggests that being within two generations of immigrating to the United States is related to scientific interest, talent, and achievement.[46] Families who recently come to the shores of the United States may well foster a particular set of values that encourages or even demands a high level achievement, whether it be in science, medicine, or business. An interesting speculation on this phenomenon is that science may be more meritocratic than most other career paths and therefore talent and achievement in and of itself is more likely to be recognized and rewarded. Another possibility involves the work ethic of recent immigrants. In the words of a 1989 foreign-born finalist of the prestigious high school Westinghouse Science Competition: "Immigrants understand the concept of hard work. In a sense we're more American than Americans. We actually believe in individualism, in going out and making our own way, like the first Americans. We had a frontier too. The frontier made America what it is." Simonton offers another possible explanation: "Individuals raised in one culture, but living in another are blessed with a heterogeneous array of mental elements, permitting combinatory variations unavailable to those who reside solely in one cultural world."[47]

### Age and Scientific Achievement over the Lifetime

Another prominent developmental question in the psychology of science concerns how scientific interest and ability change over the course of the lifetime. More specifically, there are accumulated literatures on the following questions:

*Does Precocious Math or Science Ability in Childhood and Adolescence Become Actual Achievement in Adulthood?* In one of the more ambitious longitudinal studies of extreme precocious talent in math, Julian Stanley argued for the predictive value of extreme mathematical precocity: "These young students [those who score 700 or above on the math SAT before their thirteenth birthday] seem to have the potential to become the nation's superstars in pure and applied mathematics, computer science, electrical engineering, physics, and other fields that depend heavily on great quantitative aptitude. Quite a few of the 292 appear well on the way toward excellence in such fields."[48] One of Stanley's students, Camilla Benbow, and her colleagues have presented data from the same data set showing that precocious ability predicts achievement in high school. Approximately 90 percent go on to get bachelor's degrees, slightly less than 40 percent get master's degrees, and about 25 percent receive doctorates. These figures are well above the base rates for such degrees (with 23 percent, 7 percent, and 1 percent being the national figures, respectively). Similarly, Feist has recently reported that in a sample of extremely talented high school students (Westinghouse finalists) more than 90 percent of the males and nearly 75 percent of the females went on to earn a doctorate. Given the base-rate of 1 percent in the population, these are extremely high figures. Another way to make the point that talented math students go into science is to look backwards on the math ability of those who in fact become scientists or mathematicians. Not surprisingly, such retrospective research shows scientists and mathematicians scoring well above the population norms (on average the 90th percentile) on math achievement tests in high school.[49]

As Farmer pointed out, however, even in the Stanley-Benbow data set only 42 percent of the male and 22 percent of the female extremely precocious students went on to choose a science or math graduate program. In short, only 25 percent of the extremely gifted math sample continue in science and math through graduate school, and even fewer are retained in science and math careers. Indeed, Benbow and her colleagues have recently reported that only about 17 percent of the mathematically precocious males and 6 percent of the mathematically precocious females were in math and computer science careers at age thirty-three. Moreover, from these same samples only about 4 percent of the males and 3 percent of the females were in natural or physical science careers in their early thirties. The modal career category was "executive and administrator" (with about 23 percent of the males and about 20 percent of females in this category career at age thirty-three). The retention rate, however, is somewhat higher for those who demonstrate precocious scientific achievement.

Rena Subotnik and Cynthia Steiner, for instance, reported that by their mid-twenties, 81 percent of the male and 66 percent of the female Westinghouse Science finalists were still on science-training or career tracks. Similarly, Feist reported in four different cohorts of Westinghouse finalists, ranging in age from twenty-four to fifty-four, that 89 percent of the men but only 56 percent of the women stayed in science, engineering, or math careers. If one chooses a more real-world, valid outcome criterion, such as actual creative achievement in math and science, the evidence is more mixed about whether the extremely talented youth go on to have truly influential careers. The evidence shows that neither grades nor aptitude tests are good predictors of creative achievement in scientific careers, but that many Westinghouse finalists go on to become the more productive and influential scientists of the next generation.[50]

Other predictors of high levels of scientific achievement later in life are simply the age that one decides one wants to be a scientist, the age one realizes one's talent for science, and the age that one starts doing science. In a recent retrospective study of members of the National Academy of Sciences (NAS)—our nation's most elite science organization—Feist reported that about 25 percent of the members knew they wanted to be a scientist by age fourteen and 50 percent knew by age eighteen. They knew they had talent for science even earlier, with 20 percent of the NAS members realized their talent by age thirteen and 50 percent by age sixteen. Finally, NAS members began doing science early, with 75 percent having participated in formal research by age twenty-one.[51]

*How Does Productivity Change with Age?* One of the oldest of the developmental psychology of science questions concerns whether age affects level of productivity. The question is unique not simply because it has been asked for such a long time, but because its answer is now rather consensually agreed upon. There is a relation between age and productivity in science (and other professions), and it is an inverted U.[52] Further, once controls are made for different ways of operationalizing output, the curve peaks at the same age (early forties) for quality and quantity of productivity. But it peaks somewhat differently for various disciplines (earlier in math and physics, later in biology and geology).

This is not to say that the topic of age and productivity has been without controversy. On the contrary, it has been replete with controversy from its inception. In particular, Harvey Lehman's seminal work has been the object of frequent criticism and rebuttal. Granted some of these criticisms are valid and justified, once many of the controls are made that Lehman failed to make, the result is still an inverted U. The peak may be a little flatter and it may occur a little later, but basically every study conducted on the relation between age and productivity has shown a curvilinear relation that peaks either in the late thirties or early forties and then drops off more gradually than it rose. Extreme creative talent in theoretical physics—like chess, pure math, and lyric poetry—is likely to peak earlier than in some other fields of study, such as biology, geology, history, and philosophy. It is not uncommon for major theoretical contributions in physics to be made in the decade of one's twenties, with Einstein being the most obvious but not only case, and the general peak of productivity occurring during the late twenties or early thirties, very seldom being before age twenty or after thirty-five. Even physicists themselves were aware of this phenomenon.[53] Paul Dirac, a wunderkind of theoretical physics (he won a Nobel Prize at age thirty-one for work he had done at age twenty-five), penned these words:

> Age is, of course, a fever chill
> That every physicist must fear
> He's better dead than living still
> When once he's past his thirtieth year.[54]

Although perhaps not quite the death knell that Dirac implies, after one's thirties, tremendous contributions to theoretical physics become rare. In this sense, theoretical physics is somewhat like sports—being the domain of the very young and peaking in the late twenties. There are three domains—pure math, chess, and music—in which cases of extreme precocity, that is true achievement, can occur even earlier, namely *before* one's twenties. The intriguing thing here is that these three areas do seem to be the only fields in which world-class achievement before one's twentieth birthday can been seen. Perhaps there is something about brain development (maximum neural connectivity and efficiency) that allows certain extremely gifted individuals to achieve at a high level for a relatively brief period of time. The phenomenon of brain development and giftedness awaits further neuroscientific investigation.

I must point out, however, that age accounts for only a relatively small percentage of the overall variability in productivity. The work of Karen Horner and colleagues illustrates this point. They sampled more than one thousand male research psychologists from four different birth cohorts and found a curvilinear relation between age and productivity, with the peak occurring in the early forties. In this sample, age accounted for 6.5 percent of the overall variance in publication rate. In short, it is clear that other individual difference and social factors (such as early levels of productivity, rewards and honors, and institutional support) have at least as strong if not stronger of a relation with productivity.[55]

If the description of the relation between age and productivity is relatively clear and agreed upon, its explanation is not, and the few attempts at explanation can be divided into two general categories: extrinsic versus intrinsic factors.[56] The primary candidates for extrinsic theories concern increase in family and administrative obligations and unfavorable work conditions, whereas the intrinsic factors are concerned with changes in physical health, motivation, experience, intelligence, and creativity. Empirical evidence at least partially supports the extrinsic theories.[57] For example, sociologists have long argued that an extrinsic factor (reward) plays an important role in maintaining high levels of productivity in some and discouraging it in others. In other words, scientists who produce the most impactful works early in their careers and who are thereby rewarded with tenured jobs at top departments, financial support, and prestigious awards are the ones who are most likely to continue producing. The main problem with this theory is that it cannot explain the single-peak curvilinear relation between age and productivity in all scientists, not only the most precociously productive.

Little longitudinal research has been conducted on the intrinsic theories, namely, developmental changes in health (aging), motivation, intelligence, and creativity across the lifespan. Most recently, and quite consistent with the age forty peak, is evidence from genetics and the aging human brain that shows that DNA damage starts to accumulate and numerous genes involved in cognitive functioning (memory, learning, and neural survival) start to slow down after age forty.[58] The empirical work conducted on change in intelligence across adulthood, however, points to a rather late and small decline with wide variability between individuals, which suggests that age-related declines in productivity may not be a result of a drop in intelligence.[59]

The more likely intrinsic candidate, namely, motivational decline, has received theoretical attention. More than 125 years ago, G. M. Beard argued that productivity is a function of changes in motivation (enthusiasm) and experience. The young are more enthusiastic, the old are more experienced, and both enthusiasm and experience are linear functions of age (enthusiasm negative and experience positive). Creative achievement is a result of the balance between youthful enthusiasm and the experience of age, and hence productivity peaks when these two intrinsic processes overlap (that is, in the late thirties or early forties). As Bernice Eiduson wrote in her classic longitudinal study of scientists: "from the standpoint of satisfaction there is some diminution of the involvement in work—some of the gratifications are beginning to pall and some of the fire, drive, and curiosity is gone."[60]

Dean Simonton has developed a more complex theoretical model that attempts to predict and explain the age-productivity relation by focusing on intrinsic factors, namely, cognitive components.[61] This model is based on his "chance-configuration" theory and consists of a few key assumptions: first, each creator starts off with a set amount of creative potential (number of contributions made over a normal, unrestricted life span); second, the actualization of creative potential can be broken down into two components, ideation and elaboration. Ideation is the rate at which potential ideas are expressed, and elaboration is the rate at which ideas are put into concrete, public form. As each creator produces a new work, she or he "uses up" some creative potential. The rate at which a creator actualizes potential and produces works is a direct function of the two cognitive transformations, ideation and elaboration. To graphically model the relation between these two elements, Simonton has developed one of his better-known differential equations, with the peak occurring roughly twenty years into one's career and thereafter slowly declining.

*Does Producing Works Early in Life Predict Later Levels of Productivity?* Again, enough work has been conducted on this question to provide a rather consensual answer: yes, early levels of high productivity regularly foreshadow continued levels of high productivity across one's lifetime. Those who are prolific early in their careers tend to continue to be productive for the longest periods of time. For example, one study reported that the most prolific group of scientists out-published medium and low publishers by more than two to one in the twenty-five to thirty-four age period, and they maintained a rate of about a paper per year advantage over both groups during each ten-year period until their mid-sixties to -seventies age period. At this age all three groups dropped to approximately a half a paper per year, but the precocious group still out-produced the other two groups.[62]

As is the case with productivity in general, sociologists tend to explain this phenomenon in terms of the "cumulative advantage" or the "Matthew effect." Those who publish frequently early in their careers and are therefore rewarded by their peers continue to garner more and more of the resources and continue to out-produce their peers because of the ever-increasing supply of financial and social support. Productivity data are inherently positively skewed with one-tenth of the scientists producing roughly one-half of all of the works. The rich get richer and the poor get poorer! Furthermore, there is some evidence that quantity of publication matters more than quality of publication when predicting who will receive the most peer recognition and prestigious honors; that is, who will become the most eminent.[63]

*Compared with Older Scientists, Do Younger Scientists Produce a Disproportionate Number of High-Quality Works?* Lehman suggested that younger scientists (below age forty) produce most of the highly cited and impactful works. Lehman's data, however, did not take absolute number of scientists at each age period into account, and therefore they may be biased toward the young simply because there are more young scientists. Ray Over examined whether older scientists were more likely to produce works of lower quality than younger scientists. He found that although it is true that a disproportionate number of high-quality works come from scientists less than ten years post-PhD, it is equally true that a disproportionate number of low-quality works come from this age group of scientists. In other words, more high-quality works are being produced by younger scientists not because of age but because of the high number of young scientists. The same holds true once longitudinal rather than cross-sectional data are examined. Longitudinal data are important because they do not confound age and cohort effects the way cross-sectional data do.[64]

*Are Older Scientists More Resistant to Scientific Revolutions Than Younger Ones?* The physicist and Nobel Prize laureate Max Planck experienced resistance to his novel ideas, which gave him "an opportunity to learn a new fact—a remarkable one, in my opinion: A new scientific truth does not triumph by convincing its opponents and making them see the light, but rather because its opponents eventually die, and a new generation grows up that is familiar with it."[65] This observation by Planck, which has come to be called "Planck's principle," fits well with Thomas Kuhn's notion of paradigm shift, in which the old and new paradigms are so different that they are incommensurable. Similarly, toward the end of *On the Origin of Species,* Darwin noted, "A few naturalists, endowed with much flexibility of mind, and who have already begun to doubt on the immutability of species, may be influenced by this volume; but I look with confidence to the future, to young and rising naturalists, who will be able to view both sides of the question with impartiality."[66] Darwin's primary defender, T. H. Huxley, went further, arguing that men of science ought to be strangled on their sixtieth birthday lest they retard scientific progress!

To test the connection between age and theory acceptance, David Hull and colleagues compared the ages of scientists who accepted and rejected the theory that species evolved in the ten years after publication of *Origin* and found that the rejecters, on the average, were ten years older—a statistically and practically significant difference. More recently, Hull has pointed out the other side of the coin: older scientists may be more resistant to all ideas, including the bad ones. In other words, their skepticism also prevents them from accepting ideas that are, in fact, later rejected, such as phrenology or cold fusion.[67] Perhaps this is why Carl Sagan argued that the best science involves both skepticism and openness.

As we saw with age and productivity, it is important to keep the magnitude of this effect in mind. Age ultimately accounts for less than 10 percent of the variance in theory acceptance and can provide moderate support at best for Planck's principle. Indeed, the historian turned psychologist Frank Sulloway published an exhaustive historical analysis of theory acceptance in science and concluded that birth order accounted for more variance than any other single variable.[68] Further, Peter Messeri studied age differences in relation to the acceptance of plate tectonics after the discovery of sea floor spreading in the early 1960s. During the period immediately following publication of this new idea, older scientists were significantly more likely to adopt plate tectonics than younger ones, exactly the reverse of what one would expect if Planck's principle were true. Later, after substantial confirmatory data had been disseminated, age no longer played a role in theory acceptance. Finally, other researchers reanalyzed the data collected by Hull and colleagues and divided it into the same time periods used by Messeri. They found no significant relation between age and theory acceptance and concluded, "No researcher to date has found substantial effects of age on the acceptance of new ideas."[69]

In sum, the progression from infant, to child, to adolescent, to adult, to scientist (expert) is a long and difficult one. Implicit scientific reasoning is seen in the automatic and intuitive scientific knowledge of infants and children: they are incipient psychologists, physicists, biologists, and mathematicians (as well as linguists, artists, athletes, and musicians). Moreover, children, adolescents, and adults are fundamentally similar in that they each automatically and intuitively develop explanatory cognitive representations (that is, theories) for how the world works and for their experiences. Indeed, these theoretical explanations tend to be constrained along domain-specific lines, namely, the psychological, physical, biological, and mathematical (as well as the linguistic, aesthetic, and musical). Theory construction is inherent to all age groups. The differences, however, between children, adults, and scientists stem from their degree of concreteness, explicitness, complexity, and metacognitive representation. Theory construction in childhood is inherently sensory-bound and implicit, but it gradually becomes more representational and explicit with development, culminating in the full-blown metacognitive, systematic, and explicit theories of scientists and philosophers.

Moving more directly to explicit scientific thought and behavior, the evidence for why certain individuals are more likely to become scientists and how scientific achievement and productivity change over the course of one's life confirmed the importance of two psychological factors (self-image and personality) and three family factors (birth order, religious background, and immigrant status). People are more likely to develop an interest in science if they have a self-image consistent with being a scientists; are open, introverted, and conscientious; and if they are firstborn, from Protestant or Jewish families, and have been in the country for less than two generations. The question of age and achievement demonstrated the importance of precocious talent and the curvilinear relation between age and productivity (with the peak occurring about twenty years into one's career). Moreover, although there is some evidence that younger scientists might be more receptive to new and revolutionary scientific theories, the effect is neither unequivocal nor strong in magnitude.

# Chapter 4 Cognitive Psychology of Science

Two quotations go right to the heart of this chapter. First is one by the cognitive psychologist of science Ryan Tweney: "Science is by its very nature a cognitive act!" Second, the historian of science Arthur I. Miller had this to say about Piaget, philosophy, cognitive science, psychology, and science: "The Swiss psychologist Jean Piaget made it abundantly clear that the basic problem faced by psychology, philosophy, and science is how knowledge emerges from sense perceptions or data. . . . Piaget wondered essentially about a paradox discussed by Plato in the *Meno:* How can new concepts emerge from ones already set into the brain? In other words, how can a system produce results that go far beyond the statements included in it? This is the problem of creativity."[1]

In this chapter, and indeed in this book, I attempt to answer the Platonic-Piagetian question about knowledge and creativity in science broadly defined. At its core, science involves observing nature, recognizing patterns, forming and testing hypotheses, solving problems creatively, and constructing theories. Each and every one of these activities is cognitive in nature and indeed has been the object of study for cognitive psychology as long as the field has been in existence. The process of doing science is inherently a cognitive activity, and as one might expect, cognitive psychology and the study of science are natural allies. In fact, the cognitive psychology of science is one of the largest and most well-developed disciplines within the psychology of science. Cognitive psychology, more than any other subdiscipline of psychology, has devoted attention to scientific problem solving. In this chapter I review the relevant findings from three broad categories of cognitive psychology of science: theory formation, theory change, and comparing expert to novice reasoning.[2]

## COGNITIVE PROCESSES INVOLVED IN THEORY FORMATION

### Model Building and Causal Maps

Model building in the history of science has been absolutely essential to the creation of new theories and to advancing scientific knowledge. Mental models can be literal representations of the external world (as they often are with visual imagery) or arbitrary representations (as they are with propositional, mathematical, or verbal models). In either case, they are explanatory or descriptive representations of the external world.

Howard Gruber in *Darwin on Man* provided a detailed cognitive analysis of the critical two-year period of Darwin's thinking after he returned from his *Beagle* voyage and creates perhaps one of the more enlightening portraits of the development of scientific thinking and theory formation in the history of science. Making use of the prolific and detailed notebooks Darwin used to organize and develop his ideas, Gruber argues that Darwin's thinking was not simply the result of one major moment of insight but rather a slow, complex reworking of many ideas, many insights, and many instances of making the implicit explicit. Gruber coined the phrase "network of enterprise" to capture the complex, "continually changing concerns and interests of a purposeful life" and as a way of "giving some unity to a complex and changing picture."[3] Of course, there was one moment of insight that sticks out above all others for Darwin: on September 28, 1838, Darwin read Malthus's essay on overpopulation and the struggle for existence, and suddenly many pieces of his thinking fell together in one very explicit insight.

Nature selects individuals or traits that are most adaptive to their given environment, and superfecundity is how nature ensures enough individuals to select from. The idea of natural selection was not new, but Darwin was really the first to apply it to the origin of species and not just use it as a conservative principle ("the weakest individuals get weeded out"). It was a significant theoretical insight to realize that variation need not be adaptive but rather is random or neutral. Because natural selection can now be the mechanism that allows individuals to adapt to their given environments—chance variation requires selection, and superfecundity is what allows great variation—nature then selects those individuals or traits that work best in a particular environment. Natural selection is captured by the phrase "survival of the fittest" and not merely the "death of the weakest." In short, Darwin was not the discoverer of natural selection but the discoverer of its significance.

What is fascinating about Gruber's account of Darwin's earliest ideas on evolution is how important theory was from the very beginning of his writing. Darwin had observed some amazing phenomena on his journey to South America, and although he apparently did not do much theorizing on the five-year voyage, he began his notebooks less than a year after returning to England. Above all else, it is fair to say, he had a need to organize and give meaning to the wonderful and perplexing array of observations he had made on his journey to the Galápagos. Interestingly, Darwin's own account of his earliest notebook portrays them as atheoretical: "My first note-book was opened in July 1837. I worked on true Baconian principles, and without any theory collected facts on a wholesale scale." Darwin continues: "In October 1838 [it was September], that is, fifteen months after I had begun my systematic enquiry, I happened to read for amusement Malthus on *Population,* and being well prepared to appreciate the struggle for existence which everywhere goes on from long-continued observation of the habits of animals and plants, it at once struck me that under these circumstances favourable variations would tend to be preserved, and unfavorable ones to be destroyed. The result of this would be the formation of new species. Here, then, I had at last got a theory by which to work."[4]

Gruber, however, points out after his completely detailed analysis of these notebooks that the notion that Darwin began with no theory is simply and factually wrong. Darwin had a theory at the outset of his first notebook in July 1837. In fact, although he had long bought into the relatively widespread idea of evolution (indeed his grandfather Erasmus put forth a variation of the idea), it was not until March to July of 1837 that "the idea of evolution [began] to govern his behavior and thought in a systematic way." When he began his first notebook (at age twenty-eight), Darwin already assumed an evolutionary perspective and developed a theory called "monadism" in which new life forms are constantly forming out of inanimate matter; organisms adapt to their changing environment; simple things necessarily become more complex; species, like individuals, are born, mature, and die; and new evolutionary lines are constantly branching off in an irregular "tree of nature." This is hardly Baconian induction and "without theory." Nevertheless, although Darwin quickly realized the fatal flaws of monadism, one crucial element survived: the analogy-metaphor of the tree of nature. In this sense, his early theory was quite useful to his latter thinking. To use Gruber's words: "Darwin certainly began the notebooks with a definite theory, and when he gave it up it was for what he thought was a better theory. True, when he gave up his second theory he remained in a theoretical limbo for some months. But even then he was always trying to solve special theoretical problems, such as those related to hybridization, and he almost *never* collected facts without some theoretical end in view. It was not simply from observation but from hard theoretical work that he was so well prepared to grasp the significance of Malthus's essay."[5]

Darwin himself hinted at what might be going on here: an attempt to present the professionally accepted view that one should observe objectively without theory, even if in reality one cannot do so. In a letter in 1863 to a botanist, Darwin wrote: "*let theory guide your observations,* but till your reputation is well established be sparing in publishing your theory. It makes persons doubt your observations."[6] To be sure, model building is inherent to the scientific enterprise, and Darwin was seldom without his own models and theoretical assumptions. In the end, he was to create one the greatest theoretical models in the history of science.

Other psychologists, such as Alison Gopnik and her philosopher colleague Clark Glymour, have argued that much of cognitive development is the process of theory formation and much of theory formation is based on causal maps, which they define as "non-egocentric, abstract, coherent representations of causal relations among objects."[7] Causal maps, folk theories, and scientific theories interpret, explain, and predict one's experiences with the physical, biological, and social worlds. Sometimes, in fact, causal maps and naive or folk theories of childhood can and do become more explicit and systematic theories of science. Indeed, in chapter 3, I reviewed much of the developmental literature on the topic of theory formation and attempted to show the similarities and differences between scientific theories formed by children, adolescents, adults, and scientists. Theory formation involves building ever more explicit cognitive models or representations of the physical, biological, and social worlds.

What is clear is that some kind of model building and theory formation is absolutely inherent to being human; whether the models are "myths," "magic," "metaphysical," or "scientific," they all do the same thing: provide a cognitive structure for interpreting and organizing sensory experience. Again, as I made most clear in chapter 2, cognitive models and theories are the inherent function of brains. Brains exist first and foremost as organizers of sensory experience and models and theories are convenient shorthand labels we apply to this process.

These cognitive models are not simply abstract pie-in-the-sky cognitive enterprises. Rather, they are directly and ultimately connected to the survivability and reproductive fitness of the individual and the species. Cognitive models/maps allow the individual, whether hyena or human, to interpret and explain what they have experienced and to predict what is likely to happen next. In short, they provide an internal representation of the external world. The better these models are at predicting what is likely to happen next, the better the individual is to adapt and survive in its environment. Humans, no doubt, are unique in the complexity and conscious nature of these mental models. As Darwin first recognized, mental structures and their corresponding processes are most certainly part of the evolutionary process and subject to the laws of natural selection. "A great stride in the development of the intellect will have followed, as soon as the half-art and half-instinct of language came into use. . . . The higher intellectual powers of man, such as those of ratiocination, abstraction, self-consciousness, etc., probably follow from the continued improvement and exercise of the other mental faculties." As implied here by Darwin, explicit, rational theories are not evolved "adaptations," but their implicit foundations may be. Most modern scholars on evolution of mind see consciousness, rationality, and science more as co-opted by-products of adaptations rather than adaptations directly.[8]

As much as we can learn about scientific thinking from analyzing the thinking and model building of nonliving scientists or of students and scientists solving artificial problems in vitro, Kevin Dunbar has made clear that studying how real scientists think as they are doing real science, in vivo, is absolutely essential for a full and naturalistic understanding of scientific thinking.[9] From systematic and long-term analysis of weekly lab meetings, Dunbar has concluded that four main cognitive processes occur: causal reasoning, analogy, categorization, and collaborative (distributed) reasoning. Causal reasoning exists whenever scientists "propose a cause, an effect and a mechanism that leads from cause to an effect." Dunbar has found, for instance, that up to 80 percent of the statements made during a lab meeting are causal reasoning statements. Causal reasoning, in fact, may subsume many other kinds of reasoning processes, such as analogy, deduction, and induction, rather than being a unitary process. In this sense, causal reasoning is of paramount importance in scientific reasoning. Analogy involves taking characteristics from a known source and mapping them onto a less well-known target. One interesting finding by Dunbar and his colleagues is that the majority of analogies made by scientists during lab meetings are "local" ("this is like that experiment that worked . . ."), rather than the more "distant" (and more attention-grabbing) analogies often written about ("natural selection is like artificial selection . . ."). Finally, collaborative or distributive reasoning is the informal reasoning and discussion that goes on behind the scenes or in the hallways. That is, it takes place between people in a social context and is the result of input from many people.

As already touched upon, Darwin amassed some of the more incredible and rich observations of nature ever made, before or since. His five-year expedition on the *Beagle* (from 1831 to 1836, when he was twenty-two to twenty-seven years old) was a phenomenally unique opportunity to observe the multitude and variety of nature. If not immediately, then within a year of his return, Darwin began the process of developing an explicit theory to explain his observations and other natural phenomena that he knew about from reading. But as Darwin himself alluded to, and as Norwood Russell Hanson and Karl Popper made clear one hundred years later, observation is meaningless and impossible without some kind of theory focusing one's attention on what to observe.[10] Pure observation in the strict sense is impossible. To quote Popper: "But in fact the belief that we can start with pure observations alone, without anything in the nature of a theory, is absurd." To paraphrase Kant's critique of Locke's empiricism: observation without theory is meaningless; theory without observation is blind. Darwin, too, never kept as distinct as he would like to portray his observations and his theory. We observe and immediately, automatically, and intuitively our brains try to give meaning to our observations. We try to figure out what we are seeing. Our brains serve this very function.

### Making the Implicit Explicit (Explication)

Demonstrating the power of two themes of the chapter—metacognition and explication—Howard Gruber summed up the development of Darwin's theory with these words: "In the growth of thought a given idea may move from an early phase in which it is implicit in the structure of an argument to a later phase in which it becomes explicit, consciously recognized, and deliberately expressed. As the thinking person becomes more aware of the idea, he can begin to use it more actively and purposefully. Eventually, his feeling of personal connection with it fades. He can look at it with detachment, see some of the unexpected possibilities the idea now generates, and some of the problems it raises."[11]

Although twenty years passed between the *Journals* of 1837–39 and the publication of *Origins* in 1859, the process of making the implicit explicit did not take quite that long; early formal yet unpublished versions of the theory of natural selection were written in 1842. Out of fear of persecution as much as anything, Darwin delayed making the formal theory known. The immediate impetus for publication, as is widely known, was a letter he received from Alfred Wallace in June of 1858, stating that Wallace had a theory of evolution via natural selection that, in Darwin's own words, "I never saw a more striking coincidence; if Wallace had my MS sketch written out in 1842, he could not have made a better short abstract!"[12]

If twenty years were not required for Darwin's implicit ideas of natural selection to become completely explicit and more fully developed, a few years were. Darwin had been working diligently since summer 1837 on developing a theory of evolution and natural selection and had even a few "false" attempts (for example, monadism), but the ideas that occasionally surfaced in various forms became truly explicit only when he read Malthus's essay on population.[13] But this was not the first time he was exposed to the ideas of overproduction or superfecundity. Before September 1838, however, he simply was not ready for them. By this fateful late September day, all the pieces were in place, priming Darwin for his "great insight." Indeed, although Wallace, too, had read Malthus's essay, apparently he had done so fourteen years before developing his own version of evolution via natural selection. Extensive incubation and pondering and working through the idea of superfecundity and struggle for existence were required for both creators of the theory of evolution by natural selection.

The distinction between implicit and explicit theory is more of degree than kind because implicit ideas can and often do become increasingly explicit over time. Everyone organizes and explains their observations, and most of the time we do so implicitly—without conscious awareness or verbal representation. These theories are what we have been referring to as "folk" theories, "naïve" theories, or "common sense." Infants, children, adolescents, and adults all do this and do so constantly. All cognitive development, not only scientific reasoning, involves the ever-increasing explicit and conscious control of mental representations. As I have made reference to in chapter 3, Annette Karmiloff-Smith, for instance, argued that ideas and knowledge always start out quite implicitly, and then with cognitive development and maturation they become increasingly explicit and under conscious control and reflection. Consistent with Karmiloff-Smith, as well as with many other current theories of cognitive and developmental science, I argue that the explicit theories used in science stem from and build upon these implicit everyday folk theories.[14]

Some writers, such as chemist turned philosopher of science Michael Polanyi, have argued that implicit (or tacit) knowledge is "more fundamental than explicit knowing: *we can know more than we can tell and we can tell nothing without relying on our awareness of things we may not be able to tell.*"[15] The most creative scientists, as well as artists, often have a well-honed intuitive sense, knowing implicitly and without knowing how they know, or in Polanyi's words "know[ing] more than [they] can tell." Explication happens when a thinker can delve into the less conscious, more tacit and implicit forms of knowledge and bring them above conscious threshold and use them as potential components of a scientific theory. Two of the greatest theoretical physicists of the twentieth century, Albert Einstein and Richard Feynman, both had unusually deep intuitions about the physical world that they were able to translate into mathematical language. Indeed, as the great French mathematician Henri Poincaré wrote, "It is by logic we prove, it is by intuition we invent."[16]

### The Importance of Analogy, Metaphor, and Visualization in Creative Theory Formation

One overarching theme to the cognitive psychology of science literature, especially those studies that focus on analogy, metaphor, and visualization, as well as on theory change and hypothesis testing, is the widespread use of various kinds of strategies, or heuristics, used often implicitly and sometimes explicitly to solve problems. Ryan Tweney defines heuristic as "the strategies that are chosen to organize the path from a starting point to some goal."[17] As we will see in this and the next section, there are many different kinds of heuristics applied in scientific problem solving and hypothesis testing that have a bearing on adapting theory to new and changing evidence. Some of the more common heuristics are the use of analogy, metaphor, and visualization.

*Analogy.* The essence of analogy is the seeing how something new (target) is like something old (source). Our mind seems to do this automatically and intuitively almost any time we are confronted with a new idea or new experience. Analogy is one of the more ubiquitous ways the brain takes sensory experience and gives it meaning. Indeed, in the history of cognitive science the concept of analogy and metaphor has been one of the more central mechanisms answering the Platonic-Piagetian question of how new knowledge is possible. The cognitive scientists Gilles Fauconnier and Mark Turner put it this way: "Analogy has traditionally been viewed as a powerful engine of discovery, for the scientist, the mathematician, the artist, and the child."[18]

Darwin's main analogy, and one that is seen in his very first attempt at an evolutionary theory in 1837, was the analogy between the branching of a tree and that seen in evolution. Species that are more closely related are closer together on a common branch. This branching analogy is used in current taxonomic systems. The branching tree analogy also makes clear that evolution is not a linear progressive process with humans at the apex, as many evolutionists at the time had argued. Rather, evolution consists of many dead ends and extinct species and is simply a process of the birth, growth, and death of species, all going back to one common origin. This was a truly powerful and revolutionary insight: *all life* on the planet is related! Darwin was one of the first scientists to really make this clear, and he was all too aware of the trouble this idea would cause scientifically, politically, and theologically. But he used this analogy unwaveringly from his late twenties until his death. Another foundational analogy used by Darwin was the comparison between natural selection and artificial selection seen in human breeding of plants and animals. Indeed, Darwin's first chapter to *Origins* begins with the well-known principles of "Variation under Domestication," which lays the foundation for using similar principles in nature in the next chapter, entitled "Variation under Nature."

There is a well-developed literature on the importance of analogy and metaphor in scientific problem solving and creativity.[19] The consensual conclusion from this literature is that analogy is a crucial problem-solving heuristic that allows scientists to apply schemas, models, and mental maps from known to unknown domains in order to solve problems. The success and richness of the analogy depend on how deep the similarity is between old and new. Associations no doubt play a critical role in analogical thinking, with the similarity being touched off via an association. In science, these analogies also often serve as the foundation for hypotheses.

Analysis of historical cases also consistently shows the most creative scientists discovering useful analogies to solve problems.[20] Dedre Gentner and Michael Jeziorski compared the way two physical scientists (Robert Boyle and Sadi Carnot) used analogies to the way alchemists employed them. They concluded that following particular rules when using analogies (for example, avoiding mixed analogies, understanding that analogy is not causation) was the key to distinguishing scientific from pseudoscientific reasoning. But Carnot and Boyle had different styles of analogical reasoning: the former relied on a single analogy, deriving principles from it, whereas the latter preferred to work with a whole family of analogies. Nancy Nersessian observed that James Clerk Maxwell used analogies iteratively, that is, he constantly modified them to fit his growing understanding of the constraints of the target domain. Michael Gorman demonstrated that Alexander Graham Bell deliberately "followed the analogy of nature" and used the human ear as a mental model for his telephone; like Maxwell, he was able to modify this analogy as he learned more about his target domain.

According to some cognitive developmental psychologists, the use of analogy is usually not arbitrary but rather is guided or constrained by the evolved and specific domains of mind, such as math, physics, psychology and biology. Analogy therefore involves translating ideas from one of domain to another. For instance, Susan Carey and Elizabeth Spelke argued that conceptual change often takes place through the successful use of analogy and thought experiment and offered examples from the history of science (Duhem, T. Kuhn, Maxwell, and so on). Translating models between the mathematical domain and objects (physics), people (psychology), and animals (biology) has been an especially useful analogy heuristic in solving many problems in the history of science. As Carey and Spelke put it: "In our terms, scientists who effect a translation from physics to mathematics are using their innately given system of knowledge of number to shed light on phenomena in the domain of their innately given system of knowledge of physics."[21]

*Metaphor.* Metaphor is closely related to analogy in that it, too, involves applying similarity from an old source to a new target and in this sense many metaphors are analogies. The essence of metaphor is an "as if" comparison—I am going to think about X *as if* it were Y. Some scholars, in fact, have argued that metaphor is the broader of the two concepts insofar as it can be used in both explanatory-predictive and expressive-affective contexts, whereas analogy is usually limited to the former.[22] By applying one phrase or idea to another different one that is not literally the same, we again make the unknown known.

As such psycholinguists as George Lakoff, Mark Johnson, and Steven Pinker have made clear, metaphors are so ubiquitous that we often do not even recognize the metaphorical nature of much of our thought. For instance, Pinker gives the following examples: "the messenger went from Paris to Istanbul," "the inheritance went to Fred," "the light went from green to red," and "the meeting went from 3:00 to 4:00." Only in the first phrase is "went" used literally to mean moving from one place to the other. In each of the other instances, "went" is used metaphorically. Our minds seem well equipped to make such connections and to transfer similarities as a way of understanding the new and unknown. Steven Mithen, as well as Gilles Fauconnier and Mark Turner, among others, has argued that metaphor is an essential and rather unique quality of the modern human mind, and it becomes more and more frequent with ever-increasing cognitive complexity, as seen both ontogenetically and phylogenetically.[23] Indeed, the language of children is more literal than adults, and early human language no doubt must have been more literal than modern human language. Metaphor, in this sense, is an indicator of cognitive complexity and flexibility.

It is no surprise, therefore, that science is replete with metaphors. Indeed, most major scientific insights have involved some kind of analogical metaphor.[24] Robin Dunbar has pointed out that metaphorical use of language is especially rampant in particle physics and evolutionary biology. Particle physicists refer to the different kinds of quarks as "top," "down," "bottom," "up," "charmed," and "strange." Such everyday common words could hardly be more literally removed from the abstract, unobserved, and probably unobservable quarks, and yet that is part of the joke or pun of the inventors of these terms. Mathematicians also commonly refer to equations as "beautiful," or "well behaved." Evolutionary biology is also littered with such metaphors: "the selfish gene," "kamikaze sperm hypothesis," or the "red-queen hypothesis" (from Alice in Wonderland), just to name a few. I would also point out that computer science is perhaps even more prone to metaphor: "user-friendly," "mouse," "crash," "boot-up," "file," "window," and so on. As Dunbar goes on to argue, however, there is good reason for such widespread use of metaphor in these fields: they are far removed from everyday experience and we simply have no words to describe the phenomena. Therefore scientists have little choice but to either use metaphors or invent new technical terms. Sometimes scientists do the latter, but more often than not they choose the former, and the terms "strike us" as "warmer," and "friendlier" than technical jargon.

Metaphor and analogy are so common in and out of science precisely because they are so useful to hypothesis and theory formation, thought experiments, creativity, and problem solving. They provide useful constraints to solutions to problems by focusing strategies and preventing random and fruitless searches for a solution. Scientists, and especially the best scientists, tend to use them more readily than novices and thereby go down fewer dead ends when trying to solve a problem. Of course, analogies and metaphors offer such cognitive advantage only if they are appropriate and useful. Often the more creative scientists have a feel for (that is, an intuition for) a good and productive analogy or metaphor. When they are useful, they make problem solving much more efficient than it would be otherwise.

*Visualization.* Another cognitive strategy or heuristic used in but not unique to scientific thought is visual imagery.[25] The historian of science Arthur I. Miller in his book *Insight of Genius* and elsewhere has written at great length about the important role of visualization and imagery in science. After detailing case after case in the history of physics, he summarized his main point about the role of visual imagery in creative scientific thinking: visual imagery plays an important role in scientific creativity (for example, Einstein's thought experiments) and in scientific advance, and it can carry truth value (for example, Feynman diagrams).

Einstein is one of the more noteworthy examples of the importance of visual thinking. The gestalt psychologist Max Wertheimer spent hours and hours interviewing Einstein in 1916 about how his theory of relativity came about, and among other things, Einstein remarked: "I very rarely think in words at all. A thought comes, and I may try to express it in words afterwards." The physical world after all is made up of objects and the ability to think in images would be of great help in solving physical problems. Visual intuition, to be sure, was crucial for all of Einstein's great insights: imagining himself traveling at the speed of light, imagining himself simultaneously jumping off of a roof and dropping a stone (which led to his insight that gravity and acceleration are relative quantities), and imagining two observers riding on either the northern or southern pole of a magnet. Most of these visual thought experiments occurred in the twenty-year period between 1895 and 1915, and by all accounts this was the period of Einstein's singular creative genius. In fact, other physicists have remarked on the lack of creativity in Einstein's later life being connected to his decline in his use of visual imagery. "[Richard] Feynman said to [Freeman] Dyson, and Dyson agreed, that Einstein's great work had sprung from physical intuition and that when Einstein stopped creating it was because 'he stopped thinking in concrete physical images and became a manipulator of equations.'"[26]

Physics as a whole became less comprehensible when it moved beyond what we observe in everyday life and can easily visualize. Quantum mechanics is the prime example because it deals with matter at a fundamentally different level than the observable Newtonian world. From about 1925 (with the death of Bohr's "atom as solar system" model) until 1949, quantum physics moved decidedly away from the use of visual images and toward purely mathematical formulations. In Niels Bohr's 1922 lecture on winning the Nobel Prize for his atomic model, the limitations of his own model were already clear to him: "We are therefore obliged to be modest in our demands and content ourselves with concepts which are formal in the sense that they do not provide a visual picture of the sort one is accustomed to require." In 1949, however, Richard Feynman was to change this by advancing his graphic "Feynman diagrams" to explain how subatomic particles interact. In Feynman's own inimitable language: "What I am really trying to do is bring birth to clarity, which is really a half-assedly thought-out pictorial semi-vision thing. I would see the jiggle-jiggle-jiggle or the wiggle of the path. Even now when I talk about the influence functional, I see the coupling and I take this turn—like as if there was a big bag of stuff—and try to collect it away and to push it. It's all visual. It's hard to explain."[27] Most recently, string theory has added concreteness to a concept that right now is orders of magnitude away from being observed, namely, "strings." Strings are an excellent example of both how visual images and metaphor are used in the construction of scientific theory.

*Creative and Novel Associations.* A crucial component to theory formation is creativity, and I would be remiss if I did not discuss the role of creativity in construction of scientific theory. Not all scientists are equally creative, but some clearly are and they in fact forever change the direction of the field. Newton, Einstein, Darwin, Faraday, and Feynman are luminaries in the history of science because they came up with solutions and ideas that were so novel and adaptive that they permanently changed their disciplines. The science after them was fundamentally different than the science before them.

Creativity is not, as many think, simply novel and original thought, but thought that also provides useful and adaptive solutions to problems.[28] In addition to the cognitive processes already discussed (analogy, metaphor, and visualization), a number of cognitive traits cluster around creative ability: remote and loose associations, overinclusive and disinhibited thinking (latent inhibition), fluency, flexibility, novelty, and originality. Sarnoff Mednick's theory of remote associations underscores the associational richness of creative thinkers. J. P. Guilford built a theory of creativity around ideational fluency, flexibility, and originality, arguing that creative thinking results from having many ideas (fluency) that cross boundaries and categories (flexibility) and that are novel and original. Hans Eysenck proposed that the defining cognitive characteristic of highly creative individuals, besides at least a moderately high level of intelligence, is their overinclusive and disinhibited thinking. In other words, creative people automatically have a wider range of associations and have difficulty inhibiting associations and focusing on a narrow range of relevant stimuli. For this reason, they score high in what Eysenck called "psychoticism." Finally, Dean Simonton, borrowing from Donald Campbell, put forth the chance configuration theory of creativity, which posits that ideational and associational fluency are the foundation for creative thought. In a very Darwinian fashion, ideas are first generated in great number (variation), get combined (chance permutation), and those permutations that are adaptive and useful get selected and reproduced (retention). Just as in biological natural selection, some ideas are adaptive solutions ("hits") and these get selected and retained, whereas others are not adaptive or useful ("misses") and these do not survive.

## COGNITIVE PROCESSES INVOLVED IN THEORY CHANGE

More than any other forces, experiment, data, and evidence are what change scientific theory. In this section, therefore, I will delve into how theory simultaneously guides and stems from observational evidence; how mental experimentation (thought experiments) and actual experimentation (hypothesis testing and/or research) often play a critical role in modifying theory; and how the cognitive heuristics of replication plus extension, confirmation bias, and confirm early–disconfirm late put useful constraints on testing and modifying theory.

### Coordinating Theory and Evidence

As soon as a theory starts to take shape, all kinds of evidence come to bear on its veracity. Some evidence is consistent with the theory, some inconsistent, and much else is irrelevant to the theory. The scientist is one who is most likely to confront inconsistent evidence and ultimately, if the evidence stands, change or modify theory to incorporate the new evidence. Occasionally, a scientist might deny and distort the evidence to fit a pre-existing theory, develop small post hoc modifications of the theory, or develop an entirely new theoretical paradigm to explain these anomalies. But one way or the other, evidence is always the foundation of scientific theory.

Gruber argues that Darwin's life's work consisted of two distinct yet related tasks, both of which are seen most clearly in *Origin.*[29] One task was to develop a comprehensive theory of evolution and how it worked in all forms of life, and the other was to amass the evidence supporting his theory, for he knew all too well his theory would be most controversial. The *Origin* is organized around these two major themes, with the first five chapters spelling out the basic theory and the rest of the book addressing the difficulties and possible criticisms of the theory and laying bare where the evidence supports the theory and where it does not. This scholarly and exhaustive theory-evidence format is no doubt one reason why Darwin's work has become so much more influential and well known than Wallace's less detailed account of the same ideas.

Evidence is not used only after the theory has been fully formed, but rather at each step in theory formation. The process of theory formation is a dynamic and complex process over time, as is easily seen in the case of Darwin. As Gruber wrote based on his analysis of Darwin: "The two kinds of task, theoretical and evidential, entail different activities, and in the long run yield distinguishable products for our consideration. But *in vivo,* in the life of the thinking person, they are thoroughly intertwined. . . . Hard work amassing the facts on a special point is guided by a logical relationship: the theory does not absolutely depend on the facts, nor could the facts ever guarantee the theory. The relation of theory and evidence is not simply logical but psychological."[30]

As we saw more systematically in chapter 3, Deanna Kuhn's research has been guided by the principle that the development of scientific reasoning is the ability to separate theory and evidence. Scientific thinkers, when asked to provide evidence for their theory can clearly provide independent empirical observations that bear on the theory's veracity. Nonscientific thinkers, in contrast, confabulate and confuse theory and evidence and often simply restate theoretical assumptions when asked for evidence. Separation between theory and evidence is therefore a requirement for their coordination. But again, being trained in science is to be trained in developing techniques and methods for testing hypotheses and theoretical assumptions, so a clear separation between theory and evidence is inherent to being a scientist. In fact, we could say that the scientific method codifies and institutionalizes metacognition by externalizing thought in verbal, visual, or mathematical form.[31] One cannot do science without thinking about one's thinking, modifying and coordinating it with evidence, and continuing to make implicit ideas and beliefs explicit. This may be at the root of why the scientific method has been so successful in developing models and theories of how the world works. It is inherently self-modifying and corrective.

### The Role of Experiment in Theory Change

*Thought Experiments.* Theories organize our observations of the world; in fact, we could not perceive without them. But in the end, for some people more than others, there comes a need to test our commonsense folk theories. Doing so is the beginning of science. One of the crucial developments in the history of modern science was when Galileo questioned Artistotlian assumptions, in particular those concerning mechanics. Galileo apparently began with thought experiments that only later graduated to the realm of empirical experiments. The most crucial thought experiment was imaging two objects of different weights connected to each other as they fall. When imagined as two connected objects, even common sense tells us that they must fall at the same rate, an outcome that quite contradicts Aristotle's commonsense assertion, dogmatically accepted for almost two thousand years, that objects of different weights fall at different rates.[32]

Einstein was Galileo's equal if not superior when it came to imagining hypothetical *Gedankenexperimenten* (thought experiments), which were to become the foundation of Einstein's theory of relativity.[33] The starting point for this theory was Einstein, at age sixteen, imagining himself traveling at the speed of light and asking: would I catch up with a beam of light that is emitted from a place moving at a different speed? According to Newton principles of adding velocities, he should be able to catch up with it, but intuitively he knew that he would not. His theory of relatively was developed in response to and as a solution to this paradox, and he was the first to realize intuitively and mathematically that light is a constant and a limit of nature. Indeed, with such visual images we often see the confluence of many of the cognitive processes discussed in this chapter: thought experiment, analogy and metaphor, intuition, and insight.

*Hypothesis Testing.* As useful and necessary as thought experiments have been in the history of scientific revolutions, laboratory experiments—and their related offshoot hypothesis testing—are the bread and butter of science. In science it is not enough to form an explicit theory. A theory must be stated in testable form, and explicit hypotheses must be put forth and put to empirical test.

Inspired by Popper's assertion that science should, and that the best science does, progress by falsifying hypotheses, not by proving them right, Peter Wason decided to find out whether students (novices) propose and test hypotheses systematically and, more specifically, whether they would seek to falsify or only confirm their theories.[34] Wason's research has since become a classic in the field of cognitive psychology of science. The task put to the students involved asking them to determine the rule governing a sequence of number triplets, given that the triplet "2–4–6" was an instance of the rule. When they felt they were ready, participants would test their hypothesis by telling the experimenter what they thought the rule was, and the experimenter would tell them whether it was the rule or not. If they did come up with the correct response, participants could continue to propose triples and make guesses. Participants typically proposed triples like "6–8–10" and "10–12–14" and guessed the rule was something like "numbers must go up by twos." In fact, it was simply "all three numbers in the triple must ascend in order of magnitude." Wason answered two important questions with this research: first, at least some students do test hypotheses, but second, those who tested their hypotheses did so more by trying to confirm rather than disconfirm their theories. In other words, their hypothesis testing was biased toward searching for only confirming evidence.

A related hypothesis-testing task was a four-card task, now known as the "Wason selection task" that tests students' ability to falsify their theories. There were a few variations of the task, but the prototype was as follows: students were asked to test the rule "if a card has a D on one side, it has a 3 on the other." They were then shown four cards and asked which cards they would need to turn over to test the validity of the rule. The four cards had either a letter (D or F) or a number (3 and 7) on it. Logically, the 3 card is irrelevant because the rule read "if a card has a D on one side, it has a 3 on the other" not "if a card has a 3 on one side, it has a D on the other." More formally, "if p then q" is violated only when p is true and q is false. The critical cards to test the rule (that is, the only ones that can falsify rather than confirm the rule) are therefore the D and 7 cards. But most students (between 75 and 95 percent) and even scientists or those who have taken logic courses get this wrong and fail to pick both of the correct cards. For instance, other cognitive researchers have shown that only between 10 to 20 percent of statisticians, physicists, biologists, sociologists, and psychologists select both of the critical cards.[35]

As Wason as well as the psychologist Leda Cosmides and the anthropologist John Tooby later aptly demonstrated, people's reasoning ability is not as poor as these classic results imply.[36] The context of the task matters, and so in a more realistic, concrete, and social context the same task (same structure, different specifics) is often solved correctly. For instance, pretend you are a bouncer at a bar and the rule is "if a person is drinking beer, he must be eighteen or older" (note the rule has same structure as "if D then 3"). Now, you are asked to check either ID or drink of four people: a beer drinker, a Coke drinker, a twenty-five-year-old, a sixteen-year-old. Who would you check? Most people (about 75 percent) get this task correct and would check the beer drinker and the sixteen-year-old. The structure of the task is identical to the more logical and abstract D-3 version above, and yet people test hypotheses much more reasonably.

Why? According to Cosmides and Tooby, their version of the task is testing an implicit social contract and humans have evolved cognitive mechanisms for detecting cheating on social contracts, not for abstract logical problems. According to Wason, the increase in rationality stems from the ability to give a unified representation to the material.

Another, perhaps related, heuristic involved in successful hypothesis testing is the ability to think simultaneously of two or more "problem spaces," where "space" denotes a set of possible alternatives. David Klahr and his colleagues asked participants to learn how a device called a "big trak" functions by conducting experiments.[37] They found that it was most useful to think metacognitively in terms of separate problem spaces, where one space contained ideas for possible experiments and another contained space for possible hypotheses. The most successful participants reacted to falsificatory evidence in the experimental space by developing new hypotheses that represented a shift in the way they thought about the function of the device, which in turn suggested new areas of the problem space to search for evidence.

*The Replication-Plus-Extension Heuristic.* One of the limitations of most abstract tasks and simulated scientific problems and hypothesis testing is that they include no possibility of error in the results, even though working scientists struggle constantly to separate "signal" (effect) from "noise" (error).[38] Michael Gorman told participants that anywhere from 0 to 20 percent of their results on an abstract task similar to the Wason 2–4–6 task might be erroneous (that is, a trial that was classified as inconsistent with the rule might be consistent and vice versa). Errors would occur at random, as determined by a random number generator on a calculator. Initially, the error rate was set at zero; participants encountered no actual errors. Gorman found that participants used "replication-plus-extension" to eliminate the possibility of error: they proposed experiments that were similar to, but not exactly the same as, previous experiments in an effort to replicate the current pattern.

But replication and replication-plus-extension are costly heuristics: they require a significant investment of laboratory time and resources while competing laboratories may be pursuing novel research.[39] The cost and complexity of replication can be increased on experimental tasks that incorporate the possibility of error. For example, when participants have to replicate an entire sequence of experiments rather than just a single result, the possibility of error encourages hypothesis preservation, or a reluctance to discard a hypothesis in the face of occasional disconfirmation. When the possibility of error is 20 percent, participants had even more difficulty using replication and replication-plus-extension to combat hypothesis preservation. Even on very simple artificial tasks, replication alone is not sufficient to isolate and eliminate errors. Obviously, scientists rely on other kinds of checks in addition to replication, for instance, refinement of procedures. Future experiments on error should use scientific problems and tasks that simulate them and also compare the performance of scientists to novices.

*Confirmation Bias.* One of the more pervasive topics in the cognitive psychology of science has been the tendency to selectively look for and latch onto evidence that confirms our theory and to deny, distort, or dismiss evidence that contradicts it. One of the first to put confirmation bias on the front burner of the cognitive psychology of science was Peter Wason. Wason and many others who have followed his lead have consistently demonstrated that when students test their hypotheses about how something works, they are very much disposed toward positive tests, that is, they only propose tests that support rather than refute the theory. The best scientists do this relatively infrequently and nonscientists or novices do it more frequently, but scientists are not immune to such biases. Michael J. Mahoney, in fact, compared a small sample of scientists working on the 2–4–6 task to a sample of Protestant ministers and found that the former were more prone to confirmation bias than the latter! Arie Kruglanski has argued that scientists are subject to some of the same cognitive biases as nonscientists, including confirmation.[40]

Falsification, however, is not impossible and confirmation bias is not inevitable in nonscientists. Michael Gorman and his colleagues, for example, found that instructions to falsify significantly improved performance on the 2–4–6 and similar tasks in various samples of college students.[41] So it appeared that confirmation was a bias that can be combated with education, and indeed training to be a scientist provides just such education.

*Confirm Early–Disconfirm Late Heuristic.* It is somewhat misleading to argue that confirmation bias is the only strategy involved through the hypothesis-testing process. Scientists, and especially the best scientists, appear to use a more complex heuristic: early in theory formation they look for confirming evidence, but once the theory is well developed they look for disconfirming (falsifying) evidence. This heuristic has been given the label "confirm early–disconfirm late" by the cognitive psychologist Ryan Tweney, and it echoes an important distinction first made in the 1930s by the philosopher Hans Reichenbach between the stage of discovery and the stage of verification. Early on during the discovery phase, the confirmation heuristic is most useful, but later during the verification phase, the most productive strategy is to seek to disconfirm. We see such changing strategy in Darwin's development of his theory. Darwin's first theory of evolution (monadism–spontaneous development of organic life from inorganic forms) had many dead ends, but it contributed one very fruitful component to his final theory: the tree branch model (theory). Darwin had some evidence that his theory did not explain things too well, but he did not give up on it immediately. He continued to work with his theory until he hit upon a better alternative.[42] One reason for Darwin's delay in publication, in fact, can be attributable to his desire to pre-empt every legitimate criticism, and, indeed, he was able to reflect at great length on the theory's weakness in many attempts over the years to disconfirm it.

Physicist Michael Faraday also used this strategy. Tweney constructed detailed problem-behavior graphs of Faraday's problem-solving processes.[43] Faraday wrote about the dangers of "inertia of the mind," by which he meant premature attachment to one's own ideas, but he also argued that it is important to ignore disconfirmatory evidence when one is dealing with a new hypothesis. In general, Faraday followed the "confirm early–disconfirm late" heuristic: confirm until you have a well-corroborated hypothesis, then try to disconfirm it. For example, his initial attempts to use magnets to induce an electric current produced apparent disconfirmations, but he ignored them—a single confirmation was more powerful than half a dozen disconfirmations, especially given the high possibility of error in his initial experiments. When he obtained a more powerful magnet, he was able to reduce the level of noise and obtain consistent confirmations.

The tendency to hold on to theories early and only later attempt to disconfirm them is seen empirically in college students as well. Clifford Mynatt, Michael Doherty, and Ryan Tweney developed an artificial universe task that bore more resemblance to real-life science than abstract problems like the 2–4–6 Wason problem.[44] Participants spent about ten hours on the most complex of these tasks, and none of them discovered the rule. The ones who made the most progress exhibited a kind of confirmation bias, but with one important qualification: confirmation bias is most effectively used early in the hypothesis-testing process. Tweney and colleagues concluded that confirmation was an effective heuristic early in the inference process; once a subject or scientist had discovered and verified a pattern, then she could switch to the search for disconfirmatory evidence.

*Multiple Hypotheses and Strong Inference.* As confirmation bias research demonstrates, even expert scientists are not immune from being biased toward evidence that supports their hypothesis. To counteract this tendency, the notion of "strong inference" as a guard against "parental affection" of one's ideas originated with the geologist T. C. Chamberlain in the late nineteenth century. Because his writing on this subject is so simple and elegant, it should be quoted at moderate length: "The moment one has offered an original explanation for a phenomenon which seems satisfactory, that moment affection for his intellectual child springs into existence; and as the explanation grows into a definite theory, his parental affections cluster about his intellectual offspring, and it grows more and more dear to him, so that, while he holds it seemingly tentative, it is still lovingly tentative, and not impartially tentative. So soon as this parental affection takes possession of the mind, there is a rapid passage to the adoption of the theory. There is an unconscious selection and magnifying of the phenomena that fall into harmony with the theory and support it, and an unconscious neglect of those that fail of coincidence."[45]

Recent research in the cognitive psychology of science has supported the idea that multiple hypothesis search is a better strategy than single hypothesis search and that scientists are more likely than novices to use such a strategy. Searching for two complementary rules (hypotheses) rather than one appears to increase successful hypothesis testing on abstract tasks. Many cognitive psychologists in the 1980s and 1990s altered Wason's card task to make it a search for two complementary rules rather than a single rule.[46] They found this change made it much easier for participants to explore the limits of their hypotheses, thereby facilitating discovery of the target rule, and that experts were better at this multiple-hypothesis testing than novices. Similarly, Eric Freedman argued that limits on working memory may explain why novices tend to focus on one hypothesis at a time. So, explicitly instructing participants to search for multiple hypotheses (that is, use strong inference strategies) did in fact enhance their ability to test hypotheses successfully.

### Computer Simulation of Scientific Discoveries

The work of cognitive psychologists of science has led to uncovering so many cognitive strategies or heuristics in human scientific reasoning that one might wonder whether one could program computers to use these heuristics and "discover" and "solve" scientific problems. Because this is a variation of the "artificial intelligence" question (whether we can get computers to really think like humans and not simply process information), the answer is "yes," some researchers have tried to simulate the scientific discovery process with computers.[47] The lead psychologist in this movement was the Nobel laureate Herbert Simon. Simon and a group of colleagues at Carnegie-Mellon developed a series of computer programs designed to emulate scientific discoveries, for example, Kepler's Law. The simplest of these, called BACON, was given columns of numbers and asked to find a relation, using heuristics like "if the terms in two adjacent columns increase together, compute their ratio." The relation turned out to be the numerical equivalent of Kepler's Law.

A more complex program developed by Depaak Kulkarni and Herbert Simon used a detailed study of Krebs's discovery of the ornithine cycle to create a computer simulation that followed Krebs's discovery process as closely as possible. KEKADA, as the program was called, relied on a dual-space search and a hierarchy of heuristics to accomplish this goal. The hierarchy included general heuristics that could have been used across a wide range of scientific problems and specific ones limited to the domain of organic chemistry. One of the conclusions from KEKADA is that experts possess both general and domain-specific heuristics, whereas novices are more likely to possess only the more general ones. There were even some heuristics possessed only by Krebs and a few others, including a tissue-slicing technique that greatly facilitated the discovery.

Computer simulations are beginning to use such strategies as visualization. Peter Cheng and Simon showed that it might have been easier for Christiann Huygens and Christopher Wren to have discovered the law of conservation of momentum using diagrams rather than deriving it from theory or by data-driven processes similar to those used by BACON. Cheng then created HUYGENS, a more general computational simulation of discovery, which used both diagrams and a kind of dual-space search: from given numerical data, HUYGENS switches to a space of diagrams in its search for regularities by looking for patterns in the diagrams. When patterns have been found, the regularities are simply transformed back into equations. The change to diagrammatic representation permits different operators, regularity spotters, and heuristics to be employed that are more effective than those used in the direct search of a space of algebraic terms. Cheng admits that we cannot be sure Christiaan Huygens used this method—but it is plausible historically, and HUYGENS demonstrates that it would have been more efficient than alternatives. Instead of claiming he developed a program that discovers, Cheng argued instead that he provided computational evidence for the importance of using diagrams in scientific discovery, evidence that could be combined with material from other sources (for example, fine-grained case studies of the way diagrams are used in actual discoveries). Cheng's goal appears to be to provide both a normative account—how diagrams should be used to discover—and a historically plausible one—this is how diagrams probably were used by Huygens.

Other cognitive psychologists have also attempted to simulate scientific discovery with computer programs.[48] Paul Thagard, for instance, proposed a theory that the scientific hypothesis with the most explanatory coherence wins in disputes. He developed a computer simulation program ECHO and applied it to the oxygen-phlogiston debate and the controversy surrounding the extinction of the dinosaurs. This simulation is directed more toward testing philosophical norms for settling controversies than for emulating the psychological processes of participants.

Proponents of the computer simulation approach argue that it can lead to better science. To use the words of David Klahr and Herbert Simon: "What researchers learn about the science of science leads to a kind of engineering of science in which—as in other areas—knowledge of a natural process can be used to create an artifact that accomplishes the same ends by improved means."[49] Such simulations, however, will always need psychologists of science to supply detailed data on human processes in similar domains. It is exactly this sort of rich and detailed data that make simulations like KEKADA, BACON, and HUYGENS so powerful. There is little doubt that computers will, if they have not already, develop the computing power to find new solutions to real scientific problems. The question remains, however, whether they will be emulating or assisting the human discoverer.

This is not to say there are no limitations to or criticisms of the computational approach. The major inherent limitations to such simulations will always involve the power of intuition, aesthetics, fuzzy logic, and even analogy and metaphor.[50] Humans are very gifted at perceiving, thinking, and reasoning about boundary, nonprototypic, nonliteral, analogous, and ambiguous categories of thought and perception. Indeed, as we have seen with the greatest scientists and mathematicians, having a strong intuitive sense of the crucial problems and their solutions is absolutely crucial for most scientific breakthroughs. Because intuition by definition is "knowing without knowing how you know," programmers stay away from such programming tasks. In the end, computers are extremely "talented" at well-defined, clear tasks (closed logical systems), but they have an impossible time "thinking outside the box," which is, after all, the essence of human creativity. The biggest limitation of all is that no computer, not even IBM's "Big Blue," which has beaten the best human chess player in the world (Gary Kasparov), will be able to do very well at more than a few specialized tasks. Big Blue, to be sure, can do only *one* thing: play chess. Ask it to fry an egg, and it would do nothing but try to figure out its first move on a chessboard. In short, artificial intelligence will not in the foreseeable future and maybe not ever be able to hold a candle to the general and changing abilities of the human mind (which are, we must remember, the result of millions of years of evolution). After all, it is our ability to adapt and change to an ever-changing environment that lies at the foundation of our success. If computers are to compete, they too must be able to "evolve" to a changing environment. Jeff Shrager and Pat Langley, in an excellent and sympathetic volume on computational simulations of scientific discovery, describe "two important aspects of intellectual activity—embedding and embodiment—that have significant bearing on science but that have not been addressed by existing computational models. Briefly, science takes place in a world that is occupied by the scientist, by the physical system under study, and by other agents, and this world has indefinite richness of physical structure and constraint. Thus the scientist is an embodied agent embedded in a physical and social world."

## COMPARING REASONING AND HEURISTICS USED BY SCIENTISTS AND NONSCIENTISTS

Although some of the research reviewed thus far in this chapter has compared and contrasted the cognitive strategies and principles used by scientists and nonscientists, there is enough residual research using this method to warrant a separate section.

### Experts More Readily Modify Hypotheses

Sometimes scientists may have biases *against* their proposed theories. Kevin Dunbar's analysis of real scientific thinking during in vivo laboratory meetings revealed an important finding relevant to (dis)confirmation bias. He found that when confronted with a disconfirmatory result, scientists typically did one of three things: they either changed a corollary assumption of the current hypothesis, attributed an anomalous result to error, or displayed a "falsification bias," discarding results that appeared to confirm a hypothesis. Dunbar speculated that this falsification bias was a protection against airing hypotheses that might later be proved wrong, a frequent experience for the senior scientists. Dunbar also explicitly compared experts (scientists) to novices (students). Experts were more willing to modify or discard hypotheses than novices. Part of this willingness came from the fact that group interaction helped scientists articulate alternate hypotheses. In scientific practice, much of the coordination between hypothesis and evidence goes on in groups.[51] Perhaps that explains the apparent difference between Tweney's and Dunbar's results: Tweney studied a detailed record of Faraday's experiments, and Dunbar focused on laboratory meetings. Here the cognitive psychology of science begins to merge with the social psychology of science—conceptual change occurs in a group setting.

### Experts Are Cognitively Complex

The study of dispositional cognitive styles, such as integrative complexity, provides a link between cognitive and personality psychology of science. Integrative complexity is a measure of complexity of thinking and is divided into two components: differentiation and integration.[52] The simple thinker makes relatively few qualifications and sees things in black-and-white terms. In contrast, the complex thinker not only makes distinctions and qualifications but integrates into a synthetic whole the opposing points of view. Only two studies have been conducted on integrative complexity in scientists. The founder of the field of integrative complexity, Peter Suedfeld, reported not only that the American Psychological Association (APA) presidents had the highest complexity means compared to all nonscientist samples, but that the most eminent psychologists gave the most complex presidential addresses. In addition, I interviewed a group of eminent scientists and, among other things, had them respond to a set of semistructured questions, which were transcribed and coded on integrative complexity. The mean levels of complexity in these physicists, chemists, and biologists were even higher than those in the Suedfeld study. To be clear, these eminent scientists were complex thinkers when it came to their research but not about other issues (such as science education).

### Novices Solve Problems and Evaluate Evidence Based More on Common Sense

In contrast to novices who tend to form commonsense representations, expert scientists form abstract representations of scientific phenomena. For example, when asked to predict whether a yo-yo on a table will roll to the left or right when one pulls on a string, nonscientists (novices) say right based on their commonsense experience with yo-yos, whereas scientists (experts) classify the problem in terms of momentum and force equilibrium and conclude the yo-yo will move to the left. Similarly, Kathleen Hogan and Mark Maglienti recently reported that the major difference between experts (scientists and technicians) and novices (students and nonscientist adults) in evaluating evidence from a hypothetical study was that scientists primarily used the criterion of how consistent the conclusion was with the data, whereas novices were more likely to evaluate it based on its consistency with their own beliefs (common sense).[53]

The history of science also provides a lesson about the development of scientific thinking in scientists and nonscientists.[54] For example, echoing an argument by Piaget, Michael McCloskey found that college students held beliefs about momentum that resembled those of Philoponus (sixth century) and Buridan (fourteenth century). These historical analogies suggest that novices form a kind of "commonsense" representation of scientific phenomena.

### Experts Use "Intuition" and Discover Analogies

Jill Larkin and her colleagues used kinematics problems from an elementary physics textbook and compared how scientists and nonscientists approached them.[55] Experts, of course, solved the problems much more rapidly and with fewer errors than novices. But the critical finding concerned the cognitive mechanisms that explain expert superiority: they used chunking and node-linked representations of large quantities of learned material to hone in quickly and efficiently on the problem solution. Indeed, their explanation for "physical intuition" resides in the expert's ability to almost immediately form schemata using patterns and node-linked representations. Similarly, experts worked forward from the information given, reasoning qualitatively until they arrived at a representation that suggested what set of equations to use. Novices, in contrast, worked backwards from the possible solution, applying equations early in the hopes of finding the values of specific variables.

In addition, the better and more creative scientists are more likely to discover useful analogies.[56] For instance, John Clement compared the way technical experts and novices solved more unusual problems, such as determining what happens when the width of the coils on a spring is doubled and the suspended weight is held constant. Experts used informal, qualitative reasoning processes; for example, they often constructed an analogous simpler case, imagining what happens if the coils were replaced by a U-shaped spring of the same length. Then they related the analogy to the case. Novices were not able to use such simple analogies. Additionally, Dedre and Donald Gentner demonstrated that novices who used a flowing waters analogy to understand electric circuits formed a mental model that was appropriate for battery problems but not for ones involving resistors. Finally, Kevin Dunbar, in his study of molecular biology laboratories, noticed that the least successful of his four laboratories used virtually no analogies, whereas the other three used local analogies to change representations and procedures. A local analogy involves drawing on a similar experiment to solve a problem with the current one. Dunbar also noted that expert scientists made more analogies than relative novices because the deep, structural features of a domain were obvious to them and they could therefore map them readily onto other domains.

In sum, scientists create scientific theories by building models that very often make use of analogy, metaphor, thought experiments, and visual imagery. They test their theories in dynamic interchange with observation and empirical evidence, making use of such heuristics as "confirm early–disconfirm late" or "replication-plus-extension." The more successful and creative scientists are metacognitive, flexible, and likely to use the strong-inference technique of testing more than one hypothesis at once. Lastly, there are four general conclusions from the literature comparing novice to expert reasoning: (1) experts are more willing to discard hypotheses than novices; (2) scientists are more integratively or cognitively complex (at least about certain domains) than nonscientists; (3) nonscientists solve problems and evaluate evidence based more on common sense than scientists; and (4) scientists are more likely use intuition and analogy.

## **Chapter 5** Personality

## Psychology of Science

Humans are not alone in the uniqueness and variability of individual members of the species. Individuals within every living species exhibit differences or variability. In fact, variation and selection are the two cornerstones of evolutionary theory. But the degree to which individual humans vary from one another, both physically and psychologically, is quite astonishing and somewhat unique among species. Some of us are quiet and introverted, others crave social contact and stimulation; some of us are calm and even-keeled, whereas others are high-strung and persistently anxious. What is the connection between our personalities and our scientific interest, ability, and talent? Moreover, what makes some people more likely to become scientists than other people?

### THE NATURE OF PERSONALITY

There is something quite specific in what psychologists mean by the word "personality." Uniqueness and individuality are one core component; if everyone acted alike and thought alike, there would be no such thing as personality. This is what is meant by "individual differences." The second major component of personality is "behavioral consistency," which is of two kinds: situational and temporal. Situational consistency is the notion that people behave consistently in different situations, and they carry who they are into most every situation. Temporal consistency, in contrast, is the extent to which people behave consistently over time. To illustrate both forms of consistency as well as individual differences in the context of personality, let us take the trait of friendliness: We would label a person as "friendly" only if we observe her behaving in a friendly manner over time, in many different situations, and in situations where other people were not friendly. In short, personality is what makes us unique and it is what is most stable about who we are.

Over the last sixty years, the field of personality psychology has debated the question of both the number and the structure of fundamental dimensions of personality.[1] That is, what are the universal dimensions of personality and how many are there? Although, as with every academic debate, there is some disagreement concerning the answers to these questions, during the last fifteen years a surprisingly clear answer has begun to emerge. It has been labeled the "Big-Five" or "Five-Factor" Model. According to this model, there are five major dimensions to human personality: anxiety (neuroticism), extroversion, openness, agreeableness, and conscientiousness. Moreover, each dimension is both bipolar and normally distributed in the population. That is, for instance, anxiety is one pole of the dimension, with emotional stability being its opposite. Everyone falls somewhere on the continuum from extremely anxious to extremely emotional stable, with most people falling in between. In addition, just like intelligence, if we were to plot everyone's score on a frequency distribution, we would get a very nice bell-shaped (normal) distribution of scores, with about two-thirds of the scores falling within one standard deviation of the mean. The same is true with extroversion-introversion, openness-closedness, agreeableness-hostility, and conscientiousness-unreliability.

### Origin and Function of Personality Traits

The approach that I will take here is an evolutionary and functional one. I assume that traits have evolved as adaptive behavioral shortcuts to fundamental problems (survival and reproduction). That is, a disposition is nothing other than a quick ready-made pattern of response to a given situation—a behavioral heuristic as it were. Certain behaviors were useful and adaptive for survival or reproductive success during early periods of human evolution, and these predispositions to certain behaviors were therefore products of either natural or sexual selection. As David Buss points out, many evolutionary psychologists have ignored individual differences and have focused instead on species-typical mechanisms. He goes on to ask why genetic variability exists because natural selection tends to eliminate genetic variability. As Geoffrey Miller persuasively argues, however, much of this neglect of individual differences and the enigma of human genetic variability result from a lack of appreciation by many evolutionary psychologists of the uniqueness and power of sexual selection. He argues that many such individual differences are more the product of sexual than natural selection. Hence, these traits may serve as fitness indicators during the competition between individuals for sexual reproduction. In short, sexual selection tends to magnify individual differences, whereas natural selection tends to eliminate individual differences. Indeed, David Buss acknowledges the importance of sexual selection in personality differences when he writes "personality characteristics such as dominance, friendliness, and emotional stability are intimately tied with sexual selection in that they are central to mate choice."[2]

Whatever the selection pressure, and whether they are more by-product or adaptation, the basic personality dimensions commonly known as the "Big-Five" (neuroticism, extroversion, conscientiousness, agreeableness, and openness) have emerged because they have proven useful for solving basic survival and mate-choice problems.[3] In that sense, we could say these dimensions are fundamental to the human condition. Neuroticism, for instance, may well have been quite adaptive in that it provides a beacon for danger and threat. "No fear" may be a macho ad campaign, but if taken too literally, it would quickly lead to the extinction of the species. By the same token, the other extreme—hypersensitivity to threat—is debilitating and disruptive to normal everyday functioning. A balance is usually struck between these two extremes, and hence personality traits show the classic bell-shaped distribution in the population, with very few people being on the extreme ends of the curve and most being in the middle.

Another function of traits is to lower behavioral thresholds, that is, being high on a given trait lowers the threshold for its expression in a given situation.[4] From a functional perspective, traits are dispositions to behave and serve as ready-made response options. They do not cause behavior but rather make particular behaviors more likely in particular circumstances for particular people. To be concrete: if a shy person were being introduced to a large group of people, that person's threshold for embarrassment and feeling awkward will probably be surpassed and she will experience quite a bit of discomfort. An extroverted person, on the other hand, will have a much higher threshold for embarrassment and feel quite comfortable in the same situation. Their optimal levels of arousal are different. Or, being agreeable and friendly means having a relatively low behavioral threshold for expressions of warmth and consideration for other people's feelings, so in a particular situation a person who is high on the trait "agreeable" will be more likely to express those kinds of behaviors. Being "anxious" means that one has a lower threshold for experiencing and expressing anxiety in a given situation. In short, traits lower behavioral thresholds.

### Genetics and Personality

There are two major categories for how biological and behavioral scientists study the relation between genetics and personality. The first and most direct method for establishing a genetic basis for personality traits is known as the "quantitative trait loci" (QTL) approach (see chapter 2). The essence of this approach is "not to find *the* gene for a particular personality trait, but rather *some* of the many genes that make contributions of varying effect sizes to the variance of the trait."[5]

QTL research has begun to uncover some of the many genetic markers involved in basic nonpathological personality traits, such as novelty- or thrill-seeking, neuroticism/anxiety, and indirect genetic markers in potentially pathological levels of aggression, sexuality, impulsivity, or lack of constraint.[6] To take just one example: thrill seeking appears to be associated with long repeating sequences of base pairs (rather than short sequences) in an allele for a dopamine receptor (DRD4) on chromosome 11. Having a short form of the allele appears to result in a more efficient method for binding dopamine, and hence having a long repeating sequence rather than short repeating sequence results in a dopamine deficiency. This deficiency in turn is associated with a greater need for thrill seeking, such as riding roller coasters or playing the stock market. Dopamine appears to be related to the experience of pleasure, and those with low baseline levels would then naturally seek experiences that give them a rush, that raise their low levels of dopamine as it were. It is important to point out, however, that QTL research is just a first step toward locating a particular region on the chromosome rather than the specific gene involved, and in that sense it is a beginning rather than an end in the process of uncovering the genetic basis for personality traits.

Yet there is a second, more traditional, method for examining the effect that genetics plays in behavior and personality, and that is through studying twins, both identical (monozygotic) and fraternal (dizygotic), who have been raised together and apart. As discussed in chapter 2, the logic of this twin-study approach is simple yet powerful. Identical twins are essentially 100 percent genetically alike, whereas fraternal twins, like all siblings, are on average 50 percent genetically alike. If genes play little role in personality, then identical twins (who are genetically identical) reared apart should be no more alike than any two people chosen at random from different environments. And yet, if environment plays little role, two identical twins separated at birth should be very similar—genes should influence personality regardless of environment. The emerging conclusion from twin-study research is that most basic personality traits have heritability estimates of between 40 and 50 percent. For instance, extroversion often correlates around .50 for monozygotic twins and around .20 to .25 for dizygotic twins, which leads to a heritability coefficient of between .50 and .60 (the simple model of heritability is calculated as twice the difference between mono- and dizygotic twin correlations). Neuroticism tends to have a heritability of about 50 percent, and conscientiousness, agreeableness, and openness slightly less, at between 40 and 50 percent. Such a figure leaves roughly 50 percent of the variance to be explained by three nongenetic sources: shared environment, unshared environment, and error. The surprising conclusion of recent researchers is that most of the environmental effects are of the unshared kind, such as birth order or different peer groups, and almost no variance is explained by shared environment. That is, environment does influence personality structure but not the environment that most people think, namely, growing up in a particular household. The environment that seems to matter most is the "unshared" environment between siblings, that is, their having different birth orders or different peer groups or even changes in parenting style and attitudes over time.[7]

These genetic and central nervous system (CNS) differences, in turn, have their most direct behavioral effect on temperament, that is, "a small number of traits that are present early in life, are biologically rooted, and relatively stable."[8] Temperament is the foundation for personality development and such differences in basic personality dimensions as extroversion, neuroticism, and conscientiousness. As Mary Rothbart and her colleagues put it: "Temperament arises from our genetic endowment. It influences and is influenced by the experience of each individual, and one of its outcomes is the adult personality."

There are a number of different theoretical approaches to temperament and they do not all agree on the dimensions that constitute temperament. There is not yet a widely accepted taxonomy of temperament, as exists in the field of personality (with the "Big-Five" Model; see above). Some temperament researchers argue for three, and other researchers argue for nine distinct categories of temperament. On the low end, Arnold Buss and Robert Plomin argue for the dimensions of activity level, sociability, and emotionality. On the high end, and perhaps one of the most influential models of temperament, Alexander Thomas and Stella Chess propose nine dimensions: activity level, approach-withdrawal, adaptability, emotional intensity, mood, persistence, distractibility, threshold, and rhythmicity.[9] For instance, some infants are generally more active than inactive; are more approaching than withdrawing in novel social situations; are more often in positive than negative mood states; and some have lower thresholds for stimulation; finally some are more regular and predictable in their feeding, eating, and eliminating cycles (rhythmicity). These are the types of things parents are likely to notice about how their children differ from their very first day of life. It is my argument that some of these temperamental and personality dispositions function to lower thresholds for interest, talent, and achievement in science. That is, they make becoming a scientist more likely and influence both the kind and quality of scientist one becomes. The question now is which specific traits lower these thresholds.

## PERSONALITY AND SCIENTIFIC INTEREST

With this background in the function of traits and behavioral genetics, what are some of the traits that make scientific interest and scientific creativity more likely? Research on the personality traits associated with scientific interest and creativity has existed for more than 125 years. In 1874 the British statistician and psychologist Francis Galton (Darwin's first cousin) published the first scientific investigation of the psychological characteristics of geniuses, including scientists.[10] In this sense Galton can be given the label the "first psychologist of science." He collected qualitative self-report data from 180 English men of science and found that they were distinguished by their high level of energy, physical health, perseverance, good memories, and remarkable need for independence.

In the early part of the twentieth century in the United States the study of genius was furthered by James McKeen Cattell (the first professor of psychology in the world, at the University of Pennsylvania), Lewis Terman (the founding figure of IQ tests in the United States), and his student and colleague Catherine Cox.[11] Under the guidance of Terman and using J. Cattell's eminent sample as participants, Catherine Cox carried out the most ambitious, systematic, and most quantitative of the early investigations into genius. Although she did not focus exclusively on scientists, she did report findings broken down by group. Cox found the traits that most clearly distinguished scientists from nonscien-

Table 5.1 Personality Traits More Salient of Scientists Than Nonscientists

| Trait Category | | |
|---|---|---|
| Social | Cognitive | Motivational |
| Dominant | Open | Driven |
| Arrogant | Flexible | Ambitious |
| Hostile | | |
| Self-confident | | |
| Autonomous | | |
| Introverted | | |

*Source:* Feist 1998.

tific eminent men were the desire to excel, originality, reason, tendency not to be changeable, determination, neatness, and accuracy of work.

Since the 1950s, however, more systematic work has focused on personality and scientists, with "scientists" being defined as any sample that consisted of either students of science, engineers, inventors, social scientists, biological scientists, or natural scientists. The literature on personality and creativity has advanced to the point that a quantitative review (meta-analysis) has now been published.[12] Meta-analysis is a quantitative, rather than qualitative review of the literature, that is, the outcomes of studies are quantified by means of their "effect size." Instead of just reading much of the existing literature and drawing one's own conclusions, in meta-analysis, one quantifies each study's result using effect size and then calculates to overall effect size for all studies combined. The conclusions listed in table 5.1 summarize results of a meta-analysis conducted on the twenty-six published studies comparing personality traits of nearly five thousand scientists or science-oriented students to nonscientists.

One way to organize the literature is by putting the dispositions into higher-order categories, such as cognitive, social, motivational, and affective. By so doing, dispositions are organized into related clusters. Whether a trait is social or not is determined by the extent to which it concerns one's attitudes toward and interactions with others. For instance, the tendencies to question social norms and to be relatively independent of group influence are social dispositions commonly found in creative people. Also, having a greater than normal desire to remove oneself from social interaction and being overstimulated by novel social situations (introversion) is frequently observed in highly creative people, including the sciences.

### Cognitive Traits

One of the more robust if not terribly surprising findings from the research on scientists' personalities is their higher levels of *conscientiousness,* that is, their desire for order, organization, and punctuality. The overall effect size on conscientiousness showed that scientists on average were about a half a standard deviation higher than nonscientists. More specifically, conscientiousness is made up of traits such as "careful," "cautious," "conventional," "disciplined," "orderly," "persevering," "reliable," and "self-controlled." Scientists tend to prefer settings that allow for individual expression and yet are structured. In short, being high on conscientiousness lowers one's threshold for being interested in or having a career in science.

The other primarily cognitive trait that distinguishes scientists from nonscientists is *openness to experience,* which is comprised of traits such as "aesthetic," "creative," "curious," "flexible," "imaginative," and "intelligent." The effect size, however, is a bit smaller than on conscientiousness, about a third of a standard deviation. The open person seeks out new experiences, is curious about the world, and is relatively flexible in his or her ideas. One defining characteristic of being a scientist is being willing to admit when you may be wrong once the evidence shows this to be the case. One cannot be too set in one's ways in science because nature has a way of humbling even the best of ideas. Additionally, one must be able to attack problems from different angles if one is to solve problems others have been unable to resolve.

### Social Traits

Scientists are more *dominant, assertive,* and maybe even *hostile* in personality than nonscientists.[13] For example, in a classic study by Raymond Van Zelst and Willard Kerr, personality self-descriptions were collected on 514 technical and scientific personnel from a research foundation and a university. Holding age constant, the researchers reported significant partial correlations between productivity and describing oneself as "argumentative," "assertive," and "self-confident." Dominance and assertion appear to be distinguishing characteristics of female as well as male scientists. For example, in one study of 116 female biologists and chemists listed in *Who's Who in America* and *Who's Who of American Women* compared them to female norms. Using R. B. Cattell's 16 Personality Factor questionnaire as the measure of personality, the researchers found that the women scientists were more dominant, confident, intelligent, radical, and adventurous than women in general. Furthermore, the personality profile of female scientists was quite consistent with that of male scientists. Two of the three studies that found scientists to be less dominant than a comparison group were based on samples of female scientists compared to female artists. The few studies that found no relation between dominance and creativity were on young student samples, so it is possible that dominance is an effect more than a cause of creative behavior. But not enough longitudinal research has been conducted to warrant such a conclusion at this time.

Scientists, relative to nonscientists, do prefer to be alone and are somewhat *less social and less affiliative.* Such a finding is somewhat more true of physical scientists and mathematicians than of social scientists. Scientists, as is true for creative people in general, have relatively low thresholds for social stimulation, and therefore solitary activity or small group interactions are ideal. As Hans Eysenck's theory and empirical research has demonstrated, introverts have lower thresholds for arousal than extroverts and therefore find social stimulation overwhelming. Moreover, Anthony Storr has argued that creative people in general simply have to be comfortable alone with their thoughts if they are to create.[14] Social stimulation can and often will interfere with reflection needed to solve problems. Science, especially science as it is practiced today, is less and less of a solitary activity and more and more of a team enterprise. More and more problems are being solved in teams. But theory, as well as much empirical work, for instance, is still primarily a solitary activity and hence more suitable to people with introverted tendencies.

### Motivational Traits

Scientists as a group also tend to be rather *achievement driven* and *ambitious.* Regardless of which personality inventory was used in a given study, almost all of the research into the personality characteristics of scientists has found relatively high levels of drive and focus compared to nonscientists. For instance, the widely used personality measure the California Psychological Inventory (CPI) has two achievement scales, namely, Achievement via Independence and Achievement via Conformance. The effect size for all the studies using the CPI showed that scientists were almost three-fourths of a standard deviation higher on Achievement via Independence and more than a half a standard deviation higher on Achievement via Conformance compared to nonscientists. According to the author of the CPI, Harrison Gough, a person who scores high on Achievement via Independence has a "strong drive to do well; [and] likes to work in settings that encourage freedom and individual initiative." The same drive is characteristic of a person who scores high on Achievement via Conformance, but he or she "likes to work in settings where tasks and expectations are clearly defined."[15] From this it can be inferred that scientists prefer settings that are structured and yet allow for individual initiative, an appealing characteristic of most scientific occupations.

### Personality and Domain-Specific Scientific Interest

I contend that one's preference and orientation toward people or things play a crucial role in the area of science that one becomes interested in, namely, physical, biological, or social. The foundation for the people-thing orientation comes from the vocational interest literature. Dale Prediger was the first to modify John Holland's hexagonal model of vocational interests onto two basic dimensions: people-things and data-ideas. The "people" end of the dimension is mapped onto Holland's "social" career types, whereas the "thing" end of the dimension is mapped onto "realistic" career types. According to Holland, the social career type prefers occupations that involve informing, training, enlightening other people. The realistic career type, in contrast, prefers careers that involve manipulating things, machines, objects, tools, and animals.[16]

Supporting this domain-specific view of scientific interest, Simon Baron-Cohen and his colleagues have found that engineers, mathematicians, and physical scientists score much higher on measures of high-functioning autism and Asperger's syndrome than nonscientists, and that physical scientists, mathematicians, and engineers are higher on a nonclinical measure of autism (Asperger's) than social scientists. In other words, physical scientists often have temperaments that orient them away from the social and toward the inanimate—their interest and ability in science is then just one expression of this orientation. Moreover, autistic children are more than twice as likely as non-autistic children to have a father or grandfather who was an engineer.[17]

The problem with the research on personality and science as it stands is that it is not specific to any specific domain of science but rather covers scientists in general. Very little if any research has compared the personality dispositions of physical, biological, and social scientists to examine whether social scientists have more sociable and extroverted personalities compared to their physical scientist peers. Of most interest would be developmental research that examined whether a preference for things is evident early in life for future physical scientists and, likewise, whether a preference for people is evident early in life for future social scientists. Similarly, cross-cultural work showing the same association between thing-orientation and physical science and social-orientation and social science the world over would be quite valuable. Therefore, the next line of research for the personality psychology of science is to explore differences in personality between physical, biological, and social scientists. Based on the evidence just cited, my prediction is that the physical scientists as a group will be more introverted and thing-oriented (that is, have more developed implicit physical intelligence) than the biological scientists, who in turn will be less sociable and extroverted than social scientists (that is, have more developed implicit social intelligence).

### Personality, Scientific Interest, and Theoretical Predilection

Do certain kinds of people become more interested in science and research than other kinds of people? More specifically, are people with certain kinds of personality traits more likely to be attracted to the life of science and research than other kinds of personalities? As we saw earlier from the meta-analytic review, the answer seems to be yes, certain personality traits do predict a career interest in science and research, even within the branch of social science of psychology. Clinical psychology, for example, is an ideal domain in which to address this question because it emphasizes two different and distinct sets of skills in the PhD program, applied-clinical skills and research and scientific skills. What is known as the "Boulder Model" was implemented in the late 1940s and is the basis for all doctoral programs in clinical psychology in the country. The heart of the model is equal emphasis on research training and clinical practice training. Yet in reality, a major concern for PhD programs in clinical psychology is the high rate of students who are not interested in research. Clinicians do tend to be more people oriented than investigative and research oriented.[18]

An important question, therefore, has become: what predicts interest in science and research in these students and can this interest be increased by particular kinds of training environments? The general conclusion from the studies on these questions is that one of the strongest predictors in interest in research (or lack thereof) is personality and that training environment plays but a modest role in increasing interest in research.[19] For example, a study by Brent Mallinckrodt and colleagues examined the impact of training environment, personality-vocational interest, and the interaction between the two on increasing research interest and found that personality-vocational interest was a stronger predictor than research environment in increasing interest in research over the course of graduate training.

In addition to the domain of interest (physical versus social science) and age, the work on personality can also shed light on theory acceptance and even theory creation. Or stated as a question: do certain personality styles predispose a scientist to create, accept, or reject certain kinds of theories? The first work on this question was in the mid-1970s by George Atwood and Silvan Tomkins, who showed through case studies how the personality of the theorist influenced his or her theory of personality. More systematic empirical investigations have expanded this work and have demonstrated that personality influences not only theories of personality, but also how quantitatively or qualitatively oriented and how productive psychologists are.[20] One general finding from these studies is that psychologists who have more objective and mechanistic theoretical orientations are more rational and extroverted than those who have more subjective and humanistic orientations. For instance, Johnson and colleagues collected personality data on four groups of psychologists (evolutionary-sociobiologists, behaviorists, personality psychologists, and developmental psychologists) and found that distinct personality profiles were evident in the different theoretical groups. That is, scientists who were more holistic, purposive, and constructivist in orientation were higher on the Empathy, Dominance, Intellectual Efficiency, and Flexibility scales of the California Psychological Inventory and the *intuition* scale of the Myers-Briggs Type Indicator (MBTI). Most of these studies, however, have been with psychologists, so answering the question of whether these results generalize to the biological and natural sciences remains a task for future psychologists of science.

## PERSONALITY IN SCIENTIFIC ACHIEVEMENT AND CREATIVITY

Becoming a scientist is one thing; making significant contributions and solving problems creatively is another. Just as I did with scientists compared to nonscientists, in the meta-analysis I also compared traits of the most creative and eminent scientists to the more average scientist. Such comparison provides clues as to which traits make scientific achievement, creativity, and eminence more likely. The main conclusion from the quantitative review of the literature is that many of the traits that lower thresholds for scientific interest are seen in a more extreme form in the most creative scientists. That is, the more creative scientists are more confident, open, dominant, independent, and introverted than their less creative peers, who are higher on these dimensions than nonscientists (see table 5.2).

Table 5.2 Personality Traits More Salient of Creative Scientists Than Less Creative Scientists

| Trait Category | | | |
|---|---|---|---|
| Social | Cognitive | Motivational | Affective |
| Dominant | Open | Driven | Aesthetic |
| Self-confident | Flexible | Ambitious | Expressive |
| Deviant | Intelligent | | |
| | Curious | | |
| | Imaginative | | |

*Source:* Feist 1998.

The traits of arrogance, hostility, and conscientiousness (or relative lack thereof) are most noteworthy of highly creative scientists. The confidence found in scientists in general seems to go one step further in the most creative scientists. For example, I interviewed more than one hundred very eminent scientists—a large portion of whom were members of the National Academy of Sciences (the highest honor for an American scientist after the Nobel Prize). They completed personality questionnaires, but observer ratings of personality were also made by having the audiotaped interviews evaluated—blindly—by research assistants. As it turns out, there was a positive relation between ratings of hostility and arrogance and scientific eminence. The most eminent were deemed the most hostile and arrogant by blind raters, suggesting either that arrogance and hostility make eminence more likely or that becoming eminent increases one's arrogance and hostility. Some people might be tempted to conclude there is a causal connection between arrogance and scientific eminence. For example, soon after the article based on this research was published in a scientific journal a journalist contacted me. As only journalists can, she got to the heart of the matter in her opening paragraph when she wrote, "Gregory Feist is a nice guy. Too bad for him!"[21]

Science, to be sure, is a competitive enterprise—every paper, every grant proposal, every job application, every fight for the top students is a competitive undertaking and not everyone survives such a winnowing process. It takes a real belief that one has something special to offer and that one has a way of doing things that is better than most others. This is most true of those at the top of their field. The competitive nature of science is not to everyone's liking nor consistent with his or her disposition, and certainly some more deferent souls leave the field because of it (and some manage to stay in despite the mismatch).

In the highly competitive world of science, especially big science, where the most productive and influential continue to be rewarded with more and more of the resources, success is more likely for those who thrive in competitive environments, that is, for the dominant, arrogant, hostile, and self-confident.

The other defining characteristic of creative scientists as compared to scientists in general is the relatively low levels of conscientiousness in the former. Conscientiousness is not a defining trait of the most distinguished scientists—at least when compared to the average scientist. Indeed, conscientiousness and creativity, although not polar opposites, are somewhat opposing forces of personality. There is no doubt some truth to the argument that some degree of conscientiousness is required to turn creative potential into actual creative products—the discipline and stick-to-itiveness to see the task to completion. But the traits of careful, cautious, conventional, disciplined, orderly, persevering, reliable, and self-controlled are less evident in the most creative of scientists.

## DIRECTIONAL INFLUENCE BETWEEN PERSONALITY AND SCIENTIFIC BEHAVIOR

As we saw with the arrogance-eminence connection, the most pressing question that begs to be addressed from the personality findings is whether these traits are causes or effects of scientific interest, talent, and achievement. To put it most simply: do smart, conscientious, introverted, driven, and controlled people become scientists or does science create smart, conscientious, introverted, driven, and controlled people? Out of logical necessity and the empirical evidence for when temperament-personality dimensions are formed, it would seem very unlikely that any of these characteristics would be nonexistent until one became a scientist, and therefore unlikely that being a scientist actually caused these traits of personality.[22] Some of them may in fact become more pronounced after one is trained as a scientist and after practicing science. As is often the case, however, the model that may best fit the relation between personality and scientific behavior is probably bidirectional, going from personality to scientific behavior as well as from scientific behavior to personality.

Of the dozen or so studies that have examined scientific behavior longitudinally, most have focused on questions of age and productivity and only two have looked at personality across time.[23] Results examining the directionality question from a recent study by Feist and Barron showed that certain personality traits, such as dominance, may become more pronounced during and after a career in science, suggesting a directional influence from career to personality. But before such an inference can be confirmed, one has to rule out alternative variable explanations. For instance, perhaps age and maturation, not scientific careers, lead to this difference in dominance. Lack of research on longitudinal personality change and stability is one of the real shortcomings of the personality psychology of science literature.

In sum, cognitive traits (conscientiousness and openness), social traits (confidence, arrogance, independence, and introversion), and motivational traits (achievement and drive) lower the threshold for scientific interest. Having such a cluster does not make scientific interest inevitable, only more likely. Interest in the physical sciences, more specifically, is probably influenced by the thing-orientation, whereas interest in the social sciences is influenced more by a people-orientation. The main traits that lower the threshold for creative achievement in science appear to be openness, flexibility, drive, ambition, introversion, arrogance, self-confidence, and hostility. Too little systematic longitudinal research has examined the question concerning whether personality is more the cause or the effect of scientific interest and talent to warrant any firm conclusions at this time.

# **Chapter 6** Social Psychology of Science

Science is unquestionably a cognitive activity, but it is also unquestionably a highly social activity, with much work being done cooperatively or competitively with other research teams. The social-cognitive and attributional perspectives, with their emphasis on cognitive heuristics, biases, and causal explanations, can complement the work I cited on cognitive psychology of science (see chapter 4). Addressing the social factors involved in science, the field of social psychology of science finds itself in an unusual situation. It is potentially one of the richest and most stimulating areas in the psychology of science, but as yet it remains more latent than actual. One can very easily apply all the major social psychological phenomena—social cognition, attribution theory, attitude and attitude change, competition, cooperation, conformity, gender, social influence and persuasion, and intergroup relations—to the study of science and scientists, but as yet much of this work has not been conducted.

According to Gordon Allport, the province of social psychology can be defined as "an attempt to understand and explain how the thought, feeling and behavior of individuals are influenced by the actual, imagined or implied presence of others." As others have noted, substituting "scientists" for "individuals" in Allport's quotation creates a good working definition of social psychology of science.[1] Social psychology of science may not be as well developed as the developmental, cognitive, or personality psychologies of science, but the edited volume *The Social Psychology of Science* suggests the field may be on the verge of becoming a viable discipline. Some of the main figures in social psychology have begun to produce work that is directly relevant to the social psychology of science, as indicated by their contributions to this volume. Because of the dichotomy between actual and potential literature, however, my review of social psychology of science will be divided into extant and potential topics of investigation.

## SOCIAL PSYCHOLOGY OF SCIENCE AS IT IS

### Effects of the Experimenter on the Experiment

Psychology is the only study of science that uses the experimental method, that is, experimenters manipulate variables and randomly assign participants to experimental conditions. It is now quite clear that who the experimenter is, how she or he behaves, and what she or he knows can each have subtle effects on the outcomes of experiments, not only in the social sciences but sometimes in the physical and biological sciences as well. The pioneer in the field of what has come to be known as "experimenter effects" is the psychologist Robert Rosenthal.[2]

There are two general classes of experimenter effects, namely, interactional and noninteractional. Interactional effects come into play when an experimenter interacts with participants, whereas noninteractional effects are mostly cognitive and perceptual and do not involve interaction between experimenter and participant. All I will say about noninteractional effects is that their influence has long been recognized; in fact, an astronomer from the eighteenth century (Friedrich Bessel) who studied errors in astronomical observations was labeled by Rosenthal as the "first student of the 'psychology of scientists.'"[3]

Research on interactional effects has demonstrated that participants' responses can be influenced by biosocial effects (age, gender, ethnicity of the experimenter), by psychosocial effects (personality and temperament of the experimenter), by situational effects (experience level of experimenters and environmental variability), by modeling effects (the experimenter's own performance while conducting the study), and lastly by self-fulfilling prophecy effects (experimenter's knowledge of the hypothesis or desired outcome can help bring about that very outcome). Regarding the biosocial interactional effects, Rosenthal and his colleagues, for example, have shown that male experimenters have exhibited more friendly behavior toward their (mostly female) participants, which in turn affected the results of the experiment. Psychosocial effects have been demonstrated by the finding that authoritarian and status-conscious experimenters obtain more conforming results from their participants than friendly and warm experimenters. Situational effects have been demonstrated by the finding that more experienced experimenters obtain different responses than less experienced experimenters. Self-fulfilling or expectancy effects are concerned with "how the investigator's expectation can come to serve as a self-fulfilling prophecy."[4] That a researcher's prior expectations can affect the observations as well as the final results has been demonstrated not only when the participants are humans but also when they are animals. In 1963, for instance, Rosenthal and Kermit Fode published a study in which twelve experimenters were each given five rats that were to be taught to run a maze with the aid of visual cues. Six of the experimenters were informed that their rats had been bred for maze dullness, while the other six experimenters were told their rats had been bred for maze brightness. Of course, the rats were randomly assigned to condition and were not systematically different. The results were quite pronounced: rats run by experimenters who thought they were especially bright did in fact learn better and faster how to navigate the maze than the rats run by experimenters who thought their rats were dull. The outcome, in short, was self-fulfilling.

What is even more relevant (and frightening) has been the replication of this same effect with teachers' perceptions and student performance, a phenomenon Rosenthal dubbed the "the Pygmalion effect" (after the Greek myth of a sculptor who carved a beautiful statue, fell in love with it, and believed it came to life and returned his love). Rosenthal and Jacobson set out to replicate the bright-dull finding from rats and maze performance in elementary school children. They first administered a nonverbal test of intelligence to students in eighteen classrooms (three at each grade, 1 through 6), a test, the teachers were told, that does an excellent job at predicting "intellectual blooming." Each class naturally consisted of students with above average, average, and below average ability levels, yet the experimenters randomly chose 20 percent from each classroom and informed the teachers that these particular children had performed on the intelligence test in such a way that "they would show surprising gains in intellectual competence during the next 8 months of school."[5] Testing on the same intelligence test eight months later revealed greater IQ gains in children teachers had expected to show gains than in those children not expected to show change. Indeed, experimenter expectancy and self-fulfilling prophecy effects are well known in behavioral and medical science research, and a fundamental principle of such research is the requirement to incorporate double-blind procedures where not only the participants are blind to the condition they have been assigned, but so too are the experimenters.

Social psychology of science has begun to shed light on these experimenter and observer effects, which will increase our understanding of the social cognitive processes involved in the creation and development of scientific knowledge. In the words of the social psychologist Arie Kruglanski: "Cognitive and motivational biases that influence scientific conclusions are fundamentally inevitable and are an integral part of how all knowledge is acquired. Rather than regarding them as impediments to truth, it may be more practical to take them into account to improve the quality and persuasiveness of one's research."[6] It would not be an exaggeration to say that the whole field of experimenter effects could be categorized as a subdiscipline of social and cognitive psychology of science. Indeed, this body of work provides a prototypic example of how social psychology has much to offer science studies and how it implicitly has been doing so for years.

### Social-Cultural Influences on Science and the Evaluation of Science

Another key figure in the social psychology of science is Dean Simonton, whose work has more explicitly explored how social structures influence the creation and maintenance of science within and across different historical time periods.[7] Theoretically, Simonton's chance configuration model provides an explanation for how an individual scientist's conceptual configurations and insights develop, are articulated, are communicated, are accepted/rejected, and become influential and how, over time, a group of like-minded individuals can form around them; how those who produce the most ideas are most likely to wield wide-ranging influence by their high-quality work; and how individual differences as well as social factors contribute to the "essential tension" between traditional knowledge and revolutionary, not-yet-accepted knowledge. Empirically, Simonton has shown through analysis of historical and archival data how mentors and role models, war, and political upheaval or stability can influence creative output in science. Using cross-lagged panel designs, Simonton examined the causal influence of war on scientific productivity in seven European countries from 1500 to 1900 and reported that war had a significant influence on productivity, rather than the other way around, but the influences were complex and inconsistent across country. Finally, Simonton reported that often the most creative contributions come from those familiar with two different cultures, suggesting that exposure to multiple cultural frames of reference is important for creative productivity in science.

Will Shadish has been another important figure in the field of social psychology of science and has written about the significance of a psychological perspective in the evaluation of quality in science. Quality evaluations are at the heart of the scientific enterprise. Such evaluations and their criteria and measurement determine who gets which job, who gets tenure, who gets which grants, and who gets which awards and honors. As I discussed in chapter 5, science is a very competitive enterprise, and resources (read money and recognition) are scarce. Of course, the question of quality in science immediately raises a few other critical—and very social psychological—questions: Whose perceptions do we use to evaluate quality? What criteria are used? How do we decide how to weigh the various criteria? Are these criteria and evaluations fair or biased against particular individuals or groups of individuals? What role does bias play in the peer review process?

Until recently, philosophy, history, and sociology may have been the disciplines most likely to address these questions, but as Shadish wrote: "Why should we think that psychology offers an important perspective on our understanding of science quality? The reason is this: The perception of quality in science probably exercises an inordinate amount of influence in scientific reward systems, and perception is largely a psychological variable." Social negotiations and self-presentation tactics involved with promoting one's own career clearly play a role in influencing the perceptions of the powers that be. Few would deny this. The real question then becomes "how much of a role does self-presentation play in career success?" The cynic may say a major role, whereas the more naive person may say no role. Rather than leave the question to one's predilection toward cynicism or gullibility, I would argue the question is fundamentally an empirical one and therefore should be examined empirically.[8]

Shadish argues that one of the "objective" measures important in quality evaluation is citation analysis—the importance of a particular scientist's opus easily and fairly reliably can be measured by counting the number of times her or his works is cited by peers. Most frequently used by sociologists, citation analysis seldom has examined the cognitive and psychological reasons authors have for citing any particular paper. Shadish and his colleagues, however, have used surveys to answer this question for a large sample of psychologists and found that oft-cited works were considered exemplars, higher in quality, published longer ago, and were often sources of methods or designs. Interestingly and unexpectedly, frequently cited articles were also perceived to be less creative. It is not clear, however, why psychologists believe highly cited papers are high in "quality" but low in "creativity," especially since Robert Sternberg and Tamara Gordeeva reported that papers were highly cited because they were "novel." Sternberg and Gordeeva also reported that importance of theoretical contribution and whether they generated research were rated as the most important reasons psychologists gave for citing a paper.[9]

Another topic worth mentioning briefly in evaluation of science is the peer review process. One cornerstone of the scientific method is peer review, the process whereby editors of journals send manuscripts to three or four experts in the field for critical evaluation. The editor then culls these evaluations and decides whether to publish it or not (or sometimes to request that the author "revise and resubmit"). Ideally this method ensures only papers that are original contributions and based in sound methodology get published. Yet every scientist will tell stories of bias and unfair editorial decisions, often about his or her most creative papers. Indeed, one of my own papers was soundly rejected by one journal yet went on to be article of the year in another journal. Michael Gorman in *Simulating Science* has written about the personal trials and tribulations of the peer review process as well as reviewed some of the psychological literature on the topic. Bias, for instance, seems especially pernicious when one attempts to report studies that falsify a claim, which runs in direct contradiction to Popper's well-known dictum that falsification is the benchmark for true science. Clearly, further research is needed in understanding the explicit and implicit reasons scientists cite works and the possible biases behind the peer review process, but social psychologists have begun to make important contributions to the psychological factors that go into evaluating quality in science.

### Mentorships/Training

*Family and Teachers.* What role do family members or teachers play in promoting and retaining scientific interests? Parents obviously can and do influence their children's career choices either explicitly, through encouragement, or implicitly, through modeling a career themselves, and this effect has been demonstrated in the science professions.[10] Modeling effects are evident from the findings that children are more likely to enter science careers if at least one parent is in science or engineering. Furthermore, a consistent and robust finding from the literature on father's education and occupation is that scientists overwhelmingly come from families of professional occupations and higher education. Achievement in science is also more likely within science families. One study, for instance, reported that college students who won awards and were high achievers in science had fathers who were scientists. Parents, however, need not be well educated or scientists but can simply have positive attitudes toward math and science and foster such positive attitudes in their children. One rather telling study of the influence of family in scientific interest was conducted by analyzing parent-child interactions on their visit to a local science museum. In samples of children ages one to eight, they found that there was a gender difference in frequency with which parents provided explanations (causal, correlational, or analogical) rather than merely descriptions of the exhibits. Parents were more likely to provide explanations to boys. The explanation rates were about 29 percent for boys and about 9 percent for girls. These results suggest that at a young age, parents may be treating boys and girls differently in how they explain science (see gender section, below). Either directly or indirectly, having well-educated parents familiar with and interested in science is predictive of an interest in science.

It is not only parents, of course, who can play critical roles in the development of scientific interest.[11] Bernice Eiduson reported that roughly half of the scientists she studied said some older person was important in their developing and maintaining an interest in science. Vera John-Steiner eloquently described the importance of apprenticeships and mentorships in stimulating creative activity in science (as well as art). Rena Subotnik and her colleagues report that having a strong mentor in high school and college predicts staying on and pursuing a scientific career. Furthermore, Feist reported that 65 percent of the elite biological and physical scientists in his sample reported having a significant mentor in high school, and 80 percent reported having one in graduate school. In high school, mentors tended to be either a teacher (29 percent) or a parent (26 percent), whereas in graduate school they were overwhelmingly one's PhD advisor (56 percent) or another professor (20 percent).

*The Role of the Eminent Mentor.* There is also the question of the nature of mentorships at the highest levels of scientific achievement. Do the "rich get richer"? That is, do the top scientists attract the top students who go on to be the top scientists of the next generation? The answer seems to be yes, having an eminent mentor appears to be a contributing factor in obtaining eminence.[12] This finding has been most clearly demonstrated in Harriet Zuckerman's work on Nobel laureates. One of her strongest findings concerned the "cumulative advantage" effect of those young scientists who train under the scientific elite (that is, Nobel Prize winners). They produce more at an early stage in their careers, are more likely to produce works of higher impact, and are more likely to win the Nobel Prize themselves than those who do not train under laureates. As Zuckerman and others have argued, however, the causal direction of this influence probably goes both ways: the best young scientists are chosen by the best scientists, which in turn feeds into the cycle of "cumulative advantage." Dean Simonton has also reported that American Psychological Association (APA) presidents were quite likely to have been mentored by an eminent psychologist.

### Small Group Processes in Science

There is a long and distinguished literature on group processes in social psychology, and yet only recently has any of it focused on variables and tasks involved in science. For example, work on small group processes in science has made use of experimental methods and provided some insight into differences between individuals and groups working on scientific problems. Michael Gorman and his colleagues found that interacting groups (that is, those whose members interacted directly) on a scientific reasoning task performed no better than the best individual in a coacting group (that is, those whose members work separately but were informed of other members' hypotheses). Moreover, disconfirmatory instructions (telling students to look for disconfirming evidence) were usually superior to confirmatory. These findings were replicated with individual participants, suggesting that groups perform about as well as the best of an equal number of individuals on these scientific reasoning tasks. For instance, one study reported that coacting groups were more prone to confirmation bias than interacting groups. Modern research teams succeed in part because they divide labor effectively among participants with different skills and resources.[13]

### Gender and Science

One of the more contentious and polemical questions in the psychology of science concerns the role that gender plays in science in general and in scientific and mathematical ability and achievement in particular. As Evelyn Fox Keller, among others, has pointed out, the history of science is replete with associations, both implicit and explicit, between science and male; with male scientists frequently trying to "tame" or "control" the feminine Mother Nature.[14] The topics of gender and science and gender differences in scientific achievement could be and have been the focus of a book, and I leave the more exhaustive review of this literature to others. But there are three questions on gender and science that I believe have accumulated enough literature in psychology to warrant attention: interest-attrition, mathematical ability, and productivity.

*Interest-Attrition.* One of the more entrenched influences on the development of scientific interest appears to be gender. The research over the last forty years does suggest there are gender differences in science or math, whether it comes in the form of explicit attitudes, implicit attitudes, performance on aptitude tests, or actual graduation and career data.[15] The general conclusion from this body of research is that men are more likely than women to view science positively, be more interested in science and math as a career, and less likely to drop out of science. Moreover, although there is no overall gender difference in intelligence, there does appear to be some systematic differences in the mathematical domain (males being higher) and in the verbal domain (females being higher).

There are, however, at least two important qualifications to these generalizations: gender differences are less apparent in childhood and adolescence than adulthood, and they are less apparent in the social sciences than in the physical sciences, with biological sciences being in the middle. For instance, in terms of courses taken, the gender gap in science is not evident at the high school or undergraduate level or in the social sciences. High school male and female students were equally likely to take advanced math courses (trigonometry and calculus) and almost as likely to take advanced science courses (biology, chemistry, and physics). In advanced science courses there was a slightly higher percentage of females taking biology and chemistry, and a slighter higher percentage of males taking physics.

As students progress through their academic careers, however, there is an increasing gender disparity in interest in science and math.[16] At the undergraduate level, the percentage of women who earned science or engineering degrees in 1995 was 46 percent (after being about 38 percent ten years earlier). At the graduate level a more obvious gender gap exists, even in the biological and social sciences, with 39 percent of the masters degrees in science and engineering and 33 percent of the doctoral degrees in science and engineering being awarded to women. In terms of career, the disparity widens even more, with only 4 to 6 percent of the full professors in science and math being women. The most extreme gender difference is seen at the most elite level. On average, only 2 percent of the members of the National Academy of Sciences are female. The most exhaustive and extensive study of PhD scientists over a twenty-two-year period (1973 to 1995) by the National Research Council has documented progress but not yet equality for women in science. When the appropriate controls (such as rank, field, and institution) are made, the gender disparity is not so extreme, but it still exists. For instance, men hold a fourteen-percentage-point advantage in holding tenure-track positions, but this difference approaches zero once career age is held constant. This suggests that the gender disparity in tenure-track science positions should continue to decline as more and more women become eligible. Also, salary differences diminish once rank is controlled for, but they do not disappear completely, suggesting that men do get paid a bit more for the same position. Moreover, marriage and family does affect men and women rank and productivity differently but not necessarily in the manner one might expect. J. Scott Long, for instance, reports that women who interrupted their careers for marriage and family in 1979 were less likely to obtain a tenure-track position, but there was no effect in 1995. For men, in contrast , the effect of getting married and having children had a positive effect on productivity and this effect increased between 1979 and 1995.

The other qualification is that not all fields of science are equally gender biased in their distributions. Unequal distributions are most striking in the physical sciences, less striking in the biological sciences, and least striking in the social sciences. Only 17 percent of the engineering degrees and 35 percent of both the mathematics and physical-earth science degrees were awarded to women, whereas nearly 50 percent of the biological and social science degrees, and 73 percent of the psychology degrees were awarded to women in 1995. Similarly, J. Scott Long's analysis of trends in national samples showed that from 1973 to 1995 women went from being 2 to 12 percent of the engineering PhD graduates and from 21 to 51 percent of the social-behavioral science PhD graduates. All other scientific fields were in between these two ends of the continuum. In addition, in a sample of mathematically precocious students who immediately after high school said they intended to major in math or science, five years later men were more likely to have received engineering and physical science degrees and women more likely to have received biological science and medical science degrees.[17]

One goal of the psychology of science, therefore, is to unpack some of the factors behind why women decide disproportionately to leave science, even those who are demonstrably among the most promising young scientists and mathematicians in the nation. Some previous work suggests a few possible explanations: number of hours worked per week, self-image, stereotype threat, parental behavior, and opting for and having greater talent for "people-ori-

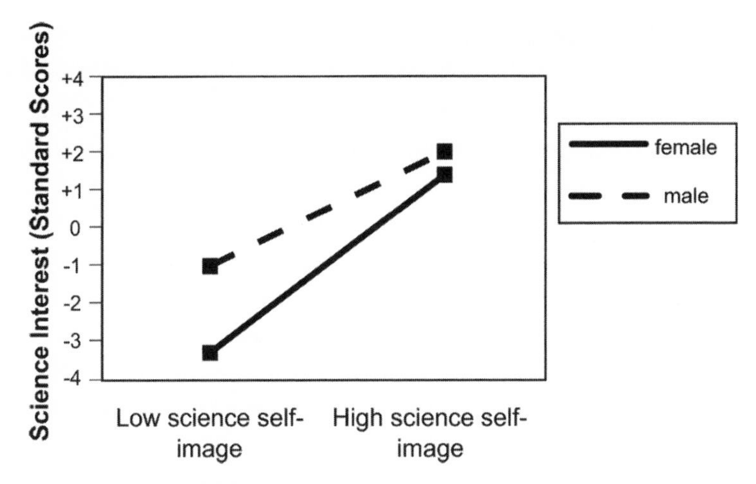


Figure 6.1. Interaction between scientific interest, gender, and self-image (after Feist, Paletz, and Weitzer, in preparation).

ented" rather than "thing-oriented" professions. For instance, Camilla Benbow and colleagues report that gender differences in math achievement dissipate once number of hours worked per week is controlled for, suggesting that workload rather than gender is the underlying cause.[18]

Another explanation focuses on self-image. One's self-perceived career identity no doubt carries a lot of weight in career choice, with people most likely to pursue those careers that match their identity and self-perceived ability. If one cannot imagine oneself in a career, there is little chance one will even attempt to pursue that career. Can one easily envision and imagine oneself as a "scientist"? For some (males) the label fits well with their projected self-image and for others (females) it does not. A recent study based on undergraduate research by Susannah Paletz reported an interaction between gender, self-image, and interest in science. In a study of 211 college students at a selective liberal arts college, men with either high or low science self-images were still slightly more interested in science than women with the same science self-image (see figure 6.1). Similarly, math skills have been seen as part of the male domain, and if one is female, there is less congruency between self-image and a career in math or science. Moreover, occupational interest research has demonstrated that congruency between talent, performance, self-perception, and drive is the best predictor of career interest. Research has demonstrated the power of self-efficacy training to increase math and science self-efficacy both for course work and career attainment, at least over a short period of time.[19]

Stereotype threat theory and research also sheds some light on the phenomenon of gender differences in science. Davies and colleagues have written of the effect of stereotype threat: "The risk of being personally reduced to a negative stereotype can elicit a disruptive state among stigmatized individuals that undermines performance and aspirations in any alleged stereotype-relevant domain—a situational predicament termed 'stereotype threat.'" Stereotype threat suggests stereotyped individuals can be adversely affected in their achievement by being reminded of negative stereotypes. Moreover, stereotype threat can both hinder and help performance through negative and positive identity. Margaret Shih and colleagues, for instance, demonstrated that when Asian women doing a math task have their ethnic identity primed, and consistent with the stereotype that Asians are good at math, they perform better than controls. When, however, they have their gender identity primed, consistent with the stereotype that women are not good at math, they perform worse.[20]

Finally, perhaps a more nativist explanation comes from various lines of research suggesting that there is a general gender effect in dispositional interest toward people and things.[21] Recall the discussion in chapter 5 (personality) that argued for the people-thing dimension to be an underlying influence on the kind of science one becomes interested in, especially physical versus social science. The "people" end of the dimension is mapped onto "social" careers that involve informing, training, or enlightening other people. In contrast, the "thing" end of the dimension involves "realistic" careers that involve manipulating things, machines, objects, tools, and animals. Building on Prediger's work, Richard Lippa reported gender ratios of roughly four or five to one of males in "thing" rather than "people" careers, and ratios of two or three to one of females in people rather than thing careers. Interestingly, there were no gender differences on the ideas-data dimension. Recall, too, the work of Simon Baron-Cohen demonstrating high proportions of high-functioning autism (Aspergers) as well as high proportions of males in the physical sciences and engineering. Simon Baron-Cohen and colleagues also have reported evidence that this gender difference in people-versus-thing orientation is present in neonates as young as three-days-old, suggesting a biological origin. The mathematical and science professions are conceptualized as being investigative but on the thing side of Holland's vocational hexagon.

*Mathematical Ability.* One of the more consistent and robust findings in the gender difference literature involves mathematical ability, with males scoring higher than females at the low, medium, and high ends of the distribution.[22] Both longitudinal and cohort data over the last twenty years suggest the gender difference is remaining constant at around one-half of a standard deviation in favor of males. But as Eleanor Maccoby and Carol Jacklin's review of the literature first made clear, there are a couple of qualifications to this generalization. First, there is little to no gender difference before adolescence, although the exact timing is somewhat disputed. Second, at least up through early adolescence, girls achieve higher grades than boys in math classes.

Recall Julian Stanley's study of extreme mathematical precocity (math scores on the SAT of 700 or above before one's thirteenth birthday). As both Stanley and his student Camilla Benbow have reported, one of the biggest surprises in collecting these data, however, was the large and consistent gender difference among the extreme scores—ultimately reaching as high as a twelve-to-one ratio in favor of males. Furthermore, Benbow's target article in *Behavioral and Brain Sciences* was commented on by more than forty experts, and although virtually none of the commentaries took issue with whether a gender difference exists, there was little agreement concerning the potential causes of this gender difference.[23] Indeed, the robust gender difference in math raises the important, if not rather unsettled, question of what its cause is.

What then are some of the possible explanations of this gender difference? Benbow reviewed the evidence for seven of the more common environmental explanations: attitudes toward math, perceived usefulness of math, confidence and self-efficacy, encouragement from parents and teachers, sex-typing, differential course-taking, and career and achievement motivation. She found that some of these environmental influences do distinguish males and females. For example, females do like math less, find it less useful for their future goals, and have less confidence in their ability than males. Furthermore, mathematics is somewhat sex-typed as a "masculine" enterprise, parents and teachers are more encouraging of male than female mathematical achievement, differences in math courses do not explain aptitude differences, and finally male career motivation is more independent of parent or teacher support than female.

These explanations, however, neither rule out nor are inconsistent with biological explanations. As Hans Eysenck's commentary pointed out, the situational findings could result from either genetic or environmental origins. Benbow also more directly addressed the biological explanations and offered four possibilities: hemispheric laterality, allergies, hormonal influences, and myopia. For instance, based on a high incidence of left-handedness in the mathematically precocious, and in particular the precocious males, and the greater bilateral or diffuse cognitive functioning of left-handed individuals, she concluded that bilateral and/or a strong right hemispheric functioning may be implicated in extreme mathematical ability. Furthermore, prenatal exposure to testosterone has been postulated to influence handedness and immune disorders and therefore could be an indirect influence on mathematical ability.[24] To quote Benbow: "In sum, the above physiological correlates, especially the possibility of prenatal testosterone exposure, lend credence to the view that sex differences in extremely high mathematical reasoning ability may be, in part, physiologically determined. . . . Of course, some of the above discussion on physiological correlates is speculative." Suffice it to say that the physiological explanations were the focus of most of the criticism in the commentaries. Many criticisms, however, did not take issue with the fact that biological explanations may play a role, but rather that their mechanisms are more complex than, and the evidence is not as solid as, Benbow's presentation. The real answer will almost certainly be found in a more complex explanation arguing that our interests and abilities, whatever they may be, require both some biological givens and some environmental training and encouragement to be fully expressed and actualized. What these precise biological and environmental mechanisms are await further investigation.

*Productivity.* Comparing publication rates of men and women has consistently shown that men produce more works than women, although the difference has declined from the 1960s to the 1990s.[25] This gender difference appears to hold for total number of publications as well as yearly average. There is some contradictory evidence regarding whether this gender difference increases or decreases across the course of one's career. Jonathon Cole reported that the gender gap on productivity increases, whereas J. Scott Long as well as Yu Xie and Kimberlee Shauman reported that it decreases over the course of one's career. For instance, Long's analysis of large national data sets in 1995 showed roughly a 30 percent greater total publication rate for men compared to women when looking across all ranks. As the data are broken down within rank, however, the male advantage drops to 13 percent at the tenure-track (assistant professor) rank and to 6 percent at the tenured rank (associate professor) and 7 percent at the full-professor rank. In short, productivity differences clearly do drop when rank is controlled for but they do not disappear entirely.

As with age and productivity, and interest in science and math, the question with gender and productivity that begs to be addressed is that of cause. Once again, and as is somewhat inevitable, explanations are more contentious and less consensual than the description of the phenomenon. Differences in marital status, family obligations, prestige of institution, rank of position, training, and motivation have each been investigated but with negative or inconsistent results. One consistent difference from the years 1935 to 1965 between male and female scientists was their marital status, with about 50 percent of female scientists and 90 percent of male scientists on average being married. Female scientists seemed to be sacrificing marriage for a career in science. This gender difference is approaching equality, having dropped to slightly more than 70 percent of the men and slightly less than 70 percent of the women scientists being married by 1995.[26]

The intuitively appealing answer that women are hindered by multiple roles of scientist, wife, and mother and are relegated to marginal departments has inconsistent empirical support. In fact, some research has shown that married women tend to slightly outproduce single women, and women with one or two children tend to outproduce women with no or more than three children. This research, however, was conducted only with women who were employed full time in academic positions. Other research has examined the more general question of the likelihood of being in the workforce, and here one sees the percentage of men above 90 percent in each of the following four categories: single (94 percent), married (96 percent), married with older children (97 percent) and married with younger children (97 percent). But for women these same four categories show a steady decline in the percentage in the workforce. For single women it is 94 percent, for married women it is 89 percent, for women with older children it is 80 percent, and for women with younger children it drops to 71 percent. By these national results, it is clear that marriage and family affect male and female scientists differently. This makes sense in light of another finding reported by Long: in 1995, 51 percent of the women who were employed part time in science said they were part time because of "family obligations," whereas only 3 percent of the part-time employed men gave that reason. Long concludes: "Overall, marriage and family are the most important factors differentiating the labor force participation of male and female scientists and engineers."[27]

Sociologists of science in general have presented evidence that gender differences in productivity cannot fully be explained by differences in type of institution (college versus university) or prestige of department. For instance, the Cole brothers reported that when both institutional and departmental variables were entered first in a regression equation and thereby held constant, the relation between gender and productivity still persisted. One explanation that does have some support is hours worked per week. For example, in research on career outcomes of the precocious math group of Julian Stanley, once hours worked per week is controlled for the gender, difference in productivity and achievement disappeared. This suggests that number of hours working rather than gender per se may explain productivity differences. An important fact to keep in mind when comparing men and women's productivity levels is that female PhD's are less likely to be employed full time.[28]

## SOCIAL PSYCHOLOGY OF SCIENCE AS IT COULD BE

The richness of social psychology is mostly an untapped and dormant resource in the study of scientific behavior, knowledge, and theory construction. Methods and theories of social psychology can be combined with those of other disciplines, such as history, to offer a unique perspective into the social nature of scientific behavior and thinking. Even more telling would be applying the experimental method so common in social psychology to the study of scientific knowledge. What follows are some possibilities for how these fields might develop.

### Possible Historical Case Studies

Relatively little actual empirical work has been carried out on the social psychology of science, but a few researchers have outlined how various methods could be applied in investigating social elements of science. Shadish and colleagues, for example, outlined a simulated experimental paradigm that would allow one to investigate issues raised by the case study of the Devonian controversy in geology as well as controversies over the existence of canals on Mars.[29] In the case of the former, the discovery of the Devonian period in geological history was not the product of a single individual; rather, it emerged out of a mix of cooperative and competitive interactions among a group of geologists—Roderick Murchison, Adam Sedgwick, and Henry Thomas De La Beche being the primary figures. As in most areas of science, there was debate and disagreement but ultimately, after dropping previous theoretical claims for lack of evidence, a single figure—Roderick Murchison—came away with the label "discoverer" of the Devonian period. Yet this is somewhat misleading for the other figures (for example, Sedgwick and De La Beche) played an essential role in presenting evidence or theoretical argument against some of Murchison's earlier claims. The processes here involved debate, evidence, theory construction, and theory change, and the label of "discoverer" is clearly partly a function of much social negotiation. Social psychology of science can help unpack these negotiations.

In addition, the Murchison case study can provide insight into the role of minority influence on majority opinion in science. As the classic work by Solomon Asch demonstrated so convincingly, a unanimous majority can cause a minority of one to conform to an erroneous position on an unambiguous perceptual task. But as the work of Serge Moscovici and Charlan Nemeth has shown, the opposite can also hold true: a consistent, determined minority can influence the judgments of a majority in an ambiguous perceptual task.[30] Moscovici and Nemeth take the view that minority influence forces the majority to look more closely at the stimuli that are the focus of argument. To return to the case of the Devonian discovery, early on in the controversy Murchison's was the novel, minority view, but there was no consistent majority opposing it. Murchison was a persuasive scientist and his consistent, determined arguments fueled a close study of those aspects of the evidence that he thought were particularly important. Gradually, Murchison's position became the majority view.

### Possible Experimental Paradigms

How can minority influence processes be studied experimentally? One could study the circumstances under which a minority can force a majority to look more carefully at the data on a scientific simulation task, such as the artificial universe used by various cognitive psychologists.[31] Such experiments could be conducted by: (1) manipulating task ambiguity by introducing different levels of error; (2) using a confederate to play the role of minority member and varying the style of argument that she/he argues; (3) manipulating the credibility of the minority (perhaps by presenting them as having had previous success with a similar task); and (4) looking at minority influence across generations, in which members of an original group are replaced one by one and each new member can consider the minority's arguments anew. In addition, Michael Gorman and Robert Rosenwein have proposed a possible quasi-experiment in which groups of individual participants try to solve problems that mimic scientific reasoning in a multifaceted environment that simulates the social negotiations found in scientific communities. Recall, too, the work of Kevin Dunbar, who has in fact investigated these processes in actual scientific laboratories.

### Other Possibilities for the Social Psychology of Science

Almost any and every topic within social psychology could be applied to the study of scientific thought and behavior, so I will outline only a few of the more obvious ones.[32] To the extent that science is becoming less and less an individual enterprise and more and more is carried out by groups and teams, the question is raised about two other fundamental social psychological phenomena, namely, competition and cooperation. As I mentioned in the last chapter on the arrogant personality of the most eminent scientists and just above with the Devonian discovery, science is a highly competitive enterprise. Competition for resources in science is a fierce and at times unpleasant enterprise. Some of the major discoveries of the twentieth century—the hunt for the structure of DNA and more recently the human genome project are prime examples—were highly competitive undertakings, with much of the stress and strain and acrimony of any other high-stakes competition. The extent to which such competition facilitates or hinders the quest for knowledge is an interesting question that needs to be explored further by social psychologists of science.

Person perception, prejudice, and even discrimination are each quite relevant to any institution, including science. And the notion of "in-group" and "out-group" can easily and quite fruitfully be applied to the study of scientific behavior, especially decisions about whom to admit as students into one's research lab, whom to bestow awards and honors on, or whom to hire for academic, government, and commercial science jobs. More specifically, when it comes to women and minorities in science, as well as decisions about honors and awards, the relatively "meritocratic" institution of science has been known to be less than purely meritocratic (awards given based purely and solely on the merit of the of work). The notion of a glass ceiling in science is also real given the consistent and robust finding that there are fewer and fewer women in science the more prestigious and elite the rank becomes. At an international conference in New Delhi, India, scholars from across the world (India, South Africa, France, Saudi Arabia, Germany, Croatia, and the United States) reported remarkably similar figures for the "inverted funnel" proportion of women in science as rank gets higher and higher.[33] Moreover, scholars at the conference discussed the phenomenon of awards and honors being vulnerable to the old-boys'-network complaint, insofar as people on panels for such decisions are naturally going to be biased toward nominees they know.

Another underappreciated topic for social psychologists of science is Harold Kelley's attribution model as it applies to scientific reasoning, that is, how people—from children to scientists—use evidence in attributing causes to effects.[34] Kelley's attribution model proposes that whether a causal attribution is internal (person) or external (situation) depends on three factors: consensus, distinctiveness, and consistency. Consensus is whether others viewing the same event come to the same conclusion; distinctiveness concerns whether the behavior/cause is unique to the person or situation in question; and consistency is concerned with whether the purported cause happens consistently over time in the same situation. When children, adolescents, adults, or scientists are attempting to test a causal hypothesis, they often implicitly use some or all of Kelley's causal criteria. To put it most generically, if everyone agrees that B follows A, if B follows A but not C, D, or E, and if B always follows A, then we have relatively solid evidence that A causes B. In short, attribution theory and the development of scientific reasoning would make perfect partners in a social-developmental psychology of science.

Finally, the applied field of Industrial-Organizational (I/O) psychology is another underdeveloped area of social psychology of science. Although its own discipline, I/O psychology is often closely allied with social psychology because I/O psychology is concerned with two different yet related topics: the industrial side addresses questions of personnel and human resource management, and the organizational side addresses questions of social and group influences on the organization. As we already saw, the question of personnel selection is of crucial importance to science: What criteria are used to select the best and most appropriate and most creative students, professors, or research scientists at colleges, universities, industry, and government science labs? How well do these criteria actually predict how well applicants do in their jobs? Is a person with high intelligence to be preferred to one with lower intelligence but more creativity? When I was interviewing scientists for my dissertation research, a chemistry professor at University of California–Berkeley told me: "We are one of the best departments in the world and our hit-rate in hiring professors is at best 50–50. If we knew in advance who would make it and who wouldn't, we would really be great."[35]

Moreover, which work environments promote and facilitate creative productivity and which hinder it? All of these questions could be better informed by a well-developed and well-established psychology of science. Knowledge of the motivational, cognitive, personality, and developmental forces behind scientific thought and behavior could help the gatekeepers of science make better informed decisions on who they want working for them and how they want to structure their work environment.

In sum, the social psychology of science has much potential and has made contributions to questions concerning the effects of experimenters on the experiment and social and cultural factors (such as war and political stability) on scientific productivity; the role of family, teachers, and eminent mentors on one's interest and achievement in science; and the role that gender plays in scientific interest, ability, and productivity. Yet the social psychology of science is unique in that much of what could clearly be has not yet become. Applying basic social psychological principles to scientific thought and behavior, such as persuasion, attitude change, influence of minority opinion on majority belief, attribution theory and scientific reasoning, person perception, prejudice and discrimination toward minorities or other out-group members, and finally human resource management and selection criteria, are tasks for the future.

# Chapter 7 The Applications and Future of Psychology of Science

Given the unique status of science and creative thinking as a hallmark capacity of our species and its dominant role in shaping modern culture, one would think that psychologists would have more systematically devoted their attention, methods, and theories to understanding the psychology of science and that other disciplines that study science would gladly make use of these contributions. Neither has been the case. The history, philosophy, and sociology of science is each well established and institutionalized. But psychology, like anthropology, is somewhere between the second and third stages (Isolation and Identification) of development (see chapter 1).

In part 1 of this book, I have summarized the current literature in the various psychologies of science—biological, developmental, cognitive, personality, and social. Much of this work, mind you, is only implicitly a psychology of science in that the scholars have not applied the label "psychology of science" to their work. As my reviews have demonstrated, however, there is a very rich and complex literature on most every topic that interests psychologists, from the genetics of intelligence to cognitive neuroscience. The literatures are rich enough that it took a chapter to summarize and integrate the empirical findings of each subdiscipline within psychology—biological-neuroscience, developmental, cognitive, personality, and social. In this chapter, the last of part I, then, I am now ready to discuss some applications of the psychology of science as well as its future possibilities.

## APPLIED PSYCHOLOGY OF SCIENCE

Industrial/Organizational (I/O) psychology investigates questions of personnel and human resource management as well as how social and group influences affect an organization. In chapter 6, I reviewed the question of social influence; personnel selection I will cover here. To be sure, selection criteria are of critical importance to science: How do we choose our best and most creative students, professors, or research scientists for positions in our college, university, industry, and government science departments and labs?[1] How well do the criteria we use actually predict real-world outcomes, such as publication rates or creative achievement? Does creativity or intelligence better predict high levels of scientific achievement? Other questions of an applied nature that are not necessarily addressed by I/O psychology are: Can we predict who will develop an interest in science at an early age? Can we predict who will choose science as a major focus of study? Or, can we predict who will stay in science after choosing that career path?

These questions address the practical importance that a psychology of science has for education, hiring, and policy making. Answering these questions by applying theories, methodologies, and findings from the psychology of science can be of tremendous practical importance, especially for math and science education and the recruitment of the best young minds into the scientific professions. The findings from an I/O psychology of science can therefore be critical in shaping educational policy at the primary school level in terms of how to best teach science and how to develop and maintain scientific interest in children. Moreover, what we know about who does science and who does the best science, as well as how science is done, should inform those who evaluate the quality of scientific research. Such evaluation occurs as early as high school science fairs, continues with undergraduate science courses, and peaks with graduate school and job applications. These are tough and important questions yet ones that a well-developed psychology of science can help answer. More specifically, parents, teachers, educators, recruiters, and policy makers want to know how we recognize, recruit and retain our most scientifically talented young people.

**Recognition of Scientific Talent**

Just as with gifted athletes, the scientifically talented start to show signs of their unique abilities by early to midadolescence. The sooner we recognize these talents, the sooner we can foster and develop them. For instance, my research on two groups of talented and elite scientists—Westinghouse finalists and members of the National Academy of Sciences—indicates that by middle to late adolescence future scientists, as well as others (for example, teachers, parents), recognized their talent. The earlier future members of the National Academy knew they wanted to be scientists and the earlier others recognized their talent, the earlier they published. Precocious scientific productivity, in turn, set them up to be the most productive scientists of their generation.[2] Precocity and its recognition are crucial in setting the scientifically gifted on the path of a productive career in science. In order to recognize scientific talent early, we must know its reliable precursors or predictors, and the strongest and most robust early signs seem to be intelligence, personality, and demographics.

High intelligence, as measured by traditional IQ tests, would appear to be a necessary but not sufficient condition for scientific thought, interest, and achievement. One's IQ predicts the kind of career one goes into, with scientists, mathematicians, engineers, medical doctors, and academics scoring higher on these standardized tests than people who go into other professions. To take one example, PhD physicists have an average IQ of 140, almost three standard deviations above the mean of the population. Such high scores for scientists are not surprising: two of the three major components of IQ tests, namely, quantitative reasoning and spatial reasoning, predict scientific interest and talent.[3] Absent these intellectual skills, one is not likely to be interested in or to become a scientist.

Although college entrance exams are not intelligence tests per se, they are similar enough to function as such. The question of what these aptitude/intelligence tests predict has become a major topic of psychological and educational research. Research generally shows that the undergraduate aptitude tests (Scholastic Aptitude Test, SAT) predict college grade point average (GPA) at a modest level (correlations generally between .20 and .50), especially grades during the first two years. For some unknown reason the SAT verbal scores do a better job of predicting grades than SAT quantitative scores. Additionally, SAT scores tend to distinguish those who graduate from those who do not. Some researchers, however, have argued that the SAT does not add enough predictive validity over and above the high school record to justify its cost.[4]

The predictive validity of entrance exams (mostly the Graduate Record Exam, GRE) at the graduate school level has been the topic of enough research, and more quantitative, or meta-analytic, reviews have been published on the topic. Some of the more narrowly focused meta-analytic reviews (particular samples or only one outcome) during the 1990s reported small but robust correlations in the .15 to .30 range with graduate school GPA.[5] Other more broadly focused studies on the predictive validity of the GRE, however, have included such outcome measures as comprehensive exam scores (comps), likelihood of attaining the degree, and publication and citation data. For instance, Robert Sternberg and Wendy Williams conducted a study on Yale University psychology PhD students between 1980 and 1991. In addition to grades, they obtained outcome ratings by graduate advisors on analytic reasoning, creativity, practical ability, research ability, and dissertation quality. They found that the GRE, especially the GRE-Subject, predicted first-year graduate school grades the best, followed by overall GPA, but second-year graduate school grades hardly at all. Of note was the fact that the effect sizes were small to medium, with correlations generally being in the .10 to .20 range. Also of note, none of the GRE scales predicted other real-world outcomes with the exception of male students' analytic scores modestly predicting almost every outcome.

In the most extensive—with more than 1,700 samples and 82,000 participants in all areas of graduate education—meta-analytic review of the literature on the predictive validity of the Graduate Record Examination, Nathan Kuncel and colleagues found that the verbal, quantitative, and subject scores had small to moderate effects on predicting first-year graduate GPA, overall graduate GPA, comprehensive exam scores, and degree attainment. Another interesting finding in this same study, and one with direct relevance to the psychology of science, concerned the ability of the GRE to predict publication and citation outcomes in 12 to 19 of the 1,752 samples, although again the effect sizes were small (.09 to .23). The subject scores of the GRE consistently were better predictors of graduate school performance than either the verbal or quantitative scales, even though graduate admissions programs often rely only on verbal and quantitative scores.

Although intelligence and aptitude tests do foreshadow scientific interests, it is important to point out what they do not portend. These tests do not predict, although people sometimes assume they do, creative achievement, job success (quality of career), well-being (happiness), and relationship satisfaction. In other words, they do not predict real-world life and career success or satisfaction variables. In all fairness, they were not designed to predict these things, but rather simply how people did in school. And at that they do reasonably well.

To better understand the relatively poor job IQ tests do at predicting creative achievement, let us examine more closely the relation between the two. Quite a bit of research has examined the relation between intelligence and creativity, and the consensual answer is that these two abilities are moderately related up to a "threshold of intelligence," around 120, and then the relation falls to essentially 0. If one were to graph this relation it would be curvilinear or, more accurately, asymptotic. In other words, a threshold of intelligence is required for creative achievement, but once one gets slightly above one standard deviation above the mean, more IQ points do not bring anything to the table. Stated differently, a person with an IQ of 160 is no more likely to be creative than someone with an IQ of 120 or 130. This seems counterintuitive only because of the common perception that intelligence is synonymous with "genius." On closer examination and reflection, it is not too difficult to understand how these processes involve different cognitive mechanisms. High intelligence is efficiency and accuracy in solving timed and structured problems that have known solutions. It is analytic and focused, or what creativity researchers have come to call "convergent thinking" because it "converges" on the known solution. High creativity, in contrast, involves finding and solving problems that are unstructured and have no known solution. Such thinking is often wide, loose, defocused, and synthetic, or what creativity researchers have come to call "divergent thinking." This form of thought requires the person to broaden and spread out their associations and ideas, to "diverge" from the usual and common.[6]

This distinction is seen in a vignette described by Yale University psychologist Robert Sternberg, who reports the case of "Alice" who came to Yale with nearly perfect GRE scores and a 4.0 undergraduate GPA.[7] As predicted by her undergraduate record and GRE, she did fine in her first year of graduate school when exams were structured and multiple choice in nature. But when the work moved away from structured course material and into less structured and creative research ideas, she did not do well and floundered. By way of contrast was "Barbara" who had relatively low GRE scores and a moderate GPA. Only her letters of recommendation hinted at her outstanding "creativity," and sure enough she ended up being one of the most creative researchers in the department.

What intelligence tests predict and do not predict is tied to one's conceptualization of intelligence, and indeed there are profound disagreements among psychologists and educators over whether intelligence is one generalized ability or specific domains of abilities. Traditionally, intelligence has been conceptualized as involving "g," which stands for "generalized intelligence," a term made famous by Charles Spearman (and in fact is sometimes referred to as "Spearman's g"). The most solid evidence for this view comes from the moderately high correlations between the verbal, quantitative, and spatial tests scores that form the core of most IQ tests.

Some psychologists, for example, Howard Gardner and Robert Sternberg, have been at the forefront of arguing for an expanded and at the same time a more specific view of intelligence. Gardner's theory of "multiple intelligences" argues for eight domains of intelligence (verbal, quantitative, spatial, musical, intrapersonal, interpersonal, natural history, and bodily-kinesthetic). The most solid evidence for this conceptualization comes from the findings concerning the criteria for domains of mind, namely, developmental automaticity, neurophysiological specificity, archeological, genetic, comparative, and precocious talent (see chapter 8). The thrust of the domain-specific argument rests on the fact that although some people may have talent in one or two of these domains, no one is really talented in all of them. More specific measures of intelligence do a better job of predicting real-world success outcomes than generalized tests because they are tied more directly to the specific talents people possess. The main problem at this point with the domain-specific view is the difficulty that psychologists and educators have had in constructing reliable and valid measures of specific domains of intelligence. Some domains have begun to be assessed and others have not. Emotional or social intelligence, for instance, has captured a lot of attention over the last ten years and shows some signs of being reliably measured by various indices.[8]

Because intelligence and aptitude tests do a relatively poor job of predicting real-world outcomes, other nonintelligence predictors have been examined. I would claim that ideally we should augment (not replace) our selection criteria with real-world predictors of creative achievement, namely, some personality, motivational, and precocity measures.[9] In the chapter on the personality psychology of science, I reviewed literature showing that cognitive personality traits (conscientiousness and openness), social traits (confidence, dominance, and introversion), motivational traits (achievement and drive), and affective traits (impulse controlled and low anxiety) were the traits most consistently related to scientific interest and creativity.

To provide one specific example for how personality predicts creative achievement over and above intellectual ability: Frank Barron and I recently published a study, begun in 1950 when the participants were twenty-seven years old, that sampled eighty male graduate students (mostly scientists) and assessed them on potential, intelligence-intellect, personality, and creativity.[10] Personality and career outcome data were collected forty-four years later (at age seventy-two). We predicted that personality would explain unique variance in creativity over and above that already explained by intellect and potential. Our results supported the prediction. For instance, observer-rated potential and intellect at age twenty-seven predicted lifetime creativity at age seventy-two, yet personality variables (such as tolerance and psychological mindedness) explained up to 20 percent of the variance over and above potential and intellect. We argued for a functional or threshold theory of personality, namely, that if traits function to lower behavioral thresholds in given situations, then the traits of self-confidence, arrogance, openness, tolerance and psychological mindedness (among others) might serve as a relatively direct link between personality and creative behavior.

Other noncognitive predictors of creative achievement include demographic factors. In chapters 3 and 6, I reviewed some of the literature demonstrating that two demographic factors are consistently related to the development of scientific interest, namely, gender and immigrant status, and briefly summarized some of the possible reasons for these associations. The question of why men are more likely than women to go into the physical sciences is a perplexing one, with no easy answer. It almost certainly has to do with a complex interplay of dispositional talent and cultural forces. The inanimate world seems to hold more fascination for boys, but why this may be remains a confounding question. The issue of immigrant status is also quite intriguing and calls for more empirical research, with each of the most likely candidates—meritocratic, work-ethic, and multicultural perspective—facilitating novel insights and perspectives. Each is plausible but all need more systematic empirical support.

A psychology of science could make a major contribution by expanding these findings and investigating more programmatically and explicitly what the robust predictors of scientific interest and talent are. The particular form of scientific interest, whether physical, biological, or social, seems very likely to be partly based on one's personality and/or talent for things versus people, or the inanimate versus the animate worlds. In addition, intelligence seems to predict interest but not talent or creative achievement, so what psychological processes and characteristics predict real scientific creative achievement? Answering this question is precisely where the applied value of the psychology of science lies.

### Recruitment of Scientific Talent

Recruitment of future scientists depends on our ability to accurately predict outcomes, and from the empirical findings just reviewed it is obvious that intelligence is but one, and maybe not the most crucial, predictor of scientific interest and talent. Intelligence seems to be better at foreshadowing interest than achievement or creativity. If this is true, then the selection criteria for recruiting future scientists would ideally include some of these other interest, personality, motivational, and demographic factors. Because there is much overlap between recognition and recruitment, my comments on recruitment will be focused on just one domain, namely, science competitions.

The traditional method of fleshing out scientific talent in this country has been through local, regional, state, and national science fairs and competitions. Beginning in 1942, the most prestigious national competition was organized by the nonprofit organization Science Service and underwritten by the Westinghouse Corporation. The Science Talent Search has been the country's premier channel through which young scientific talent has been recognized and recruited. Until 1998, when Intel Corporation took over sponsorship, the competition more commonly became known as the Westinghouse. The charge of Science Service was and still is to popularize science and make scientific information more accessible to the public. Among others, Joseph Berger, an education writer for the *New York Times,* has written a book, entitled *The Young Scientists: America's Future and the Winning of the Westinghouse,* about this talent search. As Berger writes: "From the start, the Westinghouse was different from a traditional science fair. Its goal was not simply to choose the best projects but to locate the best potential scientists."[11] Most fascinating in the history of the Science Talent Search's competition is its close association with the New York City high schools, and how since its inception in the early 1940s it has fueled a lively competition among these high schools to see who can produce the most finalists. School officials have even scouted and developed potential talent with the goal of training would-be Westinghouse finalists, much the way that many high schools might develop their football talent. The Bronx Science High School has become the New York Yankees of the science competition, having had 118 finalists from 1942 to 1990, with Stuyvesant next with 70, followed by Forest Hills High School with 42. Here we have one of the few instances in

American culture where academic and scientific talent is fostered, nurtured, and encouraged as much as athletic talent.

**Retention of Scientific Talent**

Once someone chooses science as a topic of study or a career, the next question becomes whether they stay in science or leave it. The question of attrition is of grave concern, especially if it is not random but rather affects certain groups more than other groups, for instance women or minorities. My research, as well as that of others, has reported gender differences in attrition from science, with women being more likely to leave than men.[12] For instance, I found that with an elite scientific sample (Westinghouse finalists), although most tend to stay in science, math, or technology careers, women were more than three times as likely as men to leave science during either their education or career. More specifically, by college age 6 percent of the male finalists left a science track, whereas 22 percent of the female finalists had. Once they reached career age and beyond, a total of 11 percent of the male and 43 percent of the female finalists left the science career track. It is important to note, however, as others have, that just because women may opt out of science-oriented careers more often than men does not mean they opt out of productive and achieving careers in general.[13]

Women, unfortunately, are more likely to leave science or work in it part time than men. If true, the question that begs an answer is "What can be done about this and by whom?" Obviously that is a difficult question with no easy answer and one that many different people, from teachers to scientists to politicians to policy makers, have attempted to tackle with varying degrees of success. At an international conference on women in science in New Delhi in 2004, the question of what to do about female attrition from science was central to many of the presentations. One participant, J. Scott Long, reported some of the institutional changes that American universities have begun implementing, such as flexible tenure tracks, parental leave for either parent, and increased and more flexible child-care facilities. Policy changes have come as well from politicians, as happened in 1980 when Congress authorized a $30-million budget for the National Science Foundation (NSF) to increase women's involvement in science. Mary Frank Fox reviewed intervention programs that aim to increase the number of women in science and categorized them into two groups: those that focus on individual characteristics and those that focus on institutional or environmental characteristics. Individually oriented programs focus on changing personal traits, such as cognitive strategies, motivation, attitudes, and aptitudes, whereas the institutionally oriented programs focus on changing the work environment, settings, mentorship programs, and tasks undertaken. To the extent that the proportion of women in science has increased over the last thirty years, it is possible that these programs have had some effect, but it is nearly impossible to suggest the precise impact these programs have had.[14]

## THE FUTURE OF THE PSYCHOLOGY OF SCIENCE

In this section I return to a theme that permeated the first section of the book and indeed has been a driving motive behind writing it: why is the psychology of science not a more developed discipline, especially when compared with the history, philosophy, and sociology of science? Indeed, William Shadish, Steve Fuller, and Michael Gorman boldly claimed in the opening sentence of their 1994 chapter in *The Social Psychology of Science,* "The psychology of science has finally arrived."[15] I am now ready to put forth my thoughts on whether the psychology of science has become a formal, autonomous discipline, on the verge of developing its own professional societies, conferences, journal, and university departments.

To return to a question elaborated on in chapter 1, "When, then, is a field a field?" Nicholas Mullins argued for four stages, which I simplified and reduced to three (Isolation, Identification, and Institutionalization). I said that psychology of science is between the Isolation and Identification stages and also presented evidence that the history and philosophy of science have been institutionalized for at least one hundred years and the sociology of science for roughly fifty years. They have not only formed societies and journals, but they have institutional support in the form of full-time faculty positions, funding for graduate students, and awarding of advanced degrees. Although its origins reach as far back as the 1870s with Francis Galton, the psychology of science, by contrast, first earned even its own name in the 1930s and had no more than a handful of scholars until the 1950s. During that decade there was an upswing in research and theory on the psychology of creativity, including scientific creativity. Especially post-Sputnik, anything that helped foster an interest in science was encouraged and relatively well funded. During the 1960s and most of the 1970s, however, there was very little systematic work done by psychologists on scientific thought, reasoning, or behavior. But by the mid to late 1980s, the field really stood at the precipice of full-fledged flight.

Or so it appeared. The 1986 conference held in Memphis and organized by William Shadish, Barry Gholson, Robert Neimeyer, and Arthur Houts was a good beginning, but little followed from it other than an edited volume and a book by Dean Simonton on scientific genius.[16] No society, regular conferences, or journal sprouted up afterwards. In the 1990s Ron Westrum at Eastern Michigan University started a newsletter ("Social Psychology of Science Newsletter"), but even that lasted but a few years.

Similar to Nicholas Mullins's proposal for the independence of a discipline, the psychologist Joseph Matarazzo proposed six criteria for establishing a psychological field as an independent discipline. They are: (1) its own national and international associations; (2) its own journals; (3) an acknowledgment by professionals in other fields in psychology that its subject matter, methods, and applications are distinct from its own; (4) postdoctoral training programs; (5) recognition from American Board of Professional Psychology; (6) and finally, recognition by the new American Psychological Association Commission on the Recognition of Specialties and Proficiencies in Professional Psychology.[17] With the possible exception of number 3, psychology of science meets none of Matarazzo's criteria.

Here then is my basic argument. The psychology of science has not arrived, but during the 1990s and now in the 2000s, the field *has* taken off, although primarily in hidden or implicit form. Every domain of psychology has active and talented researchers working on many different questions fundamental to understanding scientific thought and scientific behavior as well as scientific interest, theory formation, and scientific talent and creativity. Such names as Susan Carey, Alison Gopnik, Paul Klaczynski, Barbara Kosloswki, Deanna Kuhn, Elizabeth Spelke, and Corinne Zimmerman, are excellent examples—just from developmental psychology—of talented thinkers doing (implicit) psychology of science. In cognitive psychology, a list more explicitly identified with the field would include such figures as William Brewer, Kevin Dunbar, Michael Gorman, Howard Gruber, David Klahr, Roger Shepard, and Ryan Tweney. Giftedness and educational psychologists consist of such standouts as Camilla Benbow, David Lubinski, Julian Stanley, and Rena Subotnik. Lastly, there are such scholars as Dean Simonton and Frank Sulloway who cut across many traditional boundaries within psychology. The point is that some of the more talented and creative minds in psychology are interested in, have developed theories of, and have conducted research in what others and I are calling the psychology of science. The unique and interesting sociological question, therefore, is why do not more of them realize that is what they are doing.

I believe part of the answer lies in the fact that—with some exceptions—they are not familiar with or aware of the term "psychology of science." Indeed, there are no codified and institutional structures with which these scholars can identify. They did not get their PhD's in the psychology of science; there is no society to join; there is no journal to publish in; and there is no regular conference to attend. It is no wonder that many psychologists, even those studying scientific interest, thinking, talent, and creativity, do not identify with the field or call themselves psychologists of science.

It is my firm belief that much of the psychology of science is dormant, latent, and implicit, and it is one of my goals to make it manifest and explicit by laying the foundation for its infrastructure. Books such as this one might begin to change that. Another significant advance would be the formation of a journal. There are thousands of journals in science today. In psychology alone there are literally hundreds of journals, many of which are ultraspecialized and focused on very narrow aspects of human behavior. For instance, just to name a few examples, there are journals in dreaming, epilepsy, psycho-oncology, hypnosis, parapsychology, transcultural psychiatry, applied sport psychology, school health, aviation, space, environmental medicine, circadian rhythms, and eating disorders. There is even a journal devoted to science education. In many cases there are multiple (at least four or five) journals in the more specialized areas, such as dreaming, circadian rhythms, parapsychology, or hypnosis. My point is not that these are overly specialized areas and do not need the journals they currently have. Rather, my point is that if we can have multiple journals in such specialized areas, then we can and should have at least *one* journal devoted to the psychology of science. Science is such a ubiquitous and all-powerful force in modern culture that we need to examine empirically and theoretically all of the psychological factors behind the development of scientific interest, talent, and achievement. Moreover, we need a scientific outlet for publishing the results of these studies in one place.

Each of these developments is intertwined: conferences of like-minded scholars would be the most likely and most feasible first step. Out of these conferences, research ideas and collaborations could form and foment further research. If scholars begin to produce enough original research, then a journal and society could follow. If these developments were to happen, perhaps then could we start talking about an actual rather than dormant psychology of science.

# Part Two **Origins and Future of the Scientific Mind**

# Chapter 8 Evolution of the Human Mind

The ecologically most significant respect in which humankind now dominates all other terrestrial species is in its scientific understanding and technological manipulation of the world.
—*Roger Shepard,* "The Genetic Basis of Human Scientific Knowledge"

How did the human mind become capable of doing science, in all of its symbolic, mathematical, and highly specialized forms? Like all things natural, something as complex as the human brain did not appear overnight; it emerged from simpler and more rudimentary structures. This is true for the scientific mind and scientific thinking as well. Modern, explicit scientific thought evolved out of the incipient, implicit folk domains of mind (common sense). Knowing the path and history of how scientific thinking came about in our species gives us a richer and fuller appreciation of its development in modern individuals. The second theme of this book—origins of the scientific mind—rests on the assumption that ontogeny and phylogeny are related processes and that understanding how scientific interest and behavior unfold in individuals informs their development at the species level and vice versa. I argue that the natural, biological, and social sciences are not arbitrary divisions but rather fall along evolved domains of mind, that is, domains most directly relevant to problems of survival and reproduction. My perspective draws from the recent movement in social science that applies Darwinian principles of natural and sexual selection to the evolution of mental processes and human behavior.[1] Such a biological and evolutionary view, to be sure, has not been common in the social sciences over the last one hundred years, where the mind has been viewed as a malleable "blank slate," written on by the specifics of the context and environment.[2]

These two views are modern-day versions of the notorious nature-nurture debate, which goes back at least as far as the ancient Greeks. One perspective has argued that the mind is a passive and blank mechanism of thought, whereas the other has taken the view that the mind is an active and structured organizer of sensory and cognitive experience. These views reduce to the notion that knowledge comes either directly, exclusively, and passively from the senses (empiricism) or that the mind orients, predisposes, and constrains attention toward certain categories of sensations before becoming knowledge (nativism). The debate on how the mind works has taken a decisive turn toward biology and evolution over the last twenty years. Although the empiricist blank slate view held court for most of the twentieth century, since the 1980s more and more evidence has accumulated from evolutionary theory and research (anthropology, archeology, cognitive neuroscience, psychology, and philosophy) that the mind is a complex and dynamic interaction between evolved mechanisms and environmental influences.[3]

## DOMAINS OF MIND

Natural and sexual selection pressures have predisposed the human mind toward certain kinds of sensations and functions. These functions are specific mechanisms and cognitive domains that solve specific problems. Domains of mind have some degree of physical-neuroanatomical status (that is, localized to particular brain regions), but they are also conceptual and heuristic entities. As defined by Rochel Gelman and Kimberly Brenneman, a domain is a "given set of principles, the rules of their application, and the entities to which they apply." That is, domain-specific principles are interrelated and operate within a specific class of problems that have been crucial for survival and reproductive success. Domains are similar to but not synonymous with modules, for the latter are encapsulated information units that process inputs (perceptions). Domains are universal and part of human nature, and they concern knowledge of

Table 8.1 Domains of Mind Proposed by Different Theorists

| | Author | | | | | | | | |
|---|---|---|---|---|---|---|---|---|---|
| Domain | Geary and Huffman | Gopnik, Meltzhoff, and Kuhl | Parker and McKinney | Carey and Spelke | Mithen | Karmiloff-Smith | Feist | Gardner | Pinker |
| Psychology | x | x | x | x | x | x | x | x | x |
| Physics | x | x | x | x | x | x | x | x | x |
| Linguistics | | x | x | x | x | x | x | x | x |
| Mathematics | | | x | x | | x | x | x | x |
| Biology | x | | | | x | | x | x | x |
| Art | | | | | | x | x | | |
| Music | | | | | | | x | x | |
| Kinesthetics | | | | | | | | x | |
| Economics | | | | | | | | | x |

*Note:* Information is from the following sources: Carey and Spelke 1994, Feist 2001, 2004a, Gardner, 1983, 1999, Geary and Huffman 2002, Gopnik et al. 1999, Karmiloff-Smith 1992, Mithen 1996, Parker and McKinney 1999, and Pinker 2002. Pinker, in fact, argues for ten separate domains (psychology, physics, engineering, spatial, language, biology, math, probability, logic, and economics), but I have combined certain categories in this table, for example, physics, engineering, and spatial are combined and math, probability, and logic are combined.

the social (people), animate (animals), and inanimate (physical objects) worlds, being able to count, quantify (number), communicate (language) our ideas about these worlds, appreciating and creating aesthetically pleasing arrangements of visual images (aesthetics), and being sensitive to and appreciative of rhythm, pitch, timing in sounds (music).[4] Because this book is concerned with science, I focus my attention on the four specific domains that underlie the sciences, namely, folk psychology, folk physics, folk biology, and folk mathematics. Various authors have proposed anywhere from three to eight domains, with four being the most common number (see table 8.1).

To call something a domain requires specific criteria; otherwise it risks being an arbitrary enterprise. In my view, a capacity must meet the majority of seven criteria if it is to count as a domain of mind. These criteria are: archeological, comparative, developmental, universal, precocious talent and giftedness, neuroscientific, and genetic. In short, nothing less than the combined interdisciplinary evidence from archeology, primatology, developmental psychology, anthropology, giftedness-education, neuroscience, and genetics is required before something can be classified as a domain. Now is neither the time nor place to present a systematic review of all the evidence for each domain.[5]

## Domains of Mind as Domains of Science

*Implicit Psychology.* One of the most fundamental aspects of being human is our reliance on others for our survival. We are constantly confronted with problems of interpersonal relationships: from sexual behavior to child rearing, from friendship alliances to kinship-based altruism, from emotion and facial recognition to deception and cooperation. Very briefly, *implicit psychology* (*social domain*) consists first and foremost of social preferences seen in newborns as young as a few hours old, namely, an intuitive and automatic preference for humans (especially the face) over other animal forms. With development, we see a number of specific abilities involving interaction between people, for instance, the ability to recognize and infer our mental and emotional state as well as those of others, even when their beliefs and emotions differ from our own. These abilities, in short, are known as a developed "theory of mind." David Premack and Guy Woodruff first coined the phrase in 1978 and defined it as imputing mental states to oneself and to others. Imitation, pretend play, false belief, deception, mental attribution, joint attention, and self-awareness are some of the specific manifestations of theory of mind. The social-psychological domain also involves self-knowledge and self-concept. What I mean by implicit psychology is very similar to what E. L. Thorndike referred to as "social intelligence," Howard Gardner referred to as the "personal intelligences" (inter- and intrapersonal), and what Peter Salovey and John Mayer more recently referred to as "emotional intelligence."[6] The social sciences (psychology, sociology, anthropology, and economics) originate from this domain of mind.

Comparative evidence is based on examining humans in relation to our primate relatives, in particular the great apes (chimps, bonobos, gorillas, orangutans), Old World monkeys (baboons and macaques), and New World monkeys (for example, cebus, rhesus, capuchins). For the domain of folk or implicit psychology, much comparative evidence has centered on theory of mind. One manifestation of theory of mind is the false-belief task; that is, understanding that members of one's species can believe something that is not true. Primatologists and comparative psychologists have generally concluded that great apes have a theory of mind (but one that stops short on the false-belief task), can demonstrate signs of intentional deception, and have mirror-self-recognition by forty months. Old and New World monkeys have little to no theory of mind; they use deception but are not aware that others are intentionally being deceptive, and they never develop mirror-self-recognition. By comparison, humans have rather developed theories of mind, can master the false-belief task by age four, have complex understanding of deception and its intentionality, have mirror-self-recognition by twenty-four months, and go on to develop metarepresentational self-awareness (that is, awareness of awareness).[7]

Universality evidence comes from demonstrating that a behavior or capacity is found in all cultures. Although these behaviors or capacities can be manifested differently, if they are to be part of an evolved domain they must be exhibited by all current human cultures. For implicit psychology, the most consistent universality evidence exists in the area of emotion expression and recognition as well as theory of mind. In terms of emotion, Paul Ekman's work on universality of facial expression of emotion is perhaps best known. Ekman demonstrated that individuals in a preliterate culture (New Guinea) who had never been exposed to Western culture recognized the same facial configuration for the same emotions as individuals in Western cultures do. The situations that elicit these emotions are culturally specific, but once elicited the emotions have the same expressions and are recognized as such. There is cultural variation in the rules for display of emotion in certain settings, but the facial expressions associated with an emotion is universal for a small set of basic emotions: anger, disgust, fear, happiness, sadness, and surprise. Furthermore, when shown photos of basic emotional expressions, people from Europe, the Americas, and Asia label the same expressions with the same word. Some emotions, however, are not as accurately recognized as others, with fear, disgust, and surprise being the least consistent. Further, Ekman takes care to point out that universality for certain emotions does not preclude cultural variability in others. Finally, people are most accurate at recognizing emotion within their own in-group or ethnic-racial group.[8]

Theory of mind also seems to exist and develop consistently across human culture. Paul Harris argues that although relatively sparse, the universality evidence for theory of mind is encouraging. For instance, the false-belief skill seems to emerge at the same age in different Western cultures, and the distinction between real and fake emotion emerges in similar form and timing in some Western and Asian cultures. Although empirical data may be scarce, it is safe to say that there is no human culture where the beliefs, feelings, and desires of others are not recognized and represented. The specifics of *what* people attribute to what others may be thinking or wanting may vary across cultures, but the fact *that* everyone represents these beliefs about beliefs, emotions, and motivations is universal in our species.[9]

Another criterion for a domain is giftedness, that is, whether a talent is manifested in extreme forms in certain individuals. Giftedness may best be viewed as simply the extreme end of another criterion, namely, the developmental criterion. The latter is simply whether a behavior or cognitive strategy develops early and automatically in all individuals, whereas giftedness is the extreme end of that normal distribution and examines certain individuals with unusual and precocious talent in a given domain.

Relatively little research has been conducted on giftedness and talent in the social-emotional domain, partly because it cannot be assessed through pen and paper methodology. Thomas Hatch has written about these skills among kindergarten-aged children at play and argues that children with interpersonal and social skills are the leaders and diplomats of the playground. They have talents for responding to the thoughts and feelings of their playmates and can regulate their own desires and impulses. These children organize groups, mediate conflict, have empathy, and are "team players." Similarly, Alain Schmitt and Karl Grammar argued that the most socially skilled and successful children are not simply the most cognitively complex ones, but rather those who know how to produce the most desired and often simplest outcomes. When people are talented at social-emotional intelligence, they may become leaders or well liked by peers, but they seldom win awards and talent recognition contests because there are none for these skills. Gifted programs are mostly geared toward language, math, science, and music, not to interpersonal talent. Some schools, however, are beginning to include social-emotional skills in their curriculum.[10] It is interesting to speculate whether many of our best psychotherapists and even our political leaders showed precocious gifts in the area of social-emotional intelligence and mediating interpersonal conflicts in early childhood.

*Implicit Physics (Physical Objects–Spatial Domain).* Physical knowledge concerns the inanimate world of physical objects (including tools); their movement, positioning, and causal relations in space; and their inner workings (machines).[11] Because tool use is a large component of physical knowledge, some archeologists refer to this domain as "technical intelligence." It consists of the ability to solve problems of tool use (wood versus stone; simple versus complex) and mental and physical manipulation of inanimate objects of different materials, as well as an implicit understanding of physics (gravity, inertia, and dynamics of objects). Moreover, spatial knowledge and skills are involved in the physical objects domain. An "implicit physics" is also seen in children's automatic sense that physical objects obey different rules than living things (inanimate versus animate rules). Inanimate objects fall to the ground and do not get up. The physical sciences (physics, astronomy, chemistry, geology) originate from this domain.

In contrast to archeological evidence for folk psychological skills, the evidence for folk physics leaves a direct record in stone tools. Stone toolmaking, and toolmaking in general, requires specialized knowledge of objects and physical material, an understanding of their causal relation (that is, "if I do this, then that will happen"), and an understanding of their representation in three-dimensional space. The first stone tools were relatively simple objects known as Omo tools, dating from two to three million years ago (mya), and are difficult to distinguish from naturally occurring rocks and are attributed to prehomo *Australopithecus.* A revolutionary advance began with the first species of *Homo,* around 2.0 to 2.5 mya, and in fact, we have named them accordingly: *H. habilis* ("handy human").[12] Their technology industry is referred to as Oldowan and it is the greatest behavioral difference between early *Homo* and *Australopithecus.* The major advance with this technology is the appearance of the first examples of flakes being removed from rocks. But even Oldowan tools were used for little more than chopping. Another advance occurs at about 1.4 mya with the first bifacial hand axes (developed Oldowan). Roughly a quarter of a million years ago, the new species (*H. erectus*) produced the first real advancement beyond Oldowan technology, namely, precise "arrowhead" flakes (the Levallois industry).

Comparative evidence for folk physics comes from tool use in primates. Although Wolfgang Köhler and Robert Yerkes demonstrated tool use in captive great apes in the 1920s, it was only after Jane Goodall's observations of chimps in the wild that it become generally recognized that humans were not the only users and makers of tools. The question then became not whether apes used tools but how and with what level of complexity. As with all other domains, there are systematic differences in complexity of tool use and tool manufacturing as we move toward increasing complexity on the phylogenetic scale, from monkeys to apes to humans. Monkeys in the wild have seldom been observed using any tools and in captivity rarely go beyond the simplest tool use, such as using a stick to get food that is visible. Chimps, in contrast, have been observed in the wild using complex tools and even metatools (tools on tools); for instance, wild chimps have used wedges to stabilize their anvils. Chimps also have been observed using compound tools or a complex sequence of tools, such as sequentially using a chisel, awl, and dipstick to extract honey. Experiments that manipulate the complexity of problem solving using captive primates have confirmed this phylogenetic increase in complexity of tool use and tool production. Complex tool use, however, is still quite rare among primates, limited mostly to chimps, and even then it is generally limited to two tool components being used at once.[13]

A third criterion for a domain is giftedness, and evidence for giftedness in the physical domain converges on the conclusion that physical scientists from very early in life have temperaments and personalities that are "thing-oriented" rather than "people-oriented." Supporting this domain-specific view of talent, Baron-Cohen and his colleagues have found that engineers, mathematicians, and physical scientists score much higher on measures of high-functioning autism (Asperger's syndrome) than nonscientists, and moreover that physical scientists, mathematicians, and engineers are higher on a nonclinical measure of autism than social scientists. These findings may be a more extreme expression of the general phenomenon, but they do suggest that physical scientists may have temperaments that orient them away from the social and toward the inanimate; their interest and ability in science is merely one expression of this orientation. Autistic children are more than twice as likely as nonautistic children to have a father or grandfather who was an engineer. Indeed, there seems to be an intriguing connection between autism and physical-technical giftedness as seen by the fact that special talents in mechanics and space are often manifested in autistic children. For instance, Alonzo Clemons, an autistic savant, can make perfect replicas of animals in wax even if he sees them for but a few seconds. In other cases, some autistic children are experts at mechanics, being able to take clocks and radios apart and put them back together without error. In at least one recorded instance, an autistic boy could determine with a high degree of accuracy the dimensions of an object, such as a room, fence, or driveway. With objects smaller than twenty feet wide, he was accurate within one-fourth of an inch.[14]

*Implicit Biology (Natural History Domain).* Biology and natural history consists of the knowledge that the world consists of animate things that differ from inanimate (the flip side to physics). Some things live, reproduce, ingest food, and move by themselves—we call them animals. Some things fall to the ground and do get up (assuming they did not fall from too great a height). Moreover, some things are alive and grow but stay in the same place (plants). Indeed, biology and physics are part of the same innate sense that the world is divided into things that move and grow and things that do not, and that these classes of things obey different rules. Implicit biology also consists of the ability to solve problems concerning natural resources, namely, acquiring food (hunting, scavenging, foraging, forming mental maps of landscape), classifying plants (for food and medicine) as well as understanding animals and their behavior, and knowing which landscapes are resource rich and fertile. The biological sciences (biology, genetics, life science, and so on) are extensions of this domain of knowledge.

The comparative evidence for the natural history/biology domain by means of naturalistic observation has documented systematic foraging and hunting strategies of various species of chimp, especially those from the Gombe and Taï regions of Africa. Indeed, chimps are the only species of nonhuman primate that hunts large quantities of meat. Wrangham conducted very elaborate and detailed observations of food gathering among the Gombe chimps and concluded that they possess excellent botanical knowledge in terms of the location and seasonal cycles of certain plants, but they could not predict the whereabouts of plants not directly known to them. In terms of hunting, Goodall described one rather graphic scene in which six male Gombe chimps stalked a mother baboon and her infant. The males surrounded the pair and three of the chimps climbed various trees to prevent arboreal escape. When the mother did attempt to escape by climbing one of the trees, she was captured and the infant seized. The mother was released and ran off a few yards, only to witness the male chimps devouring her infant.

As summarized by Parker and McKinney, the Gombe chimps hunt thirteen species of primate, four species of ungulate, three species of rodents, and one species each of insectivore, carnivore, and hyrax, but 80 percent of their total prey comes from one species of monkey (red colobus). It is important to point out that meat eating is still a relatively rare behavior among chimps, accounting for only about 3 percent of their eating time and occurring largely during the dry season.[15] In comparing the Gombe and Taï chimps, Christophe and Hedwige Boesch reported that the Taï are more intentional, more successful, and more cooperative than the Gombe chimps. In other words, chimps possess cultural differences in the natural history domain.

If chimps exhibit such complex omnivorous hunting and food-gathering skills, then cladistically these behaviors must have originated in the common ancestor at least six million years ago. Cladistics is a branch of science where comparative psychologists and anthropologists make inferences about the origins of a behavior by seeing which species possess that behavior and where the common branch (clade) is. The logic of cladistics is simple: if many species do something, then that behavior must be much older than a behavior performed by only a few species.

By logical, if not empirical, necessity every human culture has developed either an implicit or explicit knowledge base involving the plants, animals, and geology of its local environment in order to survive. Implicit biology is therefore a human universal. Anthropologists have been the main scholars to document the similarities and differences among cultures in the folk biologies; perhaps the most prominent of such anthropologists was Claude Lévi-Strauss. In 1966 he reviewed much of the anthropological literature from the 1930s to 1950s and argued that many preindustrial cultures' knowledge of the world was concrete (rather than abstract) and sometimes overlapped with current scientific taxonomies. The Navajo, for instance, divide the animate world into two: those with and those without speech. Humans belonged to the first category and all plants and animals to the second. Animals, in turn are divided into three by method of locomotion (running, flying, and crawling). Each of these categories, moreover, is further divided into land- or sea-based travelers and night and day travelers. "The division into species obtained by this means is not always the same as that of zoology." Navajo botany uses similar obvious and nonobvious characteristics, such as sex, medicinal properties, and visual-tactile appearance (such as prickly or sticky) to classify plants. Lévi-Strauss went on to acknowledge, however, that "native classifications are not only methodical and based on carefully built up theoretical knowledge. They are also at times comparable from a formal point of view, to those still in use in zoology and botany."[16] In fact, one theme of Lévi-Strauss's writings, despite its pejorative use of the word "savage" in the book's title, is how much commonality and overlap exists between much of native natural science (botany and zoology) and their modern scientific counterparts. Such an emphasis was quite a change from the anthropology of the 1920s and 1930s, which often explicitly referred to preindustrial native cultures as "inferior."

Some of the anthropologists from the 1990s have continued the commonality theme and have focused on universals of biological knowledge. For instance, Scott Atran has documented the commonalities between certain principles of modern scientific taxonomy (that is, Linnaean taxonomy) and folk taxonomy.[17] He argues that "common sense," (that is, folk theories) rather than being detrimental to the development of science instead must be the beginning of all systematic, scientific thought, including natural history taxonomies. There is no other place to start. The development of natural history began with folk theories, moved to semiscientific Aristotelian classification, then to the more "natural system" of the sixteenth and seventeenth centuries with the recognition of the importance of self-reproduction and the concept of "genus" superordinate to species.

Giftedness in the natural history or biological science domain shows a unique orientation compared to social or physical science talent. Gifted people gravitate toward, show interest in, and have well-developed knowledge of plant and animal classification, plant and animal behavior, and their habitats. Those with talent for the natural world are less thing-oriented than physical scientists and less people-oriented than social scientists. As Howard Gardner wrote: "Biographies of biologists routinely document an early fascination with plants and animals and a drive to identify, classify, and interact with them; Darwin, Gould, and Wilson are only the most visible members of this cohort. Interestingly, these patterns are not echoed in the lives of physical scientists who, as children, more often explored the visible manifestations of invisible forces (like gravity or electricity) or played with mechanical or chemical systems." In other words, children who are talented in the biological domain spend much of their time in nature observing, collecting, and classifying—that is, creating taxonomies. One of the more interesting observations about Darwin is that by all accounts, including his own, he was not intellectually precocious—he was a most unremarkable child intellectually. The only distinguishing talent he seemed to manifest by adolescence was for identifying natural objects.[18]

*Implicit Mathematics (Numerosity and Seriation).* I believe that there is compelling evidence for including numerical or quantitative capacities as a unique domain of mind.[19] We all have a sense of numerosity; we intuitively and automatically know that there are three or five of something (one-to-one correspondence) and that one series is larger or smaller than another; we are born with this ability, but it is supplemented of course with cultural knowledge. Numerosity allows us to automatically use (add and subtract) positive whole numbers. Formal math ultimately stems from this implicit domain.

The evidence for the universality of math comes from the anthropology of number. Thomas Crump argues that few anthropologists have addressed this topic, and he went on to demonstrate that the explicit concept of number takes on a very different form and very different levels of abstraction depending on the culture. Some cultures have very few explicit words for many different numbers. For instance, some non-Western cultures seldom have words for numbers greater than two. After two often comes "many." Although this may be true, people of most cultures can easily understand the concept of four, five, and six when presented with it, and almost every human culture has had implicit mathematical skills involving seriation, numerosity, and rudimentary manipulation of quantities. Finally, the importance and ubiquity of number in cultures varies widely: "at one extreme there are, for instance, the Balinese who seem not to be able to do anything without numbers, while at the other, there are the Bemba of Zambia, who would readily dispense with them altogether."[20] Although this may be true, it is again important to distinguish explicit numbers and number words from implicit numerosity. Every culture possesses an ability to divine at a glance whether there are two or four discrete things in a series. In short, the implicit processes of distinguishing numbers of things and seriation may well be rather universal, whereas the more explicit use of numeric operations and verbal representation of number is culturally specific. Many cultures have not, of course, developed formal mathematics.

Evidence for math as a specific domain also comes from the fact that certain individuals are extremely gifted and precocious in math. Some children begin to display incredible mathematical computational and reasoning skills as early as two or three years old and by ten years of age are already performing complex mathematical calculations. The list of historical examples of inherent precocious mathematical genius is long and impressive: Pascal, Newton, Leibniz, Laplace, Gauss, Boole, Wiener, Ramanujan, and Feynman, to name but a few of the truly outstanding examples.[21]

Although I will not elaborate on the nonscience domains (see table 8.1) here, I want to at least define them. The first is *implicit linguistics* (language), which is the ability to use meaningful sounds to communicate with others and to understand abstract symbols. As many linguists have pointed out, language is composed of three components: phonology (physical expression of language, usually through sounds), morphology (words and word parts), and syntax (the combination of words and phrases). Humans the world over acquire skills in each domain of language automatically, intuitively, and without formal training. *Implicit music* is the capacity to produce, perceive, and appreciate rhythmic, melodic sounds that evoke an emotional response in oneself and others. It can also be defined as the ability to perceive changes in pitch as harmonically or rhythmically related. As one ethnomusicologist wrote recently, "All of us are born with the capacity to apprehend emotion and meaning in music, regardless of whether we understand music theory or read musical notation." There is much evidence from brain specificity, universality, ethnomusicology, the development of musical preference and ability, and archeology to argue for music being an evolved domain of the human mind. Finally, *implicit aesthetics* (art) is the sensitivity to, production of, and appreciation and preference for particular visual forms, figures, and color combinations over others. Aesthetics inherently involve emotional responses of like-dislike, which provide signals for our well-being. Indeed, a sense of aesthetics is an inevitable outcome of our sense of safety, order, and well-being.[22]

The human mind is not a blank slate, but rather it has a certain built-in structure that predisposes it toward some functions and experiences and away from others. Constraints narrow the infinite options available in solving problems and allow the individual to hone in on reasonable solutions a priori. Evolution has done with the human brain what it does with all organisms and organs: produced a biological structure that solves particular survival and reproductive problems effectively. The human brain clearly did not evolve to do science (or art, religion, or philosophy for that matter), but once it developed the amount of cortex, frontal lobes, and language systems that it did, science, although still not inevitable, was a fortunate by-product.[23]

Furthermore, some have recently argued that human intellect, creativity, and wit have resulted more from sexual than natural selection processes. That is, these traits are attractive to members of the opposite sex and are the kinds of traits we implicitly want our children to have, and therefore we are most likely to mate with people who display these characteristics. As I have argued elsewhere, I believe both natural and sexual selection play a role in the evolution of human intelligence and creativity.[24] The more applied forms of creativity—technology, engineering, and toolmaking—are probably determined more by natural selection pressures in that they have direct implications for survivability, whereas the more ornate and aesthetic forms of creativity—music, wit, and art—are probably more under sexual selection pressures, insofar as they implicitly (unconsciously) signal an individual's genetic, physical, and mental fitness.

## THREE PHASES OF EVOLUTION OF MIND

Looking back over the vastness of evolutionary time, one realizes that the only constant is change. Every form of life changes, both within the life span of its individuals and within the life span of its species. The capacities of the human mind have changed in much the same way that cognitive capacities with children change with age. The existence of distinct phases of mind over the course of evolution is hard to dispute. The question is not one of whether phases or categorical stages of mental development exist but rather of where the transitions are. On this question there is relative agreement: the major cognitive shifts occurred around 4.5 mya, 2.0 mya, and 200,000 years ago (kya).[25] These "breaks" can be seen both in major morphological change (in brain and body) and in cultural-behavioral change. In my model of cognitive evolution, I borrow from and add to other theoretical attempts to describe phases of human cognitive evolution, primarily those of Merlin Donald, Richard Klein, Steven Mithen, and Sue Parker and Michael McKinney. Even the relatively atheoretical Richard Klein places the temporal and species breaks in much the same places as the more theoretical writers (for example, Donald, Mithen), but he does not place descriptive labels on these phases. My goal is to provide a straightforward yet not overly simplistic model for the phases of hominid cognitive evolution by modifying and clarifying existing models and theories rather than fundamentally changing them.

### Phase 1: Prerepresentational Thought (6 to 1.6 mya)

The most common date for starting the human lineage is approximately six million years ago (mya), which is the estimated time that ancestral humans diverged from ancestral apes. Because this first stage of human evolution will contain three taxa (great apes, *Australopithecus,* and *Homo habilis*), we are dealing with both a very long time frame (approximately 4.5 million years) and a rather diverse set of species. There are meaningful differences in terms of time and taxa, but for heuristic purposes there is enough similarity to warrant folding them all under one phase, with distinct subphases.

The great apes provide the best comparative and cladistic picture of what our earliest human ancestors were like both morphologically and behaviorally. Be-

Table 8.2 Piagetian Stages in Hominid Cognitive Evolution

| Phase | Taxon | Piagetian Stage |
|---|---|---|
| Sensory | Monkeys | ESM LSM |
| 1. Pre-representational | Great apes | ESM LSM EPO |
| | *Australopithecus* | *ESM LSM EPO LPO?* |
| | *Homo habilis-rudolfensis* | *ESM LSM EPO LPO* |
| 2. Representational | *Homo ergaster-erectus* | *ESM LSM EPO LPO ECO* |
| | *Archaic Homo sapiens* | *ESM LSM EPO LPO ECO LCO* |
| | *Homo neanderthalensis* | *ESM LSM EPO LPO ECO LCO* |
| 3. Meta-representational | Modern Humans | ESM LSM EPO LPO ECO LCO FO |

*Source:* Feist 2004b.

*Note:* E = Early, L = Late; SM = Sensorimotor, PO = Preoperational, CO = Concrete Operational, FO = Formal Operational. Taxa in italics are based on inferences from the archeological record and are hypotheses and conjectures as much as description of fact (cf. Donald 1991; Mithen 1996; Parker and McKinney 1999, p. 279).

cause fossil finds from the period predating five million years are rare to nonexistent, a comparative perspective that includes the great apes is quite useful for an analysis of this time period. Parker and McKinney systematically detail the research on monkeys, apes, and humans (children).[26] A primary conclusion is that monkeys (macaques and cebus) never reach most of the six substages of Piaget's sensorimotor intelligence, but great apes (gorillas, chimps, and orangutans) do to varying degrees by anywhere from ten to forty months of age, and human children do by eighteen months of age. This is generally true of physical knowledge, social knowledge, logical-mathematical knowledge, and linguistic knowledge (see table 8.2).

Although there are at least three separate genera of australopithecine—*Ardipithecus, Australopithecus* (graciles), and *Paranthropus* (robust)—many scholars use the umbrella term *Australopithecus* to refer to them more broadly defined. The oldest known genus and species, *Ardipithecus ramidus,* was only first uncovered in 1994, and so currently we know next to nothing of its size, cranial capacity, or artifacts, although we do know that it was at least partially bipedal. The graciles were relatively small: 100–150 cm tall and averaged about 30–45 kg, with cranial capacities of approximately 400–500 cc. The robust form was also relatively small (110–140 cm) and (32–50 kg) but had larger cranial capacities than the graciles (500–550 cc) and larger cheek teeth and smaller canines.[27]

The major behavioral advance taken by the australopithecines was their bipedalism, not their enlarged brains or new industries or technology. But with bipedalism came a host of correlated changes: a smaller rib cage, precision grip, loss of prehensile feet, and most probably a change in social structure (with the infant no longer clinging to the mother's back and therefore an increased use of nonverbal communication and theory of mind skills). At its most advanced, australopithecine industry consisted of slightly reshaped stones, hard to differentiate from naturally occurring rocks that have been given the label "Omo industry."[28]

There is some disagreement as to which species constitute the first members of the genus *Homo,* but it is becoming more generally accepted that both *H. habilis* and *H. rudolfensis* are the defining members. I follow the lead of Klein and Mithen and include both *habilis* and *rudolfensis* in this group, but I use the *habilis* name more broadly to include *rudolfensis. Habilis* more broadly defined lived from roughly 2.4 to 1.5 mya and had cranial capacities of around 510–750 cc, which puts in midway between *Australopithecus* (400–550 cc) and *H. erectus* (approximately 1,000 cc). Other distinguishing traits of the habilines compared to the australopithecines were expanded frontal and parietal regions of the brain, a smaller and less protruding jaw, smaller cheek teeth, increasing but still not complete bipedalism, and larger consumption of meat.[29]

Recall from earlier in the chapter that *habilis* were the first stone toolmakers and therefore responsible for one of the more monumental hominid innovations. Their industry has been dubbed the Oldowan, and their tools, for the first time, could not be confused with naturally occurring rocks. As innovative as the Oldowan industry was, these tools were nonetheless relatively primitive and static by later standards. They were generally single faced, simple, and rather uniform over a long period of time (one million years).[30] No bifaced axes, for instance, are to be found in the Oldowan industry (see table 8.3).

In addition to stone tools, there were other behavioral-cognitive innovations of *H. habilis.*[31] There is archeological (bone) evidence of increased meat consumption, which implies increased hunting or scavenging, although the amount of hunting or scavenging is open to debate. Hunting almost certainly increased, and Mithen argues that this was due to the habiline's better and more complex understanding of the natural world and possibly even their ability to form hypotheses about plant life and animal behavior. There might have been some control of fire, but the evidence is ambiguous. What is less controversial is that there was an increase in social complexity, based on regression analyses from brain size and likely sharing of large meat quantities. The last behavioral-cognitive advance is also controversial: *H. habilis* may have possessed incipient

Table 8.3 Major Creative Achievements of Prehistoric Stone Age Humans (Genus *Homo*)

| Most Likely Species | Industry/Culture | Prototypic Achievement | Date (millions of years) |
|---|---|---|---|
| **Lower Paleolithic/Early Stone Age Tool Tradition** | | | 2.400–1.800 |
| *H. habilis-rudolfensis* | Oldowan | Stone Tools | 2.400 |
| *H. ergaster-erectus* | Acheulean | Hand-axes | 1.800 |
| | | Protolanguage | 1.800 |
| | | Bifaced axes | 1.400 |
| | | Fire | .790 |
| *Archaic H. sapiens?* | | Primitive bone tools | .750 |
| *H. ergaster-erectus* | Late Acheulean | First Levallois stone tools | .400 |
| | | Wooden spears | .400 |
| | | Patterned rock structures | .350 |
| **Middle Paleolithic /Middle Stone Age Tool Tradition** | | | .200 |
| *H. neanderthalensis* | Mousterian | Composite tools | .150? |
| | | Advanced Levallois stone tools | .150? |
| | | Burial sites | .090 |
| | | Music (bone flute) | .082–.043 |
| *H. sapiens sapiens* | | Language (?) | .100 |
| | | Bone harpoons–barbed points | .090 |
| | | Art (beads) | .077 |
| | | Ochre body painting | .075 |
| **Upper Paleolithic/Late Stone Age Tool Tradition** | | | .012–.045 |
| *H. sapiens sapiens* (Cro-Magnon) | Aurignacian | Semipermanent settlements | .045 |
| | | Art (figurines) | .032 |
| | | Lunar calendar | .032 |
| | | Art (Lascaux cave paintings) | .030 |
| | Souletrean | Earliest recorded numbers/tallies | .025 |
| | | Fired ceramics and sewing needle | .025 |
| | Magdalenian | Art (Altamira cave paintings) | .018 |
| | | Petroglyphs | .015 |
| **Mesolithic Tool Tradition** | | | .009–.013 |
| *H. sapiens sapiens* | Azilian | Microliths | .013 |
| | Maglemosian | Bone and antler tools | .013 |
| | Natufian | Settlements | .012 |
| **Neolithic Tool Tradition** | | | .006–.012 |
| *H. sapiens sapiens* | Natufian | Domestication of plants and animals | .011 |
| | | Townships | .010 |
| | | Astronomical inscriptions | .010 |

(*continued*)

Table 8.3 (*Continued*)

| Most Likely Species | Industry/Culture | Prototypic Achievement | Date (millions of years) |
|---|---|---|---|
| | Egyptian | Early calendars | .009 |
| | | Wheel | .008 |
| **Copper and Bronze Ages** | | | .005–.008 |
| *H. sapiens sapiens* | | Copper | .008 |
| | | Bronze | .006 |
| | | Writing | .005 |

*Sources:* The primary source for this table is Klein 1999. Secondary sources include Deacon 1997, Donald 1991, Enard et al. 2002, Falk 2000, Goren-Inbar et al. 2004, Hellemans and Bunch 1991, Henshilwood et al. 2004, Mithen 1996, Pfeiffer 1982, and Tattersall 1997. Also see http://www.chass.utoronto.ca/~banning/ANT%20200/200timelines.htm (accessed June 20, 2004).

*Note:* These dates, species, and events are not all universally agreed upon and the time periods varied somewhat geographically. I often used either the most frequent date or if there was no such date I would take an average date. Times and dates are more for when these technologies first appeared and not when they were universal. For these reasons, the dates in this table should be interpreted with some degree of caution and latitude. The table is meant simply to be a rough guide to cultural achievements.

language, based on analyses of endocranial casts showing possible increased folds in the Broca's area, a somewhat more angled basicranial flexion, as well as putative decrease in grooming and by implication an increase in language. To be sure, even if *habilis* possessed language, it would have been very rudimentary, presyntactical protolanguage.

Other theorists have written about these phases of human cognitive evolution. They all agree on the species and time periods involved (apes, *Australopithecus,* and *H. habilis*) and on the respective general cognitive abilities, but they provide different labels and different orientations. I borrow from these writers but offer the label "prerepresentational thought" for this phase of human cognitive evolution. Representation is a basic cognitive ability of "re-presenting" an idea or concept mentally, either visually or verbally, *once the object is no longer being directly sensed.* This stands in contrast to prerepresentational thought, which is tied to the here and now of what is being sensed. The essence of this stage is behavioral knowledge without any real conscious or developed cognitive representation, that is, what neuroscientists and cognitive psychologists more generally call "implicit cognition." It can be demonstrated (through cognitive tasks and neuropsychological exams) that this form of knowledge influences behavior without the person being aware of it, and indeed the current conclusion is that approximately 95 percent of mental processing occurs without awareness and without intentional control. An evolved domain is not necessarily consciously represented but can be seen and inferred from behavior. Some scholars have used the word "intuitive" to describe these incipient abilities, but I prefer the word "prerepresentational" or "implicit" because intuitive implies a kind of "I-know-but-don't-know-how-I-know" form of knowledge (which I reserve for the next phase of cognitive evolution). Prerepresentational-implicit thought, on the other hand, is a "I-don't-know-but-I-behave-as-though-I-do" form of knowledge. It is as if "my behavior knows but my conscious mind does not." In other words, knowledge gets expressed behaviorally more than cognitively. The real difference is that intuition has conscious mental representation whereas implicit thought does not. To be clear, some scholars, anthropologists, and psychologists have used the word "folk" to refer to this form of cognition, and that is pretty much what I mean by prerepresentational.[32]

The reason I apply the label "prerepresentational" or "immediate" to this phase comes from the fact that the thought processes of our ancestors during this early stage of hominid evolution were much more tied to thinking about immediately perceptible and directly sensed events and experiences than were the thought processes of their descendants. Moreover, the capacity for reflection and consciously represented beliefs were not yet possible, chiefly because language did not exist in any form that we know it. As Mithen argued, learning and problem solving consisted mostly of Pavlovian conditioned and associationistic processes, whereby something is learned because it is consistently associated with and close in time to a given outcome. Such associations form the basis for cause-and-effect thinking.[33] Learning is based on concrete reinforcements, but there is little to no ability to represent these associations. Ideas are more sensory based than conceptually based, which is what I mean by "prerepresentational."

### Phase 2: Representational Thought (1.8 mya to 30 kya)

By all accounts, 1.8 mya a major shift happened in human evolution: species appeared that for the first time did not differ morphologically from modern humans.[34] This phase of human evolution actually includes at least four species from the genus *Homo: ergaster, erectus, archaic homo sapiens,* and *neanderthalensis* and covers nearly two million years (from 1.8 mya to about 30 kya). In terms of cranial morphology, compared to *H. habilis* these more recent species of *Homo* generally had smaller canines and cheek teeth and larger brains. Cranial capacities now averaged about 900 cc in *ergaster,* approximately 1,000 cc in *erectus,* approximately 1,200 cc in *archaic H. sapiens,* and approximately 1500 cc in *H. neanderthalensis.*[35] In other words, brains during this stage grew by roughly 50 to 100 percent compared to *H. habilis.* There were some morphological differences between these species, however, with early *H. sapiens* and *H. neanderthalensis* having less sloped frontal lobes and larger cranial capacities than *H. ergaster-erectus.* In addition, unique postcranial features in general include a reduction in sexual dimorphism, a narrowing of the pelvis and a less ballooning rib cage (meaning a smaller gut and different respiration), and exclusive bipedalism for the first time. These species were essentially modern but varied in stature, with *ergaster* averaging perhaps 170 cm and *erectus* being stockier and shorter than *ergaster. Neanderthalensis* (and to a lesser extent *erectus* and early *sapiens*) was notoriously stocky, muscular, and robust by modern standards, implying regular strenuous physical activity.

The tool technology used during this period is divided into two categories: Acheulean for *H. ergaster-erectus* and Mousterian for *H. neanderthalensis* and *archaic H. sapiens* (see table 8.3).[36] The fundamental advance and prototypic Acheulean artifact is the bifaced and often symmetrical hand ax from about 1.5 mya. Interestingly, Asian *erectus* seems to have lacked hand axes and made choppers and flakes instead. The hallmark of the late Mousterian (Neanderthal) tradition (250 kya) is the Levallois flake, which is produced by carefully working the stone core. In some later Middle Stone Age (MSA) sites, the Levallois technique for the first time resulted in blade production (they were twice as long as wide). The complexity of the Mousterian-Levallois stone technologies required sophisticated knowledge of the mechanics of stone flaking, and indeed this stone technology has never been surpassed by any other group of humans, including modern *H. sapiens.* Few humans alive today can produce good Levallois points.

As was true for the prerepresentational phase, others, such as Donald, Mithen, and Parker and McKinney, provide their own labels and descriptions for the second phase of human cognitive evolution, and again I build upon them. From these theoretical descriptions, it is clear that representation is the essential and crucial element, and therefore I apply the label "representational" to this phase of human cognitive evolution. Representation, again, is the ability to "re-present" an idea once the object is no longer being directly sensed. In all likelihood, visual representations preceded verbal ones phylogenetically, but language ultimately confers the advantage of cementing the visual representations in memory. These species of hominids no doubt were capable of some verbal and propositional representation, but without fully developed grammatical language they were probably more likely to represent ideas and concepts visually rather than verbally. Gestures and nonverbal expressions were crucial to communication, and therefore people would require some degree of advanced shared or joint attention with others or (theory of mind) to communicate effectively. Another way to think about this form of thought is as first-order rather than second-order thought. First-order thinking is explicit and conscious thought that is tied to concrete ideas. It is not yet capable of reflecting back on itself and thinking about thinking (metacognition). Similarly, as Annette Karmiloff-Smith argues, such cognitive abilities are the first stage of explicit rather than implicit representation.[37]

Mithen argues that the development of explicit domains is quite clear in erectine-neanderthal mentality, with knowledge of social, physical, biological, and quantitative concepts being represented somewhat in isolation (with little fluidity between domains).[38] The limitation of this erectine-neanderthal stage is that it is not yet capable of representing representations (metacognition) or using arbitrary symbols to represent ideas. Representations would probably be rather concrete, using nonarbitrary images and utterances to describe concepts. This is one reason why images would be more common than words, because images more directly represent objects than words. Words are arbitrary sounds associated with ideas and would not obtain full development until the arrival of the only surviving species of *Homo.*

### Phase 3: Meta-Representational Thought (150 kya to present)

The last phase of hominid cognitive evolution is represented by but one species, *H. sapiens sapiens,* which had a common ancestor as early as 200 kya, started to appear anatomically around 100 kya but behaviorally only about 50 kya.[39] In terms of morphology, the chief cranial change in *H. sapiens sapiens* is not size, because Neanderthal (1500 cc) and early *H. sapiens* (1200 cc) had cranial capacities similar to or even exceeding modern humans (1400 cc). The most obvious unique cranial characteristics include a steeply rising frontal lobe, a pronounced chin, and a relatively flat face. Postcranium traits include less robustness than earlier humans; longer limbs; and narrower, shorter, and thicker pubis bones. Body size is taller than any other human species, averaging about 175cm for males and about 160 cm for females, with corresponding weights of about 65 kg and 54 kg respectively.

The major creative advances of modern humans are too numerous to list, so only some of the more noteworthy examples will be itemized (see tables 8.3 and 8.4).[40] Beginning with the first blade technology and the working and polishing of bones and antlers from approximately 90 kya and art beads from 77 kya, we see the "creative explosion" begins around 50 kya: fully grammatical language, art, music, religion, magic, and totemism. Here we also see the first signs of an incipient math and science. In short, truly symbolic and abstract thought appears in the fossil record for the first time, and with it one of the cornerstones for explicit science has been laid.

Although some scholars have argued that the origins of language date back as far as 2.0 mya, others argue for a much more recent origin of linguistic capacity. Perhaps the best-known representatives of the recent view have been Lieberman and Crelin, who reconstructed soft tissue models from Neanderthal fossil remains and concluded that Neanderthal had limited vocal capacity, especially in certain vowel (*a, i, u*) and consonant (*k, g*) sounds and were therefore not truly linguistic. Language, therefore, is unique to our species of *Homo* and probably is no more than 100,000 years old. A recent report examining the evolutionary genetics of linguistic ability places the origins of complex language capacity most likely to be within the last 120,000 years. Those who argue against earlier expression of language are either being more restrictive in what they call "language" or are arguing against a full expression of vowels. Few would argue, however, that a full range of vowels is necessary for some form of grammatical language, for one can speak rather well with just one vowel, *e*. Perhaps the most reasonable conclusion is that the evolution of language began around two million years ago with protolanguage, but that it was not fully syntactical, grammatical, and symbolic until around 100,000 to 120,000 years ago when *H. sapiens sapiens* emerged.[41]

There can be little doubt that language, in its more developed grammatical form, is the sine qua non of cultural innovation and creativity. With language one can teach and inform others what one knows, which can be passed on from generation to generation, sometimes being preserved in its original form and other times being modified and changed. Because knowledge becomes more cumulative, language speeds up cultural innovation. By logical deduction, the creative explosion of around 50 kya must have been precipitated by some new development in language. In addition, language provides a medium for expressing theretofore ineffable ideas, which can be externalized and eventually become part of cultural knowledge. Words and syntax facilitate the growth and expression of knowledge. The human need to understand the physical, biolog-

Table 8.4 Major Creative Technological-Scientific Highlights of Ancient History

| Invention | Approximate Date (thousands of years ago) | Culture |
|---|---|---|
| Metallurgy (copper) | 8,500 | Turkish (Anatolian) |
| First calendars and gnomon (sundials) | 6,000 | Mesopotamian,[a] Egyptian |
| Multiplication and square tables | 5,500 | Mesopotamian |
| Ideographic writing | 5,500–5,000 | Mesopotamian, Egyptian |
| Logographic writing (cuneiform, hieroglyphics, Harappan, Chinese) | 5,500–3,200 | Mesopotamian, Egyptian Indian, Chinese |
| Ziggurat | 5,500 | Mesopotamian |
| Papyrus | 5,200 | Egyptian |
| Geometry | 5,000 | Egyptian |
| Great pyramids (engineering) | 4,800 | Egyptian |
| Astrology, astronomy | 4,900 | Mesopotamian, Egyptian |
| Medicine, anatomy | 4,500 | Egyptian |
| Legal units of length, weight, capacity | 4,500 | Mesopotamian |
| Sexagesimal, place, decimal systems | 4,400 | Mesopotamian |
| Units of time and sundial | 4,200 | Egyptian, Mesopotamian |
| Early algebra | 4,000 | Mesopotamian |
| Consonantal "alphabet" (Abjad) | 3,500 | Phoenician, Arabic |
| Alphabet | 3,100 | Greek |
| Silk | 2,800 | Chinese |
| Ionian philosophy (natural explanation) | 2,600 | Greek |
| Ancient math | 2,500 | Greek, Indian |
| Athenian philosophy | 2,400 | Greek |
| Aristotelian science (physics, biology) | 2,300 | Greek |
| Indian-Arabic numerals | 2,200 | Indian |
| Ptolemy's geocentric astronomy | 1,850 | Egyptian |

*Sources:* The primary sources are Hellmans and Bunch 1991 and Sarton 1952a. Other sources include Butterfield 1960, I. Cohen 1985, Crump 2002, Durant 1926/1961, and S. Stumpf 1975.
[a] "Mesopotamian" is the general label that includes the Sumerian, Assyrian, Accadian, Chaldaean, and Babylonian cultures.

ical, and social worlds requires a developed language to explain the origins of its species. Once language is in place, myths and cosmologies can be passed on intergenerationally. Explicit models and theories of how the world began and how it functions originated with grammatical language. Written language (around 5 kya) takes this process one step further and truly codifies cultural knowledge by putting ideas in a concrete and semipermanent medium, making knowledge all the more cumulative. The development of pictorial language and later alphabetic language were monumental advances in the evolution of human thought. Ideas, once written, facilitate thinking about thinking (metacognition) because thought now takes an externalized, objective form. Science is simply the codification of this externalized, cumulative, and metacognitive development of knowledge.

Other symbolic, cognitively fluid, and meta-representational capacities can be seen in burial of the dead, cave paintings, formation of townships, and, finally, writing systems. Neanderthals buried their dead, but modern humans (Cro-Magnons) were the first to include important symbolic objects with their burials, implying the belief in an afterlife and possibly religion. From burials with symbolic objects one might infer the belief that death was seen as a transition to a nonphysical life, at least for the most important individuals in the community. The cave paintings provide another and the first concrete evidence of nonphysical spiritual beliefs, namely, anthropomorphism (the belief that animals have human elements) and totemism (the belief that humans have animal elements). With the end of the last ice age around 10–12 kya, we again see some monumental Neolithic changes: the first townships form, the first systematic domestication of plants (agriculture) and the invention of the wheel. The Neolithic age ends with the invention of metallurgy (first copper, then bronze, then iron). The break between "prehistory" and "history" then is seen with the first clay tablets (hieroglyphs in Egypt and cuneiform in Sumeria) around 5,200 years ago, only slightly before the first use of multiplication and square tables (see table 8.4). People first wrote on papyrus in Egypt around 5,000 years ago. Early writing was very utilitarian and was used more for recording market transactions than for expression of ideas (that would have to wait for pictorial and then alphabetic writing).[42]

All of these innovations reflect an ability not just to represent ideas, but to think about thinking. Once one can externalize one's thought in symbols, one can more readily reflect upon one's ideas, state them more explicitly, and decide whether to modify them. One expresses an idea in symbolic form, first using representational (concrete) images and then abstract ones. If one were to graph the rate of cognitive-cultural discovery, it would clearly be exponential—and one can only shudder to think what the rate over the last one hundred or even just fifty years has been! How do we explain this remarkable explosion of creative ability?

Human morphology did indeed change in the third phase. With *H. sapiens sapiens* the major cranial change was an enlarged frontal lobe. The frontal lobe, in particular the orbitofrontal region of the prefrontal cortex, allows for the level of abstraction needed to cross domains, and it also allows for the development of novel and creative relationships, and metaphorical and symbolic thought that we see only in the cognitively fluid mind. With the growth of the frontal lobes comes impulse control; the growth in intelligence is really the ability to reflect and think before acting rather than responding instinctively. The frontal lobes allow just such a break between impulse and behavior.[43]

The problem with trying to explain these changes in behavior through changes in brain size and shape is that the changes in the brain led to only relatively small technological advances (bones and antlers, pronged tools, beads) for the first 40,000 to 50,000 years. The morphological change in brain shape and size may have been a necessary cause of the creative explosion, but it was not sufficient. Richard Klein and a handful of others (such as Pinker) argue that it was not until a neural reorganization occurred in the brain that the creative explosion of around 50 kya was possible. It is not size of the brain per se that matters most but the complexity of neural organization and connectivity. Indeed, there is evidence that the human brain is more connected per cubic centimeter than that of any other primate. If neural Darwinism and plasticity are as salient in brain development as they seem to be, such growth in connectivity over 40 kya is quite possible. Unfortunately, as Klein points out, the neural connectivity hypothesis is not testable at this time because of its invisibility in the fossil record. Although it is not yet completely testable, I find the argument of neural complexity personally persuasive, given what we now know about neurogenesis (birth and growth of neurons), neuroplasticity, as well as neural processing and intelligence and creativity. The conclusion can also be made from sound logical inference: if there was no growth or change in overall brain size and morphology from 120 to 50 kya (true), and yet significant behavioral changes occurred only after 72 to 50 kya (true), then internal reorganization rather than size or morphology must be the reason.[44]

Mithen argues that modern human mentality is distinguished by its "cognitive fluidity," Donald by its "mythic ability," and Parker and McKinney by its "declarative abilities." I believe that each of these abilities is important and therefore my model incorporates each of these coevolving mechanisms. I apply the label "meta-representational" as the umbrella term for this third stage in human cognitive evolution. Many other terms may have been appropriate, such as "integrative," "fluid," or "explicit," but I chose meta-representational to emphasize the continuity with and growth out of the earlier stages in which ideas are first immediate, implicit, and prerepresentational and then capable of being represented mentally, concretely, and intuitively.[45]

Mithen puts forth the notion of horizontal fluidity between domains as one reason certain individuals are inordinately creative, I believe that a "vertical" fluidity between evolutionary distinct modes of thought, between the "older" more intuitive and implicit modes of thought and the newer more verbal, conscious, explicit modes of thought also contributes to their creativity. Recent neuroscientific research supports the notion of unusually complex and fluid neural connectivity among our most creative individuals. Moreover, many creative scientists have consistently described the importance of "intuition" in the creative problem-solving processes. Artists, of course, are known for their intuitive abilities. One could argue that the capacity to make associations between different modes of thought is analogous to different phases of brain evolution. In terms of being able to consistently arrive at creative (novel and useful) solutions to important problems, moving horizontally between ideas is as important as moving vertically between them.[46]

A related implication is to give Freud some credit for gaining insight into two distinct modes of thought, one the primary, preverbal, global, unconscious mode and the other the secondary, analytic, verbal, and conscious mode that almost certainly are vestiges of two phases of human cognitive evolution. More recently, the personality psychologist Seymour Epstein has argued for two distinct modes of processing information, namely, the experiential mode and the rational mode, with the former being intuitive, automatic, holistic, and based in emotions and "gut feelings" or "vibes" and the latter being analytical, conscious, logical, and verbal. Finally, recent neuroscience evidence would suggest that the two distinct processes are somewhat localized in the two hemispheres: the right hemisphere processes information globally, diffusely, and inferentially, and the left hemisphere processes it analytically, narrowly, and deductively. More important than just noting these two distinct ways of knowing, much evidence and theory would suggest that highly creative people are most facile at having associations between and within these distinct modes of thought and that this broad associational network is a prerequisite for consistent creative insight.[47]

Whatever the theoretical explanation—cognitive fluidity, externalization of symbols and ideas, the development of complex syntactical language, metarepresentation, prefrontal growth, neural complexity, or some combination of all of these explanations—it is undisputable that something truly unique happened to human cognitive evolution around 50 kya. It is at this time that we also see the true foundations for science and scientific reasoning: metacognition, explicit pattern recognition in the natural and social worlds, and hypothesis testing. The purpose of this excursion into human cognitive evolution is to demonstrate the ancient origins of the human mind in its attempt to understand the physical, biological, and social worlds, which is all that science is. To be sure, these evolved higher-order cognitive capacities are not limited to or unique to science; they are expressed in artistic, musical, philosophical, spiritual realms as well. They nevertheless are the origins of the scientific mind, if at the same time they are also the origins of the artistic, musical, philosophical and spiritual mind. Science and scientific thinking starts to distinguish itself from these other uniquely human qualities of mind, especially by around 2,500 years ago.

# **Chapter 9** Origins of the Scientific Thinking

There are two ways of acquiring knowledge, one through reason, the other by experiment. . . . Argument is not enough, but experience is.
—*Roger Bacon,* On Experimental Science

Insight, untested and unsupported, is an insufficient guarantee of truth, in spite of the fact that much of the most important truth is first suggested by its means.
—*Bertrand Russell,* Mysticism and Logic

When we look for the origins of science, and how and when it began to emerge as a distinct intellectual process, we must focus not just on the very recent expression of pure science and modern science as codified in the "scientific method." That is but the most elaborate and codified expression of ancient cognitive and epistemological processes. Different forms of science have been around not for hundreds or even thousands of years, but rather for millions of years. Obviously the "science" of *H. erectus* or *H. neanderthalensis* was a very different kind of science from what we know today, and we therefore have to change our notion of science a bit to apply it over such a long time period. Just as there is folk art, so too is there folk science. It is when we fail to distinguish the earliest forms of "science" and believe it to exist only in its modern mathematical explicit sense that we fail to understand the origins of science. One point to be clear on, however, is that I am conceptualizing science here as a method of acquiring knowledge and not as a body of knowledge per se. In the long run, the method has resulted in a large body of knowledge, for it is both a process and a product. But in order to understand its origins in our evolutionary past, we must understand that science in its barest form involves pattern recognition, causal thinking, hypothesis formation, and hypothesis testing. All humans in all cultures do these things and have in some form for millions of years.

Without stretching our definition of science too thinly, we can talk about two continuous dimensions for defining science: the preverbal-verbal and the applied-pure. These dimensions, in turn, allow one to create four distinct phases of science over the course of human evolution, namely, the preverbal phase, the verbal phase, the applied phase, and finally the pure-discovery phase. Mind you, these are not historically mutually exclusive phases as is readily seen by the fact that the last three exist in modern human culture. Each successive stage does not eradicate its predecessors but rather supplements them.

I also want to make quite clear at the outset that I eschew any notion of "progression" in human thought and epistemology, that younger forms of thought (for example, science) are inherently better than older forms (for example, mythology). They each are what they are for their time frame and context, namely, the ways that humans think about and make sense of their world. A major theme in the history of anthropology has been inquiry into the nature of "primitive" mentality and whether it is the same or different from "civilized" mentality. During the latter part of the nineteenth and early part of the twentieth centuries, a common position was to view preliterate cultures as "savages" and inferior to the civilized cultures of Europe. This mentality was seen most clearly in James Frazer's *The Golden Bough:*

> We shall perhaps be disposed to conclude that the movement of the higher thought, so far as we can trace it, has on the whole been from magic through religion to science. . . . Thus in the acuter minds magic is gradually superseded by religion. . . . But as time goes on this explanation [religion] in its turn proves to be unsatisfactory. . . . Thus the keener minds, still pressing forward to a deeper solution of the mysteries of the universe, come to reject the religious theory of nature as inadequate. . . . In short, religion, regarded as an explanation of nature, is displaced by science. . . . It is probably not too much to say that the hope of progress—moral and

intellectual as well as material—in the future is bound up with the fortunes of science, and that every obstacle placed in the way of scientific discovery is a wrong to humanity.[1]

Such a linear and pejorative line of thinking about cultural progress is archaic and misguided. One important conclusion from physical anthropology and, more recently, cognitive anthropology is that humans are one species with more fundamental similarity in genetics, morphology, and cognition than any superficial differences may lead us to believe. Whenever one is talking of (modern) human thought and belief systems, the notions of "primitive" and "advanced" are, from an evolutionary perspective, misguided.

But once we expand beyond our own species and examine the history of hominid thought in general, extending back to *H. erectus* if not earlier, then we can and must see some form of development, much like one sees when one examines the development of thinking in infants, children, adolescents, and adults. The current analysis of the origins of science takes the latter view and extends its purview over the entire time frame of hominid evolution, not just modern humans. Although not historically mutually exclusive, these four phases nevertheless do exhibit a historical developmental trend: "preverbal science" (folk science or ethnoscience) originated as early as two million years ago, "verbal science" as early as 100 kya, "applied science" as early as 30 kya, and "pure science" as early as 2,600 years ago. The last three phases each reside within the *H. sapiens* species, and I do distinguish them as phases but not in terms of one being better or more advanced than another. Rather, I distinguish them in terms of being younger and more or less explicit (meta-representational). It may well be true, as we see in cognitive development within an individual, that cognitive development tends to move from the relatively simple to the relatively complex, as long as we are clear that complexity is merely an increase in number of elements manipulated.[2]

## PHASES OF SCIENTIFIC THINKING

I argue for four phases of the evolution and history of science, each with at least five core components (see table 9.1).[3] With each successive phase at least one new component gets added. The five core components are observation, categorization, pattern recognition, hypothesis formation (prediction)/hypothesis testing, and causal thinking. These components themselves have a developmental path, with each one developing out of the previous one: observation →

Table 9.1 Phases of Scientific Thinking and Their Key Components, Forms of Thought, and Age

| | Phase | | | |
|---|---|---|---|---|
| | Preverbal | Verbal | Applied | Pure |
| Components | Observation | Observation | Observation | Observation |
| | Categorization | Categorization | Categorization | Categorization |
| | Pattern recognition | Pattern recognition | Pattern recognition | Pattern recognition |
| | Hypothesis testing | Hypothesis testing | Hypothesis testing | Hypothesis testing |
| | Causal thinking | Causal thinking | Causal thinking | Causal thinking |
| | | Explanation/theory | Explanation/theory | Explanation/theory |
| | | Control (magic) | Control | Control |
| | | | Measurement | Measurement |
| | | | Incipient math | Developed math |
| | | | | Controlled experimentation |
| Forms of thought | Implicit | Implicit | Implicit-explicit | Explicit |
| | Representational | Meta-representational | Meta-representational | Meta-representational |
| Age | 1.8 million years | 100 thousand years | 30 thousand years | 2.6 thousand years |

categorization → pattern recognition → prediction (hypothesis testing) → causal thinking. But the reality of it all is not linear but rather circular, with changes in each component affecting changes in the other components. Further, just as is true in science, the last component circles back and starts the cycle all over again. Another general point is that we see a development from the most implicit and sensory-bound to the most explicit and metacognitive thought across the four phases.

### Phase 1: Preverbal Science

The first vestiges of science that can be seen in the evolution of the human mind are what have to be called "implicit science" and they are preverbal—namely, the process of developing systematic knowledge of how the world works yet not knowing that one has done so. In chapter 8, I began with the earliest hominids around 6 mya, but when discussing "science" I'll start with a slightly more restrictive definition and argue that representational knowledge is required. Therefore my discussion of science begins with *H. erectus* slightly less than 2 mya. Knowledge is represented neither consciously nor symbolically but comes about behaviorally through experience with the world. As learning theorists have long argued, learning is defined as long-term change in behavior *or* knowledge due to experience, with most learning being behavioral rather than cognitive. The earliest glimmering of science is seen in bodily knowledge, meaning that it manifests itself more in what individuals are doing than in what they are consciously thinking. Moreover, this earliest phase is preverbal in that it predates the onset of grammatical-syntactical language.

During this first phase of scientific thought there are five main components at work: observation, categorization, pattern recognition (covariance), hypothesis formation (prediction)/hypothesis testing, and causal thinking (causation). The first component to science, namely, *observation,* is both an obvious and yet often overlooked starting point. Observing the world and taking in sensory input is the necessary foundation for all science, and by this criterion, many species of animal have keen powers of observation. To be clear, I should point out that observation is inclusive and encompasses all sensory modalities—hearing, tasting, feeling, smelling, not just seeing.

The second component of science is *categorization,* that is, classifying the incoming information into meaningful systems. The principles of neuroscience and sensation and perception are relevant here: central nervous systems function to organize and interpret incoming sensations, and they do this first and foremost by categorizing sensations based on perceived similarities or differences. The first patterns of perception are categories, which are of three fundamentally distinct kinds: physical things (objects), natural things (plants and animals), and social things (people). The underlying dimension for categorizing sensations are whether or not things move, live, die, or reproduce. If they do not, they are physical objects, and if they do, they are animate objects (either plants or animals). The world of stone, rock, and earth is inherently different from the world of plant and animal, and in the preverbal stages understanding consists of physical knowledge of space and causality and motion of things. If things are alive (grow and die) but do not move on their own volition, they are plants.[4] If things are alive, reproduce, eat, and move on their own volition, they are animals (that is, animate). Finally, if things live, reproduce, eat, die and if they take the same form as ourselves, they are human. In short, the central nervous system and its perceptual properties serve their most basic function when they are organizing our sensory experiences, and the most fundamental way this is done is classification. Classification is an inherent function of the brain in that it serves as a primary means of bringing order to the chaos of incoming sensory experience. In a very real sense the brain has evolved to be just such a tool: an organizer and interpreter of sensory experience.

Not just neuroscience and sensation and perception tell us that finding regularities is an inherent function of the brain. So too do philosophy and humanistic psychology. The following quote by Albert Einstein exemplifies the philosophical perspective: "The very fact that the totality of our sense experience is such that by means of thinking (operations with concepts, and the creation and use of definite functional relations between them, and the coordination of our sense experiences to these concepts) it can be put in order, this fact is one which leaves us in awe, but which we shall never understand. One may say 'the eternal mystery of the world is its comprehensibility.' It is one of the great realizations of Immanuel Kant that the setting up of a real external world would be senseless without this comprehensibility."[5] Here Einstein and Kant are saying from a philosophical perspective what I am trying to say from an evolutionary and psychological perspective: the brain in general and the scientific brain in particular function to bring sense and order to the chaos of the senses.

The need to categorize and classify was ultimately facilitated by the development of language. Likewise, the need to categorize must have been one of the driving forces for the development of language, or at least vocabulary.[6] Being social animals and having well-developed and sophisticated implicit psychologies, humans acquired over the course of their early evolution the anatomical hardware necessary for phonology (an arched cranial base, an enlarged pharynx, and a lowered larynx) and then the cognitive capacity to start making specific sounds for specific things, plants, animals, and humans (developed Broca's region and expanded frontal lobes). Instead of simply forming the concept preverbally, early protolanguage began to associate consistent sounds with particular things. This may well be how a first vocabulary developed. Connecting particular sounds to particular things helps keep them distinct and allows one to communicate this distinction, that is, to categorize them. If "ig" means "tree" and "ag" means "rock," we have not only language and communication but categorization. To name something is to categorize it. Single words are all that is needed. Not much grammar or syntax would be required (these help with more complex relations between things as well as between words). Language and categorization are mutually reinforcing—language facilitates categorization and categorization facilitates language development.

The third component of scientific thinking is *pattern recognition.* Once one has developed some way or distinguishing and classifying the world of things, one can see regular patterns in the relations between different things, plants, animals, or humans. With experience and because the brain also functions to store sensory information, we start to see consistent patterns between events—some event Y consistently follows event X. Pattern recognition is noticing that two events happen to repeatedly co-occur or covary. Finding such regularities or patterns lays the foundation for expecting the world to behave in a certain way, which is the beginning of hypothesis formation as well as causal thought. The philosopher of science Karl Popper, in *Conjectures and Refutations,* put it this way: "Thus we are born with expectations; with 'knowledge' which, although not *valid a priori, is psychologically or genetically a priori,* i.e., prior to all observational experience. One of the most important of these expectations is the expectation of finding a regularity. It is connected with an inborn propensity to look out for regularities, or with a *need* to *find* regularities, as we may see from the pleasure of the child who satisfies this need."[7]

In the fourth component, once one sees patterns, one expects them to continue, and these expectations become *hypotheses.* Observation, categorization, and pattern recognition together lead to an ability to wonder whether Y really will follow X. We implicitly and behaviorally test hypotheses when we do something to see what effect will follow, that is, we intuitively test our ideas and beliefs of how the world works. We manipulate the world in order to test it. Just as a toddler will keep throwing her food off her plate and marvel at the sound and sight of it hitting the floor ("Yes, it really does splatter each time it hits to floor, and yes Mommy and Daddy really do react each time!"), our early ancestors must have regularly tested their ideas against nature. The connection between cooking meat and its taste would be one such example. The idea of cooking meat is far from intuitive. No nonhuman animal does so and its benefits are not at all obvious. But at some point, probably after a piece of meat had fallen into a campfire (or possibly after an animal died after a forest fire), early humans realized cooked meat tastes better. This would lead to the following idea (in behavioral and not verbal form): "If I cook this meat, will it really taste better?" But by observing what preceded the outcome, a pattern was intuited (good taste follows fire). Thereafter, a more systematic testing of this connection may have followed. Once established, this knowledge would initially be transmitted behaviorally from generation to generation through imitation and observation, but once language developed the subtleties could be transmitted verbally. It is precisely such individual and cultural knowledge that we can refer to as "implicit theories." In short, perception is inherently connected with expectation and hypothesis because once a concept is formed, the world is categorized and categories lead to expectations. From expectations, hypotheses follow. The cognitive archeologist Steven Mithen, in fact, argues that pattern recognition and hypothesis testing are perhaps the essential features of scientific thinking.[8]

The fifth component of preverbal science is *cause-and-effect thinking*. If one perceives consistent covariation between two events, one is likely to infer that the earlier event caused the later event. Once a hypothesis is verified and becomes knowledge, we automatically think in terms of cause. "Fire causes better tasting meat." Causal thinking and hypothesis testing go hand in hand. One of the things that brains in general, and the human brain in particular, do well is learn cause-and-effect relations. When we experience the same event following a particular circumstance, we can hardly help but think there may be a cause. Implicit science is the accumulation of such cultural and traditional associations that become all the more readily passed on once language (storytelling) has developed. But in this first phase of scientific knowledge, such accumulated implicit knowledge is transmitted behaviorally through observation and imitation. Until any real language developed, the visual system (observation) would be the primary mode of transmitting knowledge.

I should point out that these five components do not really develop so linearly but rather much more dynamically. For instance, causal thought can be closely intertwined with observation, and hypotheses testing, through trial-and-error learning, feeds back and into pattern recognition.[9] Observation cannot be divorced from theoretical (causal) expectation. The two are impossible to separate. The main point here is that the five core components of scientific thought are separable and stand in complex and dynamic relation with one another.

In sum, scientific knowledge before language was implicit, immediate, sensory-bound, and did not accumulate in the species very rapidly. But it was an origin of science. At some point, the cognitive need to move beyond single- or double-word utterances with simple or no verb tenses gave way to syntactical and grammatical language. And once syntax formed we could do more than point and shout—we could explain. And explain we did!

### Phase 2: Verbal Science

If many species of hominid and mammal make causal connections and transmit this knowledge nonverbally or preverbally through observation, then in the next phase of scientific thought we move into the realm of one species from one genus, namely, *H. sapiens sapiens.* The development of complex language was of course a monumental innovation and is in fact limited to but one species of hominid.

It took millions of years of human evolution for language to go from simple words to the developed syntactical language that was in place by anywhere from 50 to 200 kya, with 120 kya being a recent estimate from evolutionary genetics. As Brian MacWhinney has argued, although morphological changes important for language began taking shape six million years ago (for instance bipedalism), and in the interim cognition and language continued to coevolve, it was probably only around 50,000 to 60,000 years ago that all of the necessary pieces of hardware were in place and could be integrated by the brain so that fully syntactical language could emerge. Language has three main components: phonology (expression of sounds), morphology (words and word parts), and syntax (rules for combining words). Although we have little direct evidence, logical inference would dictate that the development of language went from phonology to morphology to syntax, from production of sounds, to single words, to multiple words, to sentences. If nothing else, we can be pretty certain that this is roughly how it develops in children, and the ontogeny of childhood language development serves as a good working model for the phylogenetic evolution of language.[10]

Just as is the case with children, vocabulary no doubt preceded syntax and grammar in our species, and, indeed, it took upwards of two million years to go from simple vocabulary to grammar and syntax. When it finally happened, human knowledge and innovation changed forever. There is much evidence that a "creative explosion" happened in human culture around 50 kya.[11] This creative explosion of 50 kya apparently co-occurred with the culmination of the development of fully syntactical language. Language, whether a cause, effect, or both of these behavioral changes, clearly allows ideas to cross domain boundaries, facilitates metacognition and cognitive fluidity, and thereby allows creative solutions to problems to be realized and represented. This is consistent with the creative explosion of 50 kya. But whenever and however syntactical sentence-based language evolved, there is little doubt that it radically altered our thought and knowledge structures.

Indeed, our knowledge of the physical, natural, and social worlds became explicit and codified in the stories we began to tell each other about how things happen and why they happen, in other words in our "myths" (not being used at all pejoratively; in fact, "stories" is a better word). Imitation, observation, and protolanguage are slow and awkward by comparison. Syntactical language allows for passing on knowledge, whether it takes mundane form or ritualized and sacred form, such as grand cosmologies and storytelling. The latter were our "theories" of how the world came to be and how it continues to work. Language allowed for the less ritualized everyday knowledge of the physical, biological, and social worlds to be transmitted, and we know from anthropological studies of native cultures that this knowledge often was incredibly detailed, sophisticated, and in many cases accurate. As Claude Lévi-Strauss made clear more than thirty-five years ago, preindustrial native cultures by any current standard have incredibly sophisticated classification systems of their natural worlds, and not merely for practical purposes.[12]

In the previous section on preverbal knowledge I outlined the five fundamental components to science—observation, classification, pattern recognition, hypothesis formation and testing, and causal inferences. During the verbal phase of science, language facilitated the addition of a few new components: *explanation, explicit theory,* and *attempts at controlling nature* (magic and shamanism). Indeed, it is because nature is an incredibly powerful force, one that is clearly beyond human control, that we have tried to harness its power with mental models and explanations. By looking at the general motive behind understanding one's world we also see the commonalities between modern explicit science and early "prescientific" explanations. If we are to understand the origins of science we must break down some conceptual barriers. Science has the same motivation and origin as mythology, magic, astrology, and animism: namely, to explain, to predict, and to control. "Causal explanation, prediction, control—myth constitutes an attempt at all three, and every aspect of life is permeated by myth."[13] Once humans became aware of their mortality and developed ideas of an afterlife (perhaps around 50 kya, perhaps earlier), they no doubt became capable of wondering about their own as well as nature's origins and searched for explanations of how both came to be. Some form of self-reflective existential awareness is therefore tied to the earliest forms of verbal science.

The earliest and clearest attempts to explain the whims of nature stem from animism, the belief that spirits reside inside nature and people and that these spirits cause their behavior. By manipulating the spirits, one can control nature and people. Recall how Merlin Donald argued that "myth" is the first systematic attempt to integrate human understanding of how the world works. "The importance of myth is that it signaled the first attempts at symbolic models of the human universe, and the first attempts at coherent historical reconstruction of the past."[14] Science, too, is aimed at explaining, predicting, and controlling the physical, biological, and social worlds. Myth, magic, animism, totemism, superstition, and science are all attempts at representing the world mentally by devising models of how the world works. They all stem from the basic need to understand and make sense of our everyday experiences. As Lévi-Strauss put it: "It is therefore better, instead of contrasting magic and science, to compare them as two parallel modes of acquiring knowledge. Their theoretical and practical results differ in value, for it is true that science is more successful than magic from this point of view, although magic foreshadows science in that it is sometimes also successful. Both science and magic however require the same sort of mental operations and they differ not so much in kind as in different types of phenomena to which they are applied."[15]

I would argue that some early forms of ethnoscience were attempts at explanation, some prediction, and yet some others control. For instance, model-building (storytelling), animism, and astrology were predominantly explanatory, although mythology also had a predictive intention as well. Magic and shamanism, on the other hand, were more attempts at control than explanation. One would perform a particular ritual to appease or pay homage to the spirits in question so that the god-spirits would make things happen according to the desires and needs of the people performing the ritual. By believing either that like causes like (law of similarity) or that physical effects continue to emanate from an object once its contact with another object has been severed (law of contact), practitioners of magic attempted to control nature. As James Frazer documented, and the following is but one out of the thousands of such examples from all over the world: in Laos elephant hunters would warn their wives not to oil their hair for that would cause the elephant to slip through the hunter's grasp.[16]

My basic thesis is not to argue for the inherent superiority of the scientific method over "prescientific" modes of thinking, but rather to argue that the origins of modern explicit science are to be found in these earlier attempts of the human mind to predict, explain, and control the physical, biological, and social worlds. Early in our modern human development, there was no separation between religion, mythology, magic, and science—they were one and the same. It was not until the next phase that these forms of knowledge became their own unique entities. One cannot understand how science came to be without understanding these early forms of mental models concerning our understanding of the social, physical, and biological worlds.

An inherent limitation of verbal transmission of knowledge is that it is somewhat subject to distortion, and without more exact categories of time and space (units of measurement), it is not terribly precise. Beginning with Cro-Magnon culture of around 40 kya, things started to change. Art began to flourish, ornamental objects were made, houses were built, and clothing appeared, and by the end writing systems had developed. The human mind was set for the next phase: invention, externalized symbolic thought, and applied science.

### Phase 3: Applied Science

As Merlin Donald argued in *Origins of the Modern Mind,* the major advance in human cognitive evolution after the "mythic" (linguistic) phase was when ideas began to be externalized in concrete form, first seen with cave paintings and bone markings around 40 kya. It is important to remind ourselves that this "progression" in thinking does not preclude earlier forms of thought but rather supplements them. Earlier forms of thinking do not die out; they are built upon. The development of symbolic thought that could be externalized ushers in a third phase in the development of scientific thinking, one where knowledge of the physical world in particular leads to the construction of more and more elaborate tools and structures. I dub this phase the "applied science" or technology phase because thinking from 30 to 3 kya was very much based in a material and applied form (technology and innovation) rather than in a storytelling form. Applied science and technology is geared toward making and inventing something. It is grounded more in necessity, with its purpose and goal being to build or make, not just to understand for understanding's sake. It is more accurate to call this phase applied science or technology rather than science, for there is little doubt that technology precedes science in terms of historical development, even if now it sometimes follows it.[17]

Before any sort of applied science can be constructed, however, three related developments must be present, each of which first took applied forms: written symbols, measurement, and math. Without some standard units of measurement and without some systematic ability to quantify and determine relations, it is very difficult to do any kind of science, whether applied or pure. Of course, implicit preverbal and verbal folk ethnoscience does not require these methods. The following is a very cursory time line (recall tables 8.3 and 8.4) for major late prehistoric inventions that set the stage for the fifth and last phase of scientific knowledge, namely, pure science from around three thousand years ago.[18]

*Units of Measurement, Astronomy, and Math.* Perhaps one of the earliest expressions of applied prehistoric science is seen from the period of 30 kya with the first evidence of external counting (math) and possible astronomical recordings. They both come from bone artifacts.[19] First, bones have been found with eleven groups of five tallies, which have been interpreted as the first explicit sign of mathematical quantities being recorded. Second, a bone pressure plaque was found in Blanchard, France, showing a serpentine row of etched engravings in a bone, and the etched figures appear to be recordings of the moon phases for a two-and-a-half-month period. It is a remarkable recording, complete with waxing, waning, new, and full moons. Indeed, the first units of measurement were often astronomical and temporal. The moon, stars (constellations), sun, and planets change with regular cycles—among the most obvious cycles in all of nature. The most obvious way to mark time, from prehistoric time on, has been with celestial cycles.

Between the time of Cro-Magnon 40 to 50 kya and the rise of civilization in Mesopotamia, Egypt, China, and Central and South America between 6 and 7 kya, the major Neolithic events were the ending of the Ice Age (12 kya), the domestication of plants and animals (around 10 kya), and the establishment of city-states (around 9 kya). By the time these major changes took place, it was clear that the measurement of time and early forms of astronomy were well under way. In addition to the obvious use of sundials to tell time, prediction of eclipses and other astronomical events required careful observation and standard units of time and math. The Babylonians in Mesopotamia, for instance, made the first formal prediction of eclipses approximately five thousand years ago, which implies they had been carefully observing celestial bodies for a long time and that they had some degree of mathematical system and standardized units of time. The first irrefutable calendars occurred only around 6 kya with the Sumerians and Babylonians of Mesopotamia and the Egyptians. First the Babylonians, then many other cultures, started the practice of beginning a new month with the first sighting of a crescent moon and, in fact, referred to months with 30 days as "full" and those with 29 days as "defective"—partly because 30 is a unit of 60, on which their number system was based. Furthermore, 30 is a unit equally divisible by the 360-day year and would result in a 12-month calendar year.[20]

With the advent of agriculture, keeping track of time became not an idle intellectual enterprise but a practical one. Planting and harvesting require some standardized units of yearly and seasonal cycles. Initially, both in Egypt and Mesopotamia, the earliest calendars were lunar, but it became quickly evident that the cycles of the moon do not make the best calendar because they fall 11 days short of the solar year. The sun was the next obvious timekeeper. In Egypt years had 360 days (36 decans or 10-day intervals), partly resulting from their sexagesimal number system (base 60). Soon it was realized that basic astronomical cycles were closer to 365 days, and the Eyptians adopted perhaps the first 365-day calendar, approximately 6.2 kya (although it may have been 1,500 years later). It still had 36 decans (10-day intervals) but added a 5-day holiday.

Years and months are adequate for long-scale measurement of time, but many events require shorter units of measurement, and phases of the moon were the first basis for these shorter time periods. For the Babylonians, the four phases of the moon (every seventh day) was the basis of the week. They did not, however, conceptualize weeks as continuous as we do, but rather each new month began a new week, which circumvented the problem of sometimes 28- and sometimes 29-day cycles between full moons. In Egypt the first notion of a week was also one-fourth of the 28-day month, or 7 days. The Egyptians certainly and Mesopotamians probably also did what almost every world culture since has done: name the 7 days of the week after the 7 major (and then known) heavenly bodies, all considered by Egyptian astrologers to be planets.[21]

The day itself is a very natural and unavoidable division of time, and in Mesopotamia the Babylonians divided the entire day (day and night) into two 12-hour time periods of 30 *gesh* each (our 4 minutes). Therefore an entire day had 360 *gesh,* a year 360 days (initially), and a circle had 360 degrees, all fitting in very nicely and harmoniously into their sexagesimal number system. Furthermore, because their number system had a base 60, the Babylonians broke the full circle down into 60 units, the first fractional place being called a "minute," and the second fractional unit being called a "second," from whence we get our two basic units of time. The first actual sundials (gnomons) were invented perhaps 6.0 kya in Egypt and broke the day up into 12 hours (10 hours of daylight and 2 hours of twilight). Similarly, around 4.6 kya the Chinese used a pole and its shadow to measure time within a day.

There were some purely intellectual, cognitive, theoretical-explanatory, and spiritual reasons for observing and recording patterns of the planets, sun, stars, and moon, and therefore of time. Not all was applied and practical. After all, in most every one of these Neolithic and ancient cultures, the names and explanations for the existence and behavior of the heavenly bodies were tied directly into stories and mythologies, with gods and goddesses dictating the major events like floods and planting and harvesting.

Many elaborate and detailed stories were made up to explain heaven and earth and the movement of celestial bodies. For instance, the Egyptian calendar year is based on the heliacal rising (right before the sun) of the brightest star in the sky, Sirius (Sothis = "dog star" is in the modern constellation of Canis Major), which in 2500 BCE would have been on the longest day of the year (the summer solstice), named for Queen Isis. In Mesopotamia, Sirius was similarly referred to as "star of the dog" by the Babylonians, as "dog of the sun" by the Assyrians, and as "the dog star that leads" by the Chaldeans.[22]

The Mesopotamians (Sumerians, in particular) had their own creation stories, derived from the opening verse from *Gilgamesh,* and these creation myths also point to the early unity between astronomy and astrology. Interestingly, and consistent with modern science, the first life force was the goddess of the sea (Nammu), and the earth was formed from the mountain in the sea made by the perfect union, or marriage, between the god of heaven (Anu) and the goddess of earth (Ki). Hence the word "universe" results from combining heaven (Anu) and earth (Ki), to get "anki." The offspring of Anu and Ki made all the plants and animals, and humans were created by the combined efforts of the sea goddess Nammu, the earth goddess Ki, and the water god Enki. Disasters on earth were often attributed to wars between the gods. In Mesopotamia there was a tendency to not distinguish living from nonliving, and as a result stones, grain, and salt took on the personalities of gods.[23]

As has always been the case, observation and explanation are hard to disentangle. Even with the spiritual and mystical elements, there were very pragmatic and applied reasons for making these observations and keeping a close record. Being able to predict time and the seasons played a role in almost all aspects of late prehistoric and ancient historic life: planting, harvesting, religious ritual, navigation.[24] Indeed, one general conclusion might be that these ancient cultures made little or no distinction between the spiritual and pragmatic-scientific—they were one and the same. My purpose in this brief foray into ancient calendars and astronomy is not to provide a detailed history of ancient timekeeping, for that can be found in many histories of ancient cultures. Rather, my point is to argue for the inherent connections between explanatory-mythic and applied and practical purposes of an early form of applied science, in particular astronomy and its related disciplines of math and measurement.

The earliest expressions of math were also quite applied in their function. In order to make some of the astronomical observations and predictions explicit and testable, a moderately sophisticated mathematics was needed. Signs of early counting exist between 15 and 30 kya, but the first recorded signs of formal math in which numeric quantities were manipulated (that is, the first arithmetic) appear shortly after the end of the Ice Age, around 10 kya, in token exchange in Mesopotamia. Initially, clay tokens were used to record trade transactions, but these eventually developed, as we saw with their calendars and units of time, into the number system with a base 60. The Sumerians of Mesopotamia developed the first positional notation system around 4.4 kya and soon thereafter were developing systems of multiplication, fractions, quadratic equations, and basic geometric calculations for area, circumference, and diameter. Indeed, Mesopotamian cultures solved quadratic equations, created multiplication tables, and knew the Pythagorean theorem by 4 kya, approximately 1,300 years before Pythagoras!

In ancient Egypt, the first recorded mathematics was also very practical and trade oriented in its nature. Before the First Dynasty of 5.4 kya there were records noting 120,000 prisoners, 400,000 oxen, and 1,422,000 goats. The Egyptians clearly had a very impressive knowledge of geometry, witness the pyramids and obelisks. Indeed, starting around 3.8 kya (Twelfth Dynasty) we find the first more formal mathematical papyri—the Moscow and Rhind. The authors of these works stated and solved various problems of fractions and their multiplication, addition, and subtraction and demonstrated, in addition, fundamental knowledge of geometry (recall table 8.4).

One of the really remarkable aspects of both Babylonian and Egyptian math is that it was done without a modern Arabic numeral system, but with either cuneiform or hieroglyphics. It was not for another 1,500 years or so (200 BCE) that Arabic numerals were developed in India (Devanagari script) and were soon adopted by Islamic mathematicians. They became known as "Arabic" numerals because they were first transmitted to the West in approximately 900 CE by the Arabs.

Even though Arabs may not have been the first to use Arabic numerals, there can be little doubt that they played a big role in the development of mathematics, with algebra being the prime example (the words for both "algebra" and "algorithm" are Arabic in origin). To be sure, ancient and early math in Egypt, the Middle East, and India was relatively rudimentary by today's standards, staying mostly at the level of basic arithmetic, geometry, and algebra, but it was the first developed math in the world and therefore of tremendous historical importance to the development of science.

*Writing and the Alphabet.* Writing systems, although not an applied science, had a truly revolutionary impact on science and technology.[25] For the first time ideas could be externalized and reflected upon by many others from different places and times. Such a capacity was just the boon that was needed to allow thinking about thinking (metacognition) to take off, and it is no coincidence that cultural evolution became truly exponential at this time.

To give a simplified overview of the general typology and history of writing systems, protowriting systems, such as cave paintings and drawings, and ideographic systems (depictions of particular ideas), such as pictograms and petroglyphs, came first. But neither of these was true writing because neither tied specific vocalizations to pictures or images; they functioned more as memory aids or storytelling devices. It is fair to say, though, that cave paintings and pictographs were at the crossroads between art and writing, and they are of monumental intellectual value: they are among the first external expressions of human thought. It was not until logography/logosyllabary (characters represent words, morphemes, and syllables) (Sumerian, Egyptian, Chinese) that the first full writing systems began. The major advance here is the "rebus principle," namely, that pictographic symbols could stand for sounds.

Initially, around 5.5 to 6.0 kya, Sumerian writing was ideographic (pictorial and semantic), meaning that pictures were drawn for particular whole words. By around 3.1 kya, and mostly due to the awkwardness of the clay medium, these evolved into cuneiform logographic symbols for word parts (syllables and morphemes) that were made by a wedge-shaped reed (*cuneus* = wedge). Eventually cuneiform evolved into a syllabary/alphabet, as first deciphered by Henry Rawlinson and Edward Hincks in the 1850s. Like cuneiform, hieroglyphics began completely pictorially at the word (logos) level around 5.0 kya but transformed into a cursive and syllabic form of hieroglyphics (hieratic) by approximately 3.9 kya. When Andrew Young, and then more completely François Champollion, deciphered the Rosetta Stone in the early 1800s, it became clear that the later form of hieroglyphics was in fact a complex mixture of logographic and phonemic script. A similar transition from pictographic to at least partial phonemic script happened with Chinese as well, beginning around 3.2 kya.

The next major advance was when the smallest units of sounds (phonemes) began to be represented with their own unique character. The earliest phonemic scripts (for example, Phoenician, Aramaic, Arabic, Hebrew) had characters usually for consonants, and therefore historically many scholars have labeled these "consonantal" scripts. More recently some scholars have begun referring to them as "abjads" (after the first four letters of the Arabic script).[26] I will refer to them with the hyphenated "consonantal-abjad." Phoenician, the language spoken in what is now northern Israel, northern Palestine, and Lebanon may well have the distinction of being the first language to develop a consonantal-abjad around 3.5 kya (1500 BCE), and it probably came from the hieratic form of hieroglyphics. The last major advance in script was the addition of vowel characters to the consonantal systems, which we call "alphabets." The Greeks, borrowing from the Phoenician, Aramaic, and other Semitic languages, first developed a full phonemic alphabet around 3.1 kya (1120 BCE). It was the first true alphabet, and the Greeks were able to name it—and they chose the first two letters of their "alpha-beta."

Although complex in details, the general picture is clear: the development of written language began with whole pictures, moved to ideas, then words, then syllables-morphemes, and finally phonemes (first without and then with vowels). In other words, the progression was from stories, to ideas, to words, to consonants, and finally included vowels. All scripts show such development from larger to smaller units of language.

The last topics of the third phase of scientific thinking discussed here—city-states, architecture-engineering, and metallurgy—are each based in technology and the developments in writing, math, and measurement and are thus examples of ancient applied science and invention (see tables 8.3 and 8.4). The developments in astronomy, mathematics, and units of measurement did not happen in a vacuum. They happened in large part because humans had started to settle down in one place and were making the transition to not being hunters and gatherers.

*Architecture.* Upper Paleolithic (Neanderthal and Cro-Magnon) settlements did not take the form of city settlements, although some constructed settlements were apparent during the last Ice Age, around 40 to 45 kya. There is evidence that the first semipermanent structures were built by Neanderthals in Germany, France, Spain, and the Ukraine.[27] For instance, in the Molodova I structure in the Ukraine, there appear to have been small "houses" made of wooden support, animal skins, mud, and animal bones, although the evidence is not unambiguous.[28] But it was not until 30,000 years after these first structures that large aggregates of people settled down in one area for long periods of time. The main areas where agricultural civilizations first flourished were: northern Africa (Eygpt), the Middle East (Mesopotamia, now Iraq), India, China, and two parts of the New World (central Mexico and the Andes). The Neolithic era had arrived.

One of the major innovations of these early city-states was their architecture, which inherently combined math, engineering, masonry, physics, art, and sometimes astronomy. We know the most about the Mesopotamian and Egyptian architecture. The so-called Neolithic Natufian settlements predated farming and existed in the Middle Eastern region of the Levant. Among the oldest of these city-states is the ancient town of Jericho in present-day Israel, directly north of the Dead Sea, which began forming around 11 kya. Initially the houses of Jericho were round and made of mud brick and wood, but later they were rectangular with plaster floors. Similarly, 'Ain Ghazal near Amman, Jordon, was settled around 9.2 kya and soon had up to two thousand inhabitants who developed very symbolic art objects, such as tokens, animal and human figurines, and modeled skulls. Soon after the establishment of the first Natufian city-states, farming of emmer wheat and barley and the domestication of animals (dogs, goats, and sheep) began.[29]

The next major architectural advances were not seen for another 5,000 years, when the Great Pyramids of Giza were built (around 4,800 years ago). There are many mysteries surrounding the construction of the pyramids, but it is clear by laws of induction that because these buildings were made, the people (or at least the architects, for instance Hemon) who made them knew something about physics, geometry, and engineering. What we know is that more than two million stones weighing an average of five thousand pounds each had to be carved and moved in order to build the biggest of the pyramids (Khu-fu, also known as Cheops) with a base of 230 meters and a height of 146 meters.[30] Created from an incredibly complex mix of engineering, astronomy, architecture, mythology, geometry, worship, and slave labor, the pyramids are truly one of the great wonders of the world. In order to build these massive structures, the Egyptians had to have well-developed engineering along with some standardized units of measurement. We know that they primarily made use of body parts for units of measurement: four digits ("finger widths," ca. 1.88 cm) in a palm, and seven palms (ca. 7.50 cm) in a cubit (52.37 cm). Slightly to the north-east in Mesopotamia, around 2,500 BCE the units for length, weight, and capacity became legally fixed.

The obelisks of ancient Egypt are, to this day, another major mystery. We simply do not know how these people could have raised monuments weighing more than one thousand tons into vertical positions without the use of machinery. The best theory argues that they moved the obelisk up a vertical slope until it passed its balancing point, then dug out the ground underneath the bottom and pulled the top up to a vertical position. Similarly, a ziggurat in Mesopotamia from around 5.5 kya (3500 BCE) demonstrates Mesopotamian knowledge of columns, domes, arches, and vaults.[31]

The importance of this shift to a settled city-state lifestyle and the development of architecture and engineering is obvious and hardly needs elaboration, but for the development of technology and science suffice it to state the obvious: without a settled lifestyle with a specialized labor force and food production, there would be no real development in math, technology, and writing, all preconditions for civilization and the "pure" science that was to follow. Being able to support people who do things other than hunt, forage, or produce food is the major transition, the sine qua non of civilization in general and science and technology in particular.

*Metallurgy.* Considering that stone had been the primary material for tools for almost two and a half million years (until bone and antler came into use around 90 kya), smelting and shaping metal into tools, weapons, jewelry, and utensils marked a revolutionary change. In addition, given their shiny and aesthetically pleasing appearance, it is easy to see why metals quickly became the most valuable and mystical objects in ancient civilization and why only kings and gods were considered worthy of possessing them. Ancient metallurgy was, in fact, a complex and exquisite mix of science, art, religion, and mysticism.

The general historical progression in metallurgy, seen independently on different continents and sometimes separated by 2,000 years, was from the Copper to the Bronze to the Iron Age, although some scholars argue that these terms are now obsolete. The first artifactual native metals (native copper, silver, and gold) are seen in rolled copper beads from Turkey by the eighth millennium BCE (or 9 kya) and Mesopotamia by seventh millennium BCE (8 kya). By around the fifth millennium BCE people discovered that these metals became pliable when heated to high temperatures. The first man-made artifacts, things like beads, pins, awls, needles, knives, axes, and spearheads, date from around 4500 BCE and were made in present-day Turkey. The Egyptians also mined copper by around 4500 BCE and produced tons of it prior to the First Dynasty (ca.

3200 BCE). Copper is quite soft and has a moderately high melting point (1,083°C or 1,981°F). Therefore when bronze was discovered, it quickly became the metal of choice. Bronze is an alloy that is produced by adding about 10 to 12 percent tin to the copper, which results in a metal that is harder, less corrosive, and has a lower melting point (about 950°C) than copper. Bronze was the first metal to change society, and the cultures that mastered it deserve the moniker "the Bronze Age." Iron, although a natural element rather than an alloy, was the last metal to be smelted, and ironing was first accomplished in Eurasia (Turkey) by the Hittites around 3 kya (1000 BCE). This was a later development mainly because iron has a higher melting point (1,500°C), but once kilns were capable of such temperatures (with the aid of bellows) and it was able to be widely mined, iron became the most common of all metals, especially for making tools and weapons. It, too, had a transformative influence on the cultures that produced it.[32]

In working with metals, experimentation was necessary—hunches and hypotheses were tested and either worked or did not. Only the successful techniques were propagated, and this required empiricism and hypothesis testing. In these metallurgy industries we see the foundations for the science of chemistry (applied and implicit though it may be). But it is also true that from their earliest use, metals were accompanied by mystical and superstitious beliefs, with alchemy being the most obvious. In Egypt, Thoth, the god of knowledge, art, and science, was also a god of alchemy. It is safe to say that like that of astronomy and architecture, the early history of metallurgy was a complex mixture of empiricism, invention, mysticism, mythology, and superstition.

Within two millennia of the end of the last Ice Age remarkable changes began to take place in human cultures all over the world and especially in northern Africa, the Middle East, the Far East, and Central America: settlements, buildings, farming, ceramics, textiles, and math. Then by around 8 or 9 kya, metals were being used to craft new tools, and around 5.5 kya, the earliest writing systems were formed to record human transactions and thought. This period began in the caves and ended with the advent of civilization and written history. We have crossed the boundary separating prehistory from history, and the stage is now set for the last major development in science: discovery and science for its own sake. As is generally true in human thought, something initiated for a strictly practical purpose over time often comes to be done for its own sake, and this takes us to the last stage in the development of science, pure science. Applied and pure science are obviously not mutually exclusive; they exist in a symbiotic relation to this day and no doubt always will. My point is that between 30,000 and 3,000 years ago, there was no pure science, only applied.

### Phase 4: Discovery and Pure Science

The development of the applied sciences in Egypt and Mesopotamia by the fourth century BCE stood at the crossroads between applied and pure science. In both Egypt and Mesopotamia, many observations and technologies were tied up with the behavior and desires of the local gods, goddesses, kings, queens, or spirits. Moreover, those who practiced "science" were most often priests, and therefore science was often inherently connected to religion and mysticism. To the extent that the Egyptians and Mesopotamians were sometimes devoid of mythical explanation, they were practical and cookbook oriented. They lacked a concern with an overarching natural explanation of how the world worked.[33] Finally, because the ancient Greeks lacked a major cosmological myth about the origin of the universe and were laymen rather than priests, they were free to look for more natural explanations concerning cosmology. In the last phase of the development of science, the fundamental focus moves from mystical to naturalistic explanation and from invention to discovery.[34]

*The Ionian School.* Beginning around 2,600 years ago, Ionia, Greece, was the center of a true revolution in the history of human thought: humankind came to see nature rather than divinity or supernatural forces as the causal agent (origin) behind things. In contrast to the earlier Egyptian and Mesopotamian scientists, the Greek philosophers were laypeople rather than priests and therefore could more readily divorce their thinking from the metaphysical. This major shift was more conceptual than technological: nature rather than spirits, gods, goddesses, or magic was the causal agent behind how things worked, and it could be understood by the laws of physics, chemistry, biology, and psychology—with heavy doses of math and logic thrown in. Such a natural view, as far as we know, was a first in human history.

I am not arguing that Greek culture was void of mythology, for they clearly had a most developed mythology, given its clearest voice in Homer's *Iliad* and *Odyssey.* What I am saying is that major thinkers of the day, the philosophers, did not agree with the current mythology and began to examine the world in its more natural form. Just as mysticism and science are practiced side by side today, by different segments of the population, so too was this the case in ancient Greece. The natural philosophy of these thinkers had its price, for as is true to this day, science and rationalism can go too far in challenging the thinking of the day. Anaxagoras was condemned and exiled on the charge of "rationalism" in 432 BCE. Socrates was put to death. Even Aristotle was attacked. Rationalism and science even in ancient Greece was too much of a challenge to the established order and could go only so far with impunity.

Pure science is conducted without the applied goal of making or building something. Its aim is discovery rather than invention. It is science out of pure curiosity—to figure out how something works. Explicit science is science in which the rules and procedures are explicitly known, accessible to consciousness, and verbally transmitted from one generation to the next. In this sense, the Greeks probably created both pure and explicit science, but perfected and widely used neither, for they stopped short of systematic experimentation. It was not until the beginning of the modern era in Europe, in the sixteenth and seventeenth centuries, that pure, explicit, and experimental science became widely used.

What the Greeks added was naturalistic explanation. Thales of Miletus (ca. 640–546 BCE) is considered the "first philosopher" and the first to argue that all of nature comes from one natural source; in his mind, water was the cause of all living things. Although Greek, Thales studied in Egypt and Mesopotamia and brought the knowledge of the Middle East back to Greece. In fact, some would argue that naturalistic explanations occurred to Thales precisely because he was at the geographic and intellectual junction of Egyptian, Mesopotamian, and even Indian thought and mythology. Simultaneously being aware of such diverse beliefs allowed Thales to strip away the contradictions, and what was left was nature itself.[35] Thales' pupil, Anaximander (610–545 BCE) was, in fact, the first to propose that human life evolved slowly (not from water but from "the boundless") and that the earth was round, whereas Democritus (460–370 BCE) was the first to propose that matter was made up of indivisible "atoms." Although not from Ionia, Pythagoras (582–500 BCE) made great strides in the development of mathematics, including geometry. In contrast to the Ionians, Pythagoras argued that number is the basic element of all things (which ultimately led to his religious concern for music and its mathematical foundations). Anaxagoras (500?–428 BCE) was the last of the Ionian thinkers and the one who perhaps brought Greek thought closest to science, mainly in the realms of what we would now call cosmology, math, and theoretical physics. His ardent rationalism did land him in trouble, however, and he was banished to exile, the first person in known history to be banished for his freethinking.

*Athenian School.* If the Ionians were the midwives, then the Athenians were the mothers of modern science. As seen most clearly in Anaxagoras, by the time of Socrates (470–399 BCE) in the fifth century BCE, the concept of rational and critical examination of ideas was ripe for development. Socrates, as the founder of the famous Athenian School, publicly advocated such critical thought, but it was a specific form of critical thought, namely, psychological. Socrates' chief legacy lay in making the subject of philosophy the inner psychological world of man, especially self-knowledge and ethical behavior.[36] His major questions were "What is human nature?" and "Who am I?" The famous oracle of Delphi ("know thyself") became his driving passion. All answers to ethical, political, and social questions rest on the answers to these critical self-examined questions and therefore must be the starting point for philosophers. There was a corollary to this psychologically critical attitude, and that was the philosophically critical agnosticism of admitting one's ignorance. His starting point for everything was the assumption, predating Descartes by some 2,000 years, that "all I know is that I know nothing." Humble skepticism is the starting point for philosophy in much the same way it is for science. Socrates had a gift, ultimately leading to his execution, of challenging and questioning every basic assumption men had of who they were and what they knew. He agitated many with his demand for precision of definition and questioning of dogma. As was the case with Anaxagoras (but with more severe punishment), Socrates' ideas threatened the elders more than the youth, and in the end Socrates' execution made him philosophy's first great martyr. His death was detailed eloquently by his most famous and influential student, Plato (427–347 BCE), in the *Phaidon*.

Plato's star pupil, Aristotle (384–322 BCE), best exemplifies the movement of Greek thought into the realm of the natural. Because of his emphasis on nature and observation, Aristotle can be considered the Greek figure in whom the ancient scientific attitude reached a peak. Although quite a few of his observations and speculations were erroneous, and required nearly 2,000 years to correct, he was the first to gather many animals together in a zoological garden for observation and the first to "prove" the earth was round by pointing to the earth's shadow on the surface of the moon during a lunar eclipse. He noticed the hierarchy and similarity of many species and contributed to the taxonomy system of his day, but he overlooked Anaximander's and Empedocles' notion of evolution and survival of the fittest.[37] Aristotle contributed to observation, classification, and description of the natural world (taxonomy), but like most of the ancient Greeks he never really thought to experiment with it.

His physics, although quite detailed, was errant in its details: he was wrong about the weight of an object being related to its speed of falling; he was wrong about rejecting Democritus's atomic theory; many of his ideas about medicine were misguided; his beliefs about the makeup of elements had an influence on the pseudoscience of alchemy for centuries to come. But perhaps Aristotle's major shortcoming was that he explicitly discounted experimentation and therefore never tried to experimentally test his ideas. It is not in the specifics of his science that Aristotle's import is to be found, but rather in the organization, description, and codification of naturalistic observations in a systematic manner. As Will Durant put it in *The Story of Philosophy:* "It is again the absence of experiment and fruitful hypothesis that leaves Aristotle's natural science a mass of undigested observations. His specialty is the collection and classification of data."[38]

The ancient Greeks were the first to develop formal logic, naturalistic explanations, and doing science for its own sake. One pattern to these contributions is their foundation in observation, mathematics, and logic rather than in hypothesis testing and experimentation. The Greeks took us to the threshold of a full-blown experimental science with its focus on natural explanations and formal logic but stopped one step short of opening the door fully to experimental science, a shortcoming seen most clearly with Aristotle's approach to science. The door to experimental science remained almost completely shut for the next 2,000 years.

There were two bridges between ancient and medieval science: Islamic translations of ancient Greek philosophers and Guttenberg's most practical invention, the printing press with movable type. Arabic numerals are Indian in origin, but it was the Islamic scholars who preserved them and transmitted the ancient Greek ideas to European medieval scholars. The Arab scholars from the ninth to twelfth centuries did more than just preserve a new number system: they translated works on astronomy and mathematics and contributed their own advances in optics and chemistry.[39] Writing is so important in cultural evolution because it facilitates thinking about one's own and others' thinking, that is, it facilitates metacognition. It is not a coincidence that the last major contribution to science, the modern experimental method, was preceded by a major innovation, the printing press. Speeding up the printing process allowed ideas to be disseminated much more readily, a precondition for metacognitive thinking.

*Modern Pure Science.* Two early outliers during the so-called (and misnamed) Dark Ages were Roger Bacon (1214?–94) and St. Thomas Aquinas (1225–74). Taking issue with St. Augustine's view that sensory experience was the lowest form of knowledge, Bacon in 1268 stressed that empirical testing and observation was needed to have a complete understanding of anything. Pure logic and reason were not enough. "There are two ways of acquiring knowledge, one through reason, the other by experiment. Argument reaches a conclusion and compels us to admit it, but it neither makes us certain nor so annihilates doubt that the mind rests calm in the intuition of truth, unless it finds this certitude by way of experience. . . . Argument is not enough, but experience is." Similarly, Aquinas was the first to explicitly argue for the primacy of sensory experience over reasoning with the phrase, rekindled by Locke 400 years later, "nothing could be in the intellect that was not first in the senses."[40] Here we see one of the classic epistemological battles laid bare: which is the more valid form of knowledge, pure logic and reason or sensory experience? Rationalists argue for the former and empiricists for the latter. But the argument for experimental observation over pure logic is absolutely essential to science, and Roger Bacon and Aquinas can and should be credited for giving birth to this idea.

The first major scientist of the modern era—Copernicus (1473–1543)—was not an experimentalist but rather a careful observer of the heavens and a mathematically oriented theoretician. He created one of history's major intellectual revolutions by systematically developing the heliocentric (sun) rather than geocentric (earth) model of the solar system. Archimedes and Aristarchus in the third century BCE in Greece were the first to propose the heliocentric view, but the idea was buried for more than fifteen hundred years and was only resurrected by Copernicus. In a very real sense, one could argue that modern science began with Copernicus's heliocentric view of the solar system. But Copernicus had some conservative tendencies and was too caught up in Platonic ideals, which used the ideal perfect circle as the path of the planets. Copernicus's model solved some problems but created others.

The best astronomical empiricist of the day was Tycho Brahe (1546–1601). A privileged aristocrat who was rather eccentric (he kept a dwarf and had a pet elk that died drunk after stumbling down a stairway) and arrogant (he had much of his nose cut off in a sword-fighting duel with a man who questioned his unsurpassed mathematical talents and had to wear a gold-plated nose the rest of his life), Brahe was inspired and transformed at age thirteen by accurate predictions of a solar eclipse. Without a doubt he had the most careful and accurate planetary observations in the world, and he greatly improved astronomical observations by as much as sevenfold accuracy by recording the positions of the planets not just at important points in their orbit, but throughout their entire orbit. The truly remarkable thing about these observations is that they were done with the naked eye.

Brahe, however, was rather guarded with his mounds of data, even when the young Johannas Kepler (1571–1630) showed real interest in his observations. Kepler had been deeply perplexed by the discrepancies between Copernicus's model and known data and attempted many unsuccessful variations to bring the heliocentric model more in line with known observations. He knew also that accurate and complete observations were in the hands of Tycho Brahe. Kepler was granted unrestricted access to Brahe's observations only after Brahe's death. The completeness and accuracy of Brahe's data formed the foundation for Kepler's three laws of planetary motion, which were among the first great mathematically based models of how the celestial bodies operate.

Moving to the realm of physics, William Gilbert (1544–1603) published a work in 1600 that was the first to report electrical properties of certain objects and to argue for magnetism as the cause of gravity, hypothesizing that the center of the earth was a giant magnet. Moreover, Gilbert was an early proponent of the idea that size and gravity were related, stating that the larger the body, the greater its pull or magnetism. He extended this view to the sun, moon, and earth's rotation, and like Copernicus, he believed that the sun was the major mover of heavenly objects because it was the largest. Gilbert's systematic work on electricity, magnetism, and gravity placed him at the forefront of the experimental method. One problem with Gilbert's view on gravity, however, was that it was particular, not yet applied to all stars or to the earth's rotation—an extension only made ninety years later by Newton.[41]

Francis Bacon (1561–1626) was not the first to argue for the importance of empirical observation over theoretical speculation, but he codified and integrated more clearly than others before him much of the thinking concerning the scientific method. One of his main inspirations for the importance of observation came from witnessing the discoveries Galileo made with the telescope. Bacon argued that empirical observation must be wedded to logical thought; most important, to logical thought based in observation. He further argued that being skeptical of earlier ideas and developing theories (induction) about one's own observations were the cornerstones of science. In this attempt to wipe the slate clean of ancient authorities, Bacon was expressing a key attitude of science.

Although living nearly 350 years after Roger Bacon and Aquinas first proposed the experimental tenet of modern science, and nearly 100 years after Copernicus's heliocentric astronomical theory, Galileo (1564–1642) was perhaps the first to integrate these principles in one coherent view. With such an integration of experimental method and scientific-mathematical theory Galileo earns the title "founder of the scientific method of inquiry."[42] By building one of the most sophisticated telescopes of his day in 1610, he was able to make such significant astronomical discoveries as Jupiter's moons, the phases of Venus, and sunspots that provided rather strong and direct support for Copernicus's heliocentric view and contradicted the views of Ptolemy and Aristotle. Furthermore, he was the first to systematically set forth the law of inertia, which later became Newton's first law of motion.

Most important for the history of science, Galileo was the first to experimentally test some of Aristotle's assumptions concerning physics, in particular the proposition that weight determined the speed and acceleration of falling bodies. Galileo purportedly performed one such test by climbing to the top of the leaning tower of Pisa, dropping two balls of different weights, and observing them hit the ground at the same time (which Aristotle would not have predicted). Whether he actually did so from the leaning tower, it is certain that over the course of his life Galileo experimented with the motion of falling objects from calibrated slopes and observed at least two things that directly contradicted Aristotle: balls of different weights reached the bottom at precisely the same time and the velocity of these balls accelerated as they went along. It took almost 2,000 years, but Aristotle's untested genius would never have the same sway over natural science again. In Galileo we find the final piece to the full-blown scientific method, the piece that was missing in ancient Greece, namely, experimental testing of hypotheses. Galileo's astronomy and physics, however, ran counter not just to Aristotle and Ptolemy, but they were counter to the Church's teachings. Unfortunately, reminiscent of the fate of Anaxagoras and Socrates, the authorities (in this case, the Church) ordered formal inquisitions into Galileo's ideas, which were ultimately declared heretical and were banned. By threat of imprisonment and even death, Galileo somewhat softened his views, but he never fully recanted.

Finally, we have Sir Isaac Newton (1642–1727). The seventeenth century had two major problems in physics: dynamics (in particular, what kept the planets moving and in their path) and gravity. In one fell swoop, Newton solved and integrated both problems (dynamics and gravity) in perhaps the most important scientific work ever published, *Principia Mathematica* in 1687. He merely invented calculus (independently of and simultaneously with Leibniz) and the concept of mass in order to solve the problems of dynamics and gravity. The solution to the problem of dynamics culminated in Newton's three well-known laws of motion.

Moreover, Newton's solution to the problem of gravity was so complete that limits to his theory were not recognized until Einstein and the development of quantum physics in the early and middle part of the twentieth century, and even then Newton's ideas were not wrong, only "limited" to the mechanical world of earthly and celestial objects. Newton's theory of gravity supplied the mechanism (force) behind Kepler's laws of planetary motions (modifying Kepler's third law in the process). The law is a very elegant formula: $F = Gm_1m_2/d^2$, where F = force of gravity, G = gravity's constant, m = mass of each body, and d = distance between the two bodies/masses. With this formula he moved beyond all of his predecessors and universalized the effects of gravity; the effects were valid for all physical bodies, from apples to planets. Indeed, the airplane and the space flight programs have been two tremendously successful applications of Newtonian physics. In *Principia* we see the first fully synthetic scientific theory based in mathematical reasoning and models and grounded in careful observation of nature and experimental hypothesis testing. Science now existed in full force. Argumentation, indeed, was not enough and with perhaps only Galileo as a rival, Newton was the scientist who put this new integrated scientific method to practice. Science would never be the same after Newton.

We can stop here in our brief history of science, because our main intent was to retrace the major evolutionary and historical ingredients of science. With Newton they are all there: careful observation, pattern recognition, hypothesis testing, causal thinking, theoretical explanation, and mathematical and experimental thinking. What we see in the late 1500s and 1600s is a concentrated development of the scientific attitude of making ideas, methods, and observations increasingly explicit, increasingly pure (that is, discovery oriented without direct application), and increasingly experimental.

## SCIENTIFIC ATTITUDE

Many have written about what makes up the scientific attitude, but for our purposes the writings of the psychologist B. F. Skinner do as good a job as any of summarizing the ingredients of the scientific mind-set.[43] Perhaps the most essential of these is the ability and disposition to reject ideas that are based on authority. Science and the scientific attitude begin where dogmatism and adherence to ideas based in authority end. Historically, one reason that science did not develop much from 200 BC to 1600 AD was the dominance of Aristotelian natural philosophy (and the Church). Science was mainly what Aristotle wrote. Galileo's test of Aristotle's mechanics was science at its most basic—seeing for oneself whether nature behaves the way we, or some authority, say it does or should. Nature itself, at least our direct and verifiable observation of it is the cornerstone of science. As Bertrand Russell's quote at the opening of this chapter so eloquently points out, insight, no matter how seemingly profound at the time, does not guarantee truth. Science begins with this awareness, namely, that empirical testing of one's ideas is required. Indeed, in the last phase of scientific thought, one comes to not only know what one knows (metacognition) but also what one does not know. It is knowing what one does not know that is the driving force before experimental investigation.

Related to a rejection of authority is a predisposition toward skepticism, for gullibility is antithetical to science. It is better to withhold belief than to prematurely latch onto an idea. In this sense, science is inherently conservative. But unbridled skepticism is also not what science is about. If one is overly skeptical and critical, then one never believes anything and misses ideas that ultimately may have great value. As Carl Sagan pointed out: "It seems to me what is called for is an exquisite balance between two conflicting needs: the most skeptical scrutiny of all hypotheses that are served up to us and at the same time a great openness to new ideas. If you are only skeptical, then no new ideas make it through to you. You never learn anything new. You become a crotchety old person convinced that nonsense is ruling the world. (There is, of course, much data to support you.) On the other hand, if you are open to the point of gullibility and have not an ounce of skeptical sense in you, then you cannot distinguish useful ideas from the worthless ones. If all ideas have equal validity then you are lost."[44]

I refer to this as "open skepticism." In science, the default attitude must be skeptical nonbelief, but once evidence is in, one has little choice but to believe it, regardless of one's wishes and preconceived notions. This results in a dialectical tension between skepticism and belief. The history of science, as Thomas Kuhn made clear, is the back and forth between doubting old ideas, developing new ideas based on evidence, having anomalies build up that lead again to new doubts, and then revolutionary new ideas building on or replacing old theories.[45]

If one is willing to let direct observation of nature be the final arbiter, one ultimately must either have or develop honesty with oneself. Another characteristic of the scientific attitude, therefore, is intellectual honesty. It is not that scientists are inherently more honest than other people—they clearly are not. It is just that when verifiable observation dictates "truth," all the logic and authority in the world will not by themselves make an idea right.

Foregoing intellectual honesty and being prematurely emotionally attached to an idea before it can be empirically verified can be embarrassing at best, career ending at worst, as the case of two scientists who claimed to have achieved cold fusion in the late 1980s demonstrated. Stanley Pons and Martin Fleishmann, two chemists from the University of Utah, claimed that with a few simple everyday pieces of equipment, they were able to obtain fusion, which results in extremely high energies, a process known as "cold fusion." Cold fusion had been a holy grail of physics for it would mean obtaining much energy from very little energy. Pons and Fleishmann set the world of physics on its head overnight by making the cold fusion claim at a press conference in March 1989. This was the first sign that something awry might be taking place because the tried-and-true method of communicating important scientific findings is and has been the peer review method. Pons and Fleishmann made a fundamental mistake by trying to bypass this quality control process. But the real sin was committed from the view of science when many labs worldwide attempted to replicate the Pons and Fleishmann study and could not get the purported results. After months of negative findings, and a handful of neutral findings, it was clear that Pons and Fleishmann had a chance finding, were sloppy with their techniques, or, worse, were fabricating data. No one claims the latter, but most physicists are convinced that Pons and Fleishmann were guilty of sloppy and bad science.[46] Intellectual honesty, to be sure, is directly tied with the predisposition to deal with facts and data rather than with one's wishes and desires. Science requires such honesty because nature cannot be fooled by our foibles and wishes.

One final driving force behind the desire to do science is a profound awe and curiosity toward nature and the world. Curiosity is just an example of the brain trying to make sense of the sensory input it receives. The world is a confusing place and we all have infinite questions. What is that? How does that work? Why did that happen? Does that always follow this? Science is simply the codification of these questions, again pointing to how science really is nothing more than an elaboration of everyday thinking. This curiosity for many people peaks at about age three, but for the scientifically minded it continues, if perhaps more funneled toward a specific domain such as physical objects, biological processes, or social and human behavior, unabated over the course of one's life. The ability to wonder and be awestruck drives scientific curiosity.

If science and the scientific attitude are a part of who we are, then an obvious question arises concerning its universality and ease of learning. Why have most human cultures historically not had "science" per se? The flip side of which is why did science develop in Mesopotamia-Egypt-Greece when China, for instance, had a similarly developed culture and writing system (although pictographic more than syllabic-phonetic)? These are complex questions beyond the scope of this book, but I offer the following hints at an answer.

There is little doubt that some ideas of science are beyond everyday thought and common sense, and it is quite true that much of modern science is difficult and counterintuitive. The reason that science and math are so difficult to learn for so many people today and are not cross-culturally universal has to do with the fact that these forms of knowledge are now rather removed from the implicit forms of knowledge they are based on (cultural evolution is much faster than biological evolution). That sponges are animals or that plants reproduce is not really obvious unless one very carefully observes nature over a long period of time. Moreover, genetics and atomic physics are based in phenomena that are completely invisible to observation and yet still very much part of nature. Indeed, scientific theories are more removed from direct sensation and perception (more abstract) and hence are more culture specific.

In short, science and math are not evolved adaptations of the human mind, but rather are co-opted by-products of evolved adaptations.[47] Evolution did not produce brains so they could do systematic and explicit math and science, but it did evolve a sophisticated central nervous system that organizes and interprets sensory information and is able to reflect upon experiences and put thought between impulse and behavior. In the process of organizing and interpreting sensory input, the brain recognizes patterns, makes causal connections, and forms expectations and predictions (hypotheses). This is what the central nervous systems does. Once certain parts of the brain evolved, especially the cortex and the frontal lobes as well as the language areas, our species began the slow but upwardly asymptotic trend toward making ancient and implicit methods of organizing sensory experience (that is, gaining knowledge) of the world more and more systematic, quantitative, explicit, and metacognitive. Science as we know it requires not just careful observation, pattern recognition, hypothesis formation, and hypothesis testing, but also written language, mathematics, precise units of measurement, tools and technology that measure these units, and finally explicit knowledge of principles discovered by prior scientists. These things first appeared together in cultures only four or five thousand years ago. Although the products of modern science are far removed from the evolved cognitive processes that make science possible, they are in fact the direct result of long evolutionary, historical, and cultural processes.

# Chapter 10 Science, Pseudoscience, and Antiscience

One thing I have learned in a long life: that all our science, measured against reality, is primitive and childlike—and yet it is the most precious thing we have.
—*Albert Einstein*

## SCIENCE AND PSEUDOSCIENCE

### The Nature of Pseudoscience

Many Americans still seem quite willing to believe things that science and skeptics would just as easily dismiss. Michael Shermer reported in his book *Why People Believe Weird Things* percentages of Americans in 1991 who believed in the following paranormal experiences: 67 percent had actually had a psychic experience; 65 percent believed in Noah's flood; 52 percent believed in astrology; 46 percent believed in extrasensory perception (ESP); and 42 percent believed in communication with the dead.[1]

Ever since the 1930s, when Karl Popper first argued for falsification as the main criterion for demarcating science from nonscience, the topic of "pseudoscience" has played an important role in the philosophy of science. Just because someone claims to be doing science or to be a scientist does not mean they are. Popper argued that if the theory did not put forth predictions that were "brittle" and potentially "falsifiable," then they were not science. Theories that can be twisted post hoc to explain any kind of experimental outcome are not science. Astrology, Marx, Freud, and Adler—according to Popper—put forth such vague, general, and nontestable theories that they do not earn the title "science." To quote Popper's view on astrology: "Astrology did not pass the test. Astrologers were greatly impressed, and misled, by what they believed to be confirming evidence—so much so that they were quite unimpressed by any unfavorable evidence. Moreover, by making their interpretations and prophecies sufficiently vague they were able to explain away anything that might have been a refutation of the theory had the theory and the prophesies been more precise. In order to escape falsification they destroyed the testability of their theory. It is a typical soothsayer's trick to predict things so vaguely that the predictions can hardly fail: that they become irrefutable."[2]

Although few philosophers of science today would adhere to Popper's rather strict criterion of falsifiability, his writings on science and pseudoscience have proven to be very important in the formation of a movement both inside and outside of philosophy of science to defend science against fakery and false science. For example, in 1976 the Committee for Scientific Investigation of Claims of the Paranormal (CSICOP) was formed to encourage "the critical investigation of paranormal and fringe-science claims from a responsible, scientific point of view and disseminate factual information about the results of such inquiries to the scientific community and the public. It also promotes science and scientific inquiry, critical thinking, science education, and the use of reason in examining important issues."[3] A number of well-known scientists, such as Isaac Asimov and Carl Sagan, were prominent founding members. In addition to holding conferences and providing a network of critical thinkers, it also publishes research that investigates the paranormal claims in its monthly journal, *Skeptical Inquirer.*

In his book *What Is Science and How It Works,* Gregory Derry lays out specific prototypic characteristics of pseudoscience.[4] A precondition for being a pseudoscience is the claim that what one is doing is science. Philosophy, art, music, and religion, for instance, cannot be pseudosciences because they do not claim to be science. The prime candidates for pseudoscience, therefore, are alchemy, creation science, perpetual motion machines, astrology, UFOlogy, ESP, mental telepathy, and possibly Christian Science, although its claims to be science may be more in name only. Without going through each candidate one by one on each criterion, I will summarize Derry's criteria and touch on how various pseudoscience candidates fit them.

First, pseudoscience lacks the cumulative progress seen in science. This does not imply that science consists only of linear and upward progress. That is clearly not so. Its progress is better conceptualized as a circular spiral upwards rather than a straight line. But there is clear progress. We know more about the way the physical, biological, and social worlds work now than we did 150 or even 25 years ago. The key point to make here is that in science the new builds upon the old and does not contradict it. The old gets folded into the new and is continuous with it. In other fields, including pseudoscience, the new is categorically different and is even cut off from the past. The past is deemed irrelevant or wrong. The shifts and changes in pseudoscience, however, are not signs of progress but rather shifts. They are lateral changes rather than progressive changes. In this way, alchemy, flat-earth theory, ESP, and UFOlogy are not fundamentally more advanced now than they were 30 or 40 years ago. They may be different but they have not progressed.

The second criterion for pseudoscience is a disregard for empirical and established facts or results and a contradiction of what is already known. Pseudoscientists sometimes propose bold new ideas and see themselves very much on the cutting edge—or in their eyes, "mavericks." For them, established science is simply not capable of seeing the truth for what it is. Moreover, pseudoscientists claim that what established science says is not relevant and can often be dismissed out of hand. Even established laws of physics and geology do not hold in pseudoscience. For example, in the 1950s Immanuel Velikovsky claimed that commonalities between world mythologies were the result of various natural disasters, the most fantastic of which was that a large piece of Jupiter was ejected and became a comet.[5] This comet made a close pass at Earth, causing floods, plagues, and other calamities. The point here is that most of Velikovsky's assumptions violate or contradict basic principles of mass, gravity, momentum, and force—in other words, Newtonian physics.

Third, there is a paucity of internal skepticism (that is, of its own assumptions) as well as an unwillingness to independently test ideas. Pseudoscientists, to be sure, are capable of criticism and exercise it quite liberally toward "established" science. But their own assumptions are self-evident and beyond doubt and empirical examination. Their claims and methods would not meet and are resistant to the peer review process so crucial to science. Put most succinctly by John D. Barrow: "Beyond refutation, it is always the last word."[6] In chapter 9, I discussed the scientific attitude and argued that its quintessential trait is open skepticism. Only when evidence using established and verifiable methods have replicated a finding is one then "required" to accept it, whether one likes the idea or not. What is now evidence soon comes under a new round of critical and empirical evaluation, and ultimately it too will become outdated and supplanted by new evidence and/or theory, which is the heart of progress. Yet, once again, in science the "new" often builds on and incorporates the "old" rather than directly contradict it. Einstein's theory of relativity, for example, incorporated and expanded upon Newtonian principles of gravity.

Fourth, only vague mechanisms are put forth for how pseudoscientists come to their conclusions. The scientific method consists of numerous components that are common to—although not completely identical to—the physical, biological, and social sciences. These methods consist of theory and observation, testable hypotheses, careful and controlled manipulation of variables (holding extraneous variables constant), random assignment of "participants" to experimental or control conditions, precise measurement (valid and reliable), data analysis (usually using inferential statistics), and public communication of the methods and results.[7] The crucial element to these techniques is their public nature. The methods used to test the hypotheses must be made transparent enough so that others can independently reproduce them. There are no secrets to conducting scientific research. Similarly, when reporting original research, scientists must make use of the peer review process. Peer review, to be sure, has its problems, but it does raise the bar of accountability and does generally weed out the bad from the good science (and no doubt sometimes the most creative science). Pseudoscientists, however, do not practice such open and independently verifiable methods. They often hide behind vague and general descriptions, if any at all, of how they collected their data and refute verifiability as being a hallmark of traditional narrow science. The methods are often evaluated only by people on "the inside" and not independently by multiple peer experts.

Fifth, an essential element added by the Greeks to the development of science was formal logic, both deductive and inductive. One's reasoning must be sound, clear, and without basic fallacies and contradictions. One must be able to draw conclusions that follow from the evidence and data and not make leaps that contradict basic principles that are well established. For instance, Derry describes the case of Ignatius Donnelly who published a book in 1882 claiming not only that the legendary continent of Atlantis existed but that it was the only viable explanation for commonalities between Egyptian and South American cultures. Even in its day, it was criticized for it gaps in logic and lack of evidence. To give the reader a sense of the mystical and spiritual connotations that

Atlantis still holds for some, here is a quote from a current Web site. It is clear from this quote that the belief in Atlantis is just that, more a matter of faith, spirituality, and longing for utopia than it is scientific fact: "What is it about Atlantis, an Atlantic continent said to have disappeared 11,600 years ago, that evokes such endless fascination in human minds and hearts? For many the allure is that of a riddle or romance or genuinely sensational detective story. For others, however, it goes deeper. . . . What follows, then, is less an effort to resolve the question of whether Atlantis existed and, if so, where it was and what it was like . . . as it is an attempt to portray, in words and images, the way in which this vanished land has deeply captured so many aspects of our human soul and psyche."[8]

Conspiracy theories are another example of loose and distorted pseudoscientific logic, with only the flimsiest ties to evidence. Ever since the terrorist attacks on the United States of September 11, 2001, there has been an explosion of conspiracy theories, the spread of which is greatly facilitated by the Internet. These are much too ubiquitous to summarize, but I will outline one or two to give a flavor of the kind of thinking and logic behind these theories. In 2004 the French anthropologist Bruno Latour published an article questioning whether, in the post-9/11 world, the postmodern retreat from reality has not gone too far. When people in his French village thought him naive to believe that terrorists and not the CIA were responsible for the attacks in New York on September 11, 2001, he wonders whether there is any difference between postmodern critique and conspiracy theories. Similarly, Latour asks: "What has critique become when a French general, no marshal of critique, Jean Baudrillard, claims in a published book that the Twin Towers destroyed themselves under their own weight, undermined by the utter nihilism of capitalism itself—as if the terrorists planes were pulled to suicide by the powerful attraction of this black hole of nothingness?"[9] For Latour, granddaddy of postmodern fantastic and nihilist claims, such constructivist conspiracy theorizing seems to have lost touch with the ground.

Closely related to these pseudosciences are domains that start with legitimate scientific techniques but then move into the realm of bad or "pathological" science. Some legitimate science starts off as fantastic and, indeed, many great discoveries have elements of being incredible. So whenever scientists make fantastic claims it behooves others in the field to replicate them. If the results do not hold up, then perhaps the original "discovery" was either chance or a result of sloppy science.

Examples of nonreplicatable science might be cold fusion (see chapter 9) or polywater. Polywater, like cold fusion, is an example of fantastic claims—pardon the pun—not holding water upon further scientific investigation.[10] In the mid-1960s some Soviet scientists claimed to have discovered a new form of water that was denser and boiled and froze at more extreme temperatures than normal water. Only after a few hundred scientific articles were published on this new "polywater" did the evidence against its veracity become clear. Microscopic analysis of the water revealed contaminations with various other substances. The new form of water turned out to be a result of poorly controlled experiments and problematic methods. In contrast to pseudoscience, however, at this point something remarkable happened: the originator of the idea admitted it did not exist, that it was a result of poor-quality research. Rather than stubbornly adhering to its legitimacy, the discoverers acknowledged the errors in their ways. In this sense, they in fact were being scientists and not pseudoscientists.

A wonderfully humorous spoof of our gullibility toward scientific-sounding explanations (pathological science) was written in the satiric newspaper *The Onion* in 1999.[11] Entitled "Revolutionary New Insoles Combine Five Forms of Pseudoscience," the piece opened with the following sentence: "Stressed and sore-footed Americans everywhere are clamoring for the exciting new MagnaSoles shoe inserts, which stimulate and soothe the wearer's feet using no fewer than five forms of pseudoscience." Almost as if he had read Derry, the writer elaborates on examples of these different elements of pseudoscience. For instance: "According to scientific-sounding literature trumpeting the new insoles, the Contour Points™ also take advantage of the semi-plausible medical technique known as reflexology. . . . 'Only MagnaSoles utilize the healing power of crystals to restimulate dead foot cells with vibrational biofeedback . . . a process similar to that by which medicine makes people better.'" Finally, using two common hallmarks of pseudoscience, anecdote and irrefutability, the article ends with quotes from two "customers," one of whom said this: "I twisted my ankle something awful a few months ago . . . but after wearing MagnaSoles for seven weeks, I've noticed a significant decrease in pain and can now walk comfortably. Just try to prove that MagnaSoles didn't heal me!"

Living in Northern California where many New Age fashions have sprouted, I must say that the satire is so funny precisely because it lampoons what does in fact happen. People are swayed by many bogus but scientific-sounding "research" claims that fall apart on more critical examination. Being poorly trained in either scientific or critical thinking, many people are easy prey to products and ideas that sound plausible and intriguing but in reality have little foundation. We need not all to be scientists, but we would be better off if we thought more critically about the flimsy nature of many commercial claims about what works and what does not and the kinds of evidence on which such claims are based. Critically evaluating evidence, whether in science, advertising, or politics, is a necessary skill that ideally each person who goes through our educational system would possess and use in everyday life. It would be nice to think that the psychology of science could contribute to the development and dissemination of sound scientific reasoning among a wider populace.

I put forth the view that pseudoscience, conspiracy theories, superstition, and gullibility—to name a few—are so common because they are part and parcel of what the brain does (and has evolved to do), namely, to recognize patterns, to think causally, and to intuitively and automatically see meaning behind events that are connected by time or space. In reality these meanings and connections exist only in our minds, not in the events themselves. Left unchecked by skepticism, such a mind-set can spin a tangled web of conspiracy theories on most any event at any time.

### Balancing Skepticism with Belief

One of the central tenets of science is balancing skepticism with belief, but the default attitude must be skepticism. The skepticism-and-belief dimension is played out any time a scientist is confronted with either a new theory or a new interpretation of evidence. Only after the evidence becomes clear do people switch over to believing a phenomenon. I coined the phrase "open skepticism" to capture this tension between skepticism and belief as a fundamental principle of the scientific attitude. Skepticism is primary, but based on evidence one must be open to all kinds of "beliefs" in science.

The fields of paleontology and archeology are perhaps the most prototypic domain for conflicts between skepticism and (scientific) belief because of the inherent ambiguity of almost all fossil evidence. Every major find in archeology is met with debate and sometimes open conflict between competing schools of interpretation. Does this find belong to that species or an entirely new species? What trait demarcates one species from another? Answers are intricately wedded to theoretical assumptions and points of view. A relatively minor example of this is seen in the debate over Neanderthal burial rituals. Some burial sites have been found littered with pollen, teeth, and bone artifacts, as well as remains placed in particular body positions (for example, arms crossed over the chest). Some archeologists propose these findings are evidence of ritualistic burials, whereas others are more skeptical and argue that these artifacts are random and not suggestive of any ritualistic behavior. One's viewpoint depends, at least in part, on one's prior convictions concerning whether Neanderthals are a subspecies of *H. sapiens* or not. The point here is not to debate the merits of each argument, but rather to make the case that conflict between skepticism (tendency to not believe and critical thought) and belief is ubiquitous in science, and dispositional skepticism plays a role in scientists' reactions to new theories and new empirical evidence.

Observation, we see once again, cannot be divorced from its theoretical framework. One scientist's belief is another's skepticism. One could argue that this back and forth between skepticism and belief is the very reason science advances. One constantly has to answer the sharp doubts and criticisms of one's peers. As the astronomer Carl Sagan put it, science requires "an exquisite balance between two conflicting needs: the most skeptical scrutiny of all hypotheses that are served up to us and at the same time a great openness to new ideas."[12]

## SCIENCE AND ANTISCIENCE

Science is a dominant force in our lives, and as such it presents many people with a very ambivalent love-hate relationship. It is also an inherently two-headed force; its potential for good is matched by its potential for destruction. The positive and constructive outcomes of science and technology are too numerous to list—medical and agricultural advances, instant worldwide exchange of information, and entertainment are just some of the more prominent examples. When advances of science and technology are used for destructive political purposes, however, as they have been most dramatically with the atomic and hydrogen bombs and biological warfare, one cannot help but acknowledge the potential for destruction that comes with knowledge. Science and technology have also led to some unwanted industrial consequences, such as chemical pollutants, massive consumption of energy, and destruction of some of the Earth's beauty and resources. Finally, advances in science and technology have opened up new ethical questions concerning genetically modified food, stem-cell research, artificially extending the life of persistently vegetative patients, "designer babies," and, of course, cloning. When and where do we as a culture decide to place limits on technological and scientific advances? Just because we can do certain kinds of research does not mean we should, and there are, in fact, serious restrictions placed on human and animal research for ethical reasons.

Many people see problems with "progress," and some even actively oppose it. A benign example of such an attitude can be seen in the following anecdote: At a recent social gathering of parents of young children, my wife informed a well-educated mother that results from a meta-analysis showed no evidence that sugar increases hyperactive behavior in children. Her retort was simply, "That's just the science. I don't believe the science." More extreme examples of an antiscience attitude exist, however. The pull of a "prescientific" and "pretechnology" utopia is strong for those who focus primarily on the negative consequences of science and technology. The case of the Unabomber (Theodore Kaczynski), who from the late 1970s to mid-1990s sent mail bombs to numerous scientists and technologists, injuring twenty-nine and killing three, is one extreme example. Kaczynski also wrote a manifesto that fundamentally opposed almost all scientific and technological advance, arguing that only by destroying technology (killing if need be) could society avoid a "collapse of civilization."[13] There are others on the extreme left wing of the environmental movement, such as the members of "Earth-First" who also advocate violent opposition to technological and economic developments that impinge on the environment. To be sure, science, technology, and the environment are not inherently opposing forces, for science and technology have been applied in recycling and conservation programs from the beginning. Knowledge in itself is neither constructive nor destructive but becomes so only when applied to the real world.

In addition to opposition to the applied outcomes of scientific and technological advance, there is also opposition to science on a more abstract and pure level, namely, as a means of gaining knowledge. This battle is being played out more within the halls of academia, especially between the "two cultures" of the humanities and the sciences. In this section I will focus on these "science wars," as some have recently come to call this conflict. The modern version of the conflict was first made explicit by C. P. Snow in his book *Two Cultures,* published in 1959, where he argued that there was a growing conflict between the cultures of the humanities and the sciences. The conflict was obvious to people in both cultures and Snow—being both a novelist and a scientist—was in as good a position as anyone to give voice to the conflict. Here is one recent example of the kind of the antiscience perspective, written by a humanist, that C. P. Snow discussed: "The things that really matter to us—the secrets of the heart, of what it means to be an individual, the depths and heights of human experience—all are accessible, if at all, only through literature and the creative arts.

Science has no purchase on them, and precious little to say about them beyond the posturings of reductionists."[14]

Scientists are not unaware of this conflict. In 1990 when I interviewed full professors of physics, chemistry, and biology at major research universities in California, one of my questions concerned C. P. Snow's analysis of two cultures and whether they saw an inherent conflict between them. Many scientists responded that to the extent there was conflict, it was not inherent but more practical, in that people do not have enough time to be knowledgeable in domains outside their area. Furthermore, science is very technical, and therefore it is much harder for artists or writers to read science than it is for scientists to read literature or enjoy or play music. Indeed, a common theme in the scientists' responses was something like "we don't dislike them, but they do dislike us." One relatively consistent explanation the scientists gave for the humanists' hostility to the sciences was that the science departments on university campuses bring in much more money, train more students, and garner more resources and attention from deans. Scholars in the arts and humanities are resentful of these resources and attention. As one physicist, who also was a near-world-class violinist, put it: "Many scientists I know are quite conversant with one or the other aspect of the humanities. Many scientists are interested in music and art. Many of them are good at foreign languages and know a lot about the world. I would say the reverse is not true. I would say that many people in the humanities are absolutely illiterate about science. And proud of it." A chemist, however, went on to argue that part of the problem that humanists have in understanding science rests partly on the shoulders of scientists themselves: "I think the sciences have made themselves in part unintelligible. So they have taken a certain amount of humanity out of what they do. That is probably our own fault. And we need to spend some time repairing that. But scientists still have the opportunity to enjoy the fruits of the arts. And I think that many scientists take full advantage of that. . . . I am not so sure that academic scientists talk to academic humanists, but they certainly can enjoy music, painting, sculpture, and I think some of them do. . . . I am pretty well versed in modern painting and modern sculpture and I enjoy that. And I listen to a fair amount of music, and I think that is important for my life."[15]

As reasoned and balanced as these views of the conflict may have been, the cultural conflict nearly became an all-out intellectual war in the late 1990s with the publication of *Higher Superstitions* by Paul Gross and Norman Levitt. The "war" was between the same two cultures that Snow had written about, but now it was couched more hostilely in terms of science and antiscience or the science wars. To quote Gross and Levitt:

> Our subject is the peculiarly troubled relationship between the natural sciences and a large and influential segment of the American academic community which, for convenience but with great misgiving, we call here 'the academic left.' . . . To put it bluntly, the academic left dislikes science. . . . Most surprisingly, there is open hostility toward the *actual content* of scientific knowledge and toward the assumption, which one might have supposed universal among educated people, that scientific knowledge is reasonably reliable and rests on a sound methodology.
>
> It is this last kind of hostility that scientists who are aware of it find most enigmatic. There is something medieval about it, in spite of the hypermodern language in which it is nowadays couched. It seems to mock the idea that, on the whole, a civilization is capable of progressing from ignorance to insight. . . . We have the sense, encountering such attitudes, that irrationality is courted and proclaimed with pride. All the more shocking is the fact that the challenge comes from a quarter that views itself as fearlessly progressive.[16]

More specifically, Gross and Levitt argued that four intellectual-political movements have been most unreasonable in their attacks on the basic principles of science, namely, cultural/social constructivism, postmodernism, elements of feminist theory, and radical environmentalism. For sake of space, I will focus on only two of these domains—social constructivism and postmodernism—here and go on to propose how an advanced and informed psychology of science can mediate these conflicts. As Gross and Levitt explicitly acknowledge in their opening chapter: "Our approach in these sections is conspective and polemical. Nothing else will get attention."[17] In other words, they attempted to start a fuss and they did. Indeed, at times their language and summaries of the positions they are critiquing are a bit heavy-handed and polemical; one is tempted to say "rhetorical."

### Cultural/Social Constructivism

The heart of the cultural constructivist viewpoint is that all knowledge systems, including science, are (mere) social constructions replete with political, social, and cultural overtones and therefore no one knowledge system deserves to be deemed inherently "more reliable" or "more valid" than any other. In other words, social constructivism is a variation on "cultural relativism." Taken literally, the view holds that there is no real distinction, for instance, to be made between science and pseudoscience. Gross and Levitt summarize the position in the following terms: "Scientific questions are decided and scientific controversies resolved in accord with the ideology that controls the society wherein the science is done. Social and political interests dictate scientific 'answers.' Thus, science is not a body of knowledge; it is, rather a parable, an allegory, that inscribes a set of social norms and encodes, however subtly, a mythic structure justifying the dominance of one class, one race, one gender over another." They go on to give an example of a Marxist theorist describing Einstein's theories in terms of the "collapse of the bourgeois ego" and "commodity relations." Gross and Levitt dismiss this viewpoint out of hand: "Such propositions have all the explanatory power of the Tooth-Fairy Hypothesis. Still, hundreds of left-wing theorists dote on them."[18]

Bruno Latour's work is perhaps the prototypic and best-known example of the relativist and constructivist position.[19] Latour is as much an anthropologist of science as a sociologist, and, indeed, one of his main arguments is that to understand science one must study it the way anthropologists study any tribe. In his first book with Steve Woolgar, *Laboratory Life,* he provides the subtitle: *The Construction of Scientific Facts.* It is of note and of importance that in his second book, *Science in Action,* he used the phrase "fabrication of scientific facts" quite intentionally, with its implication of storytelling and myth-making. Truth and facts, in the end, are inherently and maybe even completely the end result of social negotiations involving power, money, class, and race. The position ultimately dissolves into a deconstructionist argument: "It is a reminder that the value and status of any text (construction, fact, claim, story, this account) depend on more than its supposedly 'inherent' qualities. As we suggested earlier, the degree of accuracy (or fiction) of an account depends on what is subsequently made of the story, not on the story itself." To demonstrate the near absurdity of their position, Latour and Woolgar apply it to their own work in a postscript to their book: "The concluding chapter of *Laboratory Life* addresses the status of our own account, the question of whether or not we are (merely) supplying a new fiction (about science) with an old. In the closing section of the original draft we declared our analysis was 'ultimately unconvincing.' We asked readers of the text not to take it seriously. But our original publishers insisted that we remove the sentence because, they said, they were not in the habit of publishing anything that 'proclaimed its own worthlessness.'"[20] Indeed, it is hard to take a position seriously that ends up arguing that it is worthless. The best one can say is "at least they are honest!"

Thomas Kuhn, clearly a leading influence on the early constructivist movement, became quite disillusioned with what the position became by the 1970s and 1980s: "What passes for scientific knowledge becomes then, simply the belief of the winners. I am among those who have found the claims of the strong [relativist-constructivist] program absurd: an example of deconstruction gone mad." To be sure, other leading sociologists of science, such as Stephen Cole, also became quite critical of the extreme conclusions drawn by the relativist-constructivist position.[21]

### Postmodernism

Indeed, meaninglessness and lack of solid reality bring us to the heart of a related movement, postmodernism. With the French philosophers Foucault and Derrida as its founding fathers, postmodernism raises nihilist desconstruction of texts to one of its central tenets. In this sense, social constructivism and postmodernism have something in common: the text itself is meaningless and without value (nihilism), and all its power resides in its deconstruction or interpretation. The author becomes meaningless and the reader all-powerful. The power of deconstructionists to destroy meaning does not stop with texts, however, for if this principle applies to all written texts, it applies to political, scientific, economic, and spiritual texts. In other words, the most basic principles of science, as well as rationalism, Enlightenment, democracy, and religion are undercut and deemed meaningless in themselves. To quote the postmodern philosopher Steven Best, "Postmodernism stresses the relativity, instability, and indeterminacy of meaning; it abandons all attempts to grasp totalities or construct Grand Theory."[22]

Postmodern scholars hold that notions of progress, science, objectivity, and validity are nothing more than a house of cards built by those in power as attempts to hold on to their power. Ironically, a number of such theorists have latched on to and co-opted such scientific and mathematical concepts as relativity, chaos, and nonlinearity. As Gross and Levitt point out ruthlessly, postmodern writing is replete with superficial understanding of science at best and misunderstanding at worst, which makes their claims to destroy its meaning all the more reprehensible. What does appear to be a major shortcoming of postmodern thought, to be less harsh than Gross and Levitt, and a bit ironic, is the lack of training and understanding of scientific principles and the touting of such ignorance with pride. As leading postmodern writer Andrew Ross, an English professor at Princeton, has written in his influential book *Strange Weather: Culture, Science and Technology in the Age of Limits:* "This book is dedicated to all of the science teachers I never had. It could only have been written without them."[23] It would be one thing if the topic of the book were English literature, but it is science and technology. It is with irony that such writers claim to provide "deep readings" of scientific texts while bragging about their scientific ignorance. In so doing they strip scientific knowledge of its meaning and its "privileged status" while relegating it to something like the "dominant paradigm of the hegemonic powers of the elite and privileged class."

To demonstrate the absurdity of the postmodern position, one need go no further than the now rather famous "Sokal's hoax." Alan Sokal, a physicist at New York University and a die-hard leftist, was inspired by reading Gross's and Levitt's book. He was inspired initially because he suspected Gross and Levitt of merely trying to advance their own rather conservative political agenda. Sokal also believed that Gross and Levitt had to be exaggerating their portrayal of postmodernism as a movement in which the quackiest and most unsubstantiated claims get passed off for profundity. In 1994 Sokal decided to test the ability of deconstructionist thinkers to detect nonsense under the guise of postmodernism by submitting an over-the-top parody of the postmodern perspective ("Transgressing the boundaries: Towards a transformative hermeneutics of quantum gravity") to the journal *Social Text.* No one but Sokal at the time knew it was a hoax, but to him its flaws were rather transparent and he was not convinced it would be accepted. It was in fact accepted for publication.

I only give the briefest of summaries of the original article, for it and its aftermath have become a cottage industry unto themselves, with accusations and counteraccusations flying back and forth mostly within and between the United States and France. It begins by making a general postmodern point about natural scientists still clinging to the "dogma imposed by the long post-Enlightenment hegemony over the Western intellectual outlook, which can be summarized as follows: that there exists an external world, whose properties are independent of any human being . . . and that human beings can obtain reliable albeit imperfect and tentative knowledge of these laws by hewing to the 'objective' procedures and epistemological strictures prescribed by the (so-called) scientific method."[24]

Sokal goes on to argue that movements in the philosophy and history of science, as well as poststructuralist and feminist theory, have undermined this "Cartesian-Newtonian metaphysics" in that they have demonstrated that "reality is at bottom a social and linguistic construct." He next moves to an overview of his purported purpose of the article: to take these "deep analyses one step further" and to apply them to "quantum gravity . . . [where] space-time manifold ceases to exist as an objective physical reality." The article was published in May

1996, and the very same day Sokal's admission of the hoax was published in *Lingua Franca.* With these two articles the science wars erupted in full force inside and outside the halls of the academy.[25]

Before summarizing some of the wartime salvos that were launched by both sides, let me first describe Sokal's reasoning for why he did what he did. The primary reason, in his words, was to "test the prevailing intellectual standards. . . . The fundamental silliness of my article lies, however, not in its numerous solecisms, but in the dubiousness of its central thesis and of the reasoning adduced to support it. Basically, I claim that quantum gravity—the still speculative theory of space and time on scales of a millionth of a billionth of a billionth of a billionth centimenter—has profound *political* implications (which, of course, are 'progressive')."[26] He wanted to see whether the standards of evaluating postmodern theory would be high enough to detect such intentional quackery and questionable reasoning. They were not.

The defense by some postmodern theorists that scholarly work rests on the ethical assumption that submitted articles must be assumed to be based on valid and honest methods does not hold here. For Sokal did not make up data (in fact, all his quotes and references were real); he merely made up interpretations that pushed the envelop of credulity and were based on intentionally flimsy reasoning. It is the job of editors and reviewers to pick up on illogical and faulty interpretation and reasoning. They did not. Should this fact lead to an indictment of all of cultural studies, especially social studies of science? Of course not. There is some perfectly solid and important work being done in science studies (I would like to think that this book is one of them), a fact that Sokal, Gross, and Levitt themselves acknowledge.

Nevertheless, some scholars normally aligned with postmodern perspective have acknowledged how the Sokal affair has undermined the legitimacy of some aspects of cultural studies. For instance, the feminist scholar of science Evelyn Fox Keller responded:

> For many scientists, this episode will only bolster their fear that postmodernism (and science studies more generally) threatens the integrity and well-being of their own disciplines. But it is not science that is threatened by the hapless publication of gibberish; it is science studies itself. And the embarrassing defense offered by Ross and Robbins (not to mention the many counter-attacks) just makes the problem worse. Scholars in science studies who have turned to postmodernism have done so out of a real need: Truth and objectivity turn out to be vastly more problematic concepts than we used to think, and neither can be measured simply by the weight of scientific authority, nor even by demonstrations of efficacy. Yet surely, the ability to distinguish

> argument from parody is a prerequisite to any attempt at understanding the complexities of truth claims, in science or elsewhere. How can we claim credibility for responsible scholarship—for the carefully reasoned and empirically founded research that makes up the bulk of science studies—if we do not recognize a problem here?[27]

A central theme of the science wars that erupted post-Sokal was whether any postmodern cultural studies scholar would actually assert the view that reality does not exist, as Sokal claims, or merely that theories of reality do not exist. Although many postmodern apologists defended themselves by claiming that no one makes the former more radical claim, the practical outcome of their theorizing has been that one can argue anything no matter how little evidence there is for the position because evidence and reality are arbitrary and meaningless. Recall the Twin Towers argument of Baudrillard.

How did we get to this extreme position? The issue of the reality of an external world and our experience of it is rooted in the annals of philosophy and most systematically dealt with by the German philosopher Immanuel Kant when he argued for an inherent gap between the "nomenal" and "phenomenal" worlds. The former is the "world in itself" and the latter is "the world as we know it." It does not require deep Kantian insight to realize these two must be separate and cannot be the same. The very function of our sensory system and brain, after all, is to process information from the outside world and give it order and meaning. That is, we will never know the world objectively and as it exists but rather only through our conceptions, theories, expectations, and beliefs. Any scientist, save the most hard-core and logical empiricists, would acknowledge this.

Whatever the gap between reality and our understanding of it, there is no justification for the solipsistic belief that reality does not exist or that all theories and representations, no matter how absurd, are equally valid (relativism). That is precisely the leap, however, that social constructivists and postmodern theorists seem to have taken. The Kantian distinction should also not repudiate the fact that science is perhaps the best method we have at narrowing the gap between the nomenal (objective) and phenomenal (subjective). Perfect and absolute scientific knowledge is not and never will be. But it is the best we have. Recall the epigraph to this chapter: "One thing I have learned in a long life: that all our science, measured against reality is primitive and childlike—and yet it is the most precious thing we have." Recall, too, from chapter 1 the discussion of realism in current philosophies of science, namely, the idea that science can come to *approximately* realistic understanding of nature itself. What scientists for the most part therefore reject is the jump that because we cannot have a direct and completely objective understanding of the natural world, either it does not exist or our epistemologies and theories about it are all relative and of equal validity.

The proof of the approximately accurate nature of scientific models of nature is seen in the fact that they work. Just to take a few of the more obvious examples: we regularly take people in machines that weigh tons (airplanes) and fly them from one part of the world to another; we have medicines that treat disease; and we have landed humans on the moon (although some conspiracy theorists continue to deny the reality of this). The point is that if these scientific models had no connection to the real world, they would not work. Our understanding of reality may be imperfect, but it is not totally arbitrary and relative. As Carl Sagan wrote about the pull of nihilism in *The Demon-Haunted World:* "If we were not aware of our limitations, though, if we were not seeking further data, if we were unwilling to perform controlled experiments, if we did not respect the evidence, we would have very little leverage in our quest for truth. Through opportunism and timidity we might then be buffeted by every ideological breeze, with nothing of lasting value to hang on to."[28]

Psychologists of science, to be sure, could have a field day analyzing the psychological forces behind such conflict between scientists and social critics—motivation, personality, open and closed cognitive processes, cognitive consistency, concept formation, rational and nonrational thought, persuasion, ingroup/out-group prejudice, and so on. Developmental psychologists, for instance, such as Deanna Kuhn and Paul Klaczynski, have found that people are motivated to find fault with evidence that is inconsistent with their beliefs and in the end their original theory is defended against and safe against evidence to the contrary. The Sokal affair is a wonderful real-world example of precisely such a psychological process. Apologists of postmodernism, for the most part, are unmoved by the hoax and continue to defend their position and decisions. Critics of postmodernism are convinced that Sokal's hoax confirms their belief that the quality of scholarship therein is a joke at best and inane and completely devoid of content or substance at worst.

Systematic science and modern scientific reasoning are clearly neither easy nor natural states of mind. Recall, however, that a major thesis of chapter 9 was that scientific thinking is rooted in and stems from ancient cognitive capacities of early humans. Similarly, Robin Dunbar wrote in *The Trouble with Science* "these [scientific] methods are, at root, simply the natural mechanisms of everyday survival." As we specifically saw with math, it is precisely because we have evolved intuitive forms of knowledge about number (numerosity) that conceptions that go beyond the intuitive and implicit are so difficult to understand. They contradict common sense and are counterintuitive. Formal math and science training in this sense can be difficult to learn initially because it requires a certain degree of "unlearning" common sense and the intuitive. The anthropologist Scott Atran has demonstrated this phenomenon very cogently in his work on folk taxonomies and their development across world cultures. Folk taxonomies have remarkable similarities with modern systematics and scientific taxonomies, but there also are some fundamental divergences because some scientific taxonomies are not at all intuitive and obvious—sponges being classified as animals is a good example. Science begins with the folk, implicit, and common sense, but as it develops it becomes more and more divorced from it.[29]

The gap between those who are trained and practice science and those who are not and do not grows and widens. In no domain is this more evident that in theoretical physics, with its current "quanta," "string theory," "quarks," "multiple universes," and "black holes." None of these has been directly observed and each is at this point quite theoretical. As science moves more and more away from the observable and commonsense (Newtonian) world, the potential for animosity between scientists and nonscientists grows. Indeed, one could argue that it is precisely such gaps that have led to the science wars. As hinted at by a scientist I interviewed, however, part of the difficulty that nonscientists have with science is scientists' own doing. Some scientists look with disdain upon any attempts to make their research "palpable" and "popular" for the masses. They eschew journalists, who they feel will do nothing but butcher, misinterpret, and bastardize the meaning of their work. Popular science is beneath them and of no concern for them. But by moving further and further away from common sense and the intuitive, modern science has begun to alienate itself from large segments of the population.

Of course, the view that pop science is qualitatively different from and incapable of capturing the essence of "real" science is based on a false premise. Real science, even difficult, complex theoretical science, can be translated meaningfully into everyday language. Even Einstein was able to come up with some relatively simple nontechnical descriptions of his theory of relativity. Fortunately, there have always been a few talented scientists who could make their ideas and findings as well as those of other scientists understood by those outside the scientific community. Three prime recent examples come immediately to mind: Carl Sagan, Stephen J. Gould, and Isaac Asimov (granted Asmiov was not a scientist, but close enough).[30]

The arrogance exhibited by those scientists who do not see the value in translating scientific findings for mass consumption does not win them friends among the nonscientists of the world and no doubt feeds the flames of antiscience sentiment. It is little wonder that some people who lack the training would become hostile toward this arrogant and aloof enterprise. What is at first a little more surprising, though not so on further reflection, is the animosity toward science and technology among fellow academics and progressive thinkers, such as the social constructivists, postmodernists, some femininists, and some radical environmentalists.

In this book I have attempted to do two things: first, to present a cogent argument for why the time is ripe for the psychology of science to mature alongside the other major studies of science; and second, to demonstrate that an understanding of how science developed in our species can enlighten our understanding of how science develops in individuals (and vice versa). The topics outlined in this book are topics the psychology of science has begun to examine and can and will continue to do so with increasing theoretical and empirical sophistication. It is time for psychologists who have these interests and who are conducting research on any form of scientific thought or behavior to organize and codify their interest by explicitly identifying with the field and developing the necessary institutional infrastructure to support such identification. It is time for the psychology of science to leave the stage of isolated scholars working on questions and for all of the latent and hidden psychologists of science to come out of the closet and publicly declare their identification with the field. Identification is the necessary step toward Institutionalization. This book will have served its purpose if it plays a role in moving the forces that be in the direction of solidifying, codifying, and establishing the foundations for the psychology of science.

# Notes

## CHAPTER 1. PSYCHOLOGY OF SCIENCE AND THE STUDIES OF SCIENCE

1. Staats 1991.
2. Mullins 1973; see Matarrazo 1987 for similar criteria for the establishment of a psychological discipline.
3. Whewell 1840, 1856.
4. Jevons 1874, Mach 1883, Pearson 1892.
5. Boyd 1991a.
6. Popper 1974.
7. Popper 1974, p. 59, emphasis in original.
8. T. Kuhn 1962.
9. Boyd 1991a; quote: Boyd 1991b, p. 196; cf. Schlick 1932–33/1991.
10. See Fuller 1988, Heyes 1989, Hull 1988, 2001, Kantorovich 1993.
11. See Feist 1995, Simonton 1995, Sulloway in preparation; quote: Hull 2001, p. 7.
12. Sarton 1952b.
13. Ibid.
14. See http://www.bshs.org.uk/links/societies.html (accessed March 13, 2003) for a listing of different societies in the history of science; see http://www.bshs.org.uk/links/journals.html (accessed March 13, 2003) for the journals. Moreover, T. Kuhn (1977, chap. 6) makes a cogent case for the history of sci-

ence being a fringe discipline in mainstream history textbooks, seldom given much attention at all and if so, only science up until 1750.

15. Hull 2001, Simonton 1995, Sulloway in preparation.
16. Sulloway 1996, p. 539.
17. Barber 1952, Merton 1945, quote from Merton 1945, p. 212.
18. See Dr. Patrick Hamlett's site at North Carolina State University's Web page http://www.ncsu.edu/chass/mds/stslinks.html (accessed June 10, 2004). See also http://echo.gmu.edu/center (accessed June 10, 2004) for a site that has links to many history of science Web sites.
19. As quoted in Merton 1973, p. 214.
20. "This book examines . . .": Cole and Cole 1973, p. ix; "that the single most important . . .": Cole and Cole 1973, p. 243.
21. Zuckerman 1996 (2nd ed.), p. xiii.
22. I will mention in passing that there is also some movement in the "anthropology of science," but because it is not as well developed as the psychology of science and has already been discussed on the natural history domain of knowledge in chapter 1, I forgo reviewing its history and key assumptions and contributions.
23. 1930s: Stevens 1936, 1939; 1950s: Roe 1952a, 1952b, 1953, R. Cattell and Drevdahl 1955; 1960s: Chambers 1964, Eiduson 1962, Gough and Woodworth 1960, Maslow 1966, Simon 1966, Taylor and Barron 1963.
24. Singer 1971, p. 1010, Fisch 1977, p. 298, Mahoney 1979.
25. Gholson, Shadish, Neimeyer, and Houts 1989.
26. Grover 1981, Shadish, Fuller, and Gorman 1994, p. 3, Shadish, Houts, Gholson, and Neimeyer 1989, p. 1.
27. Houts 1989, Shadish, Houts, Gholson, and Neimeyer 1989.
28. Campbell 1960.
29. Sulloway in preparation.
30. Historians: Holton 1973, A. Miller 1996; psychological and historical perspectives: Elms 1994, Gruber 1981, Runyan 1982, Simonton 2002, Sulloway 1996.
31. Popper 1959, p. 31.
32. T. Kuhn 1970, p. 21.
33. Popper 1970, pp. 51–58, emphasis in original.
34. Popper 1974, p. 33.
35. Fuller 1988, Hull 1988, Kantorovich 1993. Likewise, some psychologists have turned toward naturalistic epistemology, e.g., Heyes 1989.
36. Hull 2001, Sulloway 1996, in preparation.
37. Cole and Cole 1973, p. 12.
38. See Collins 1981, Knorr-Cetina 1981, Latour and Woolgar 1986.
39. Kluckholm and Murray 1953.
40. cf. Eysenck 1995, Feist 1993, Feist and Gorman 1998, Helmreich et al. 1980.
41. For reviews of the literature on genetics and talent, see Eysenck 1995, Karlson 1991, Katzko and Monks 1995, Reznikoff, Domino, Bridges, and Honeyman 1973, Saklofske and Zeidner 1995, Vernon 1989.

### CHAPTER 2. BIOLOGICAL PSYCHOLOGY OF SCIENCE

1. Lake 1998, Mithen 1996.
2. The overview on the principles of genetics, with an emphasis on behavorial genetics, is based on the work of Clark and Grunstein 2000, Hamer and Copeland 1998, and Plomin and Caspi 1999.
3. See Lederberg 2001.
4. Hamer and Copeland 1998.
5. See Gray and Thompson 2004 for a review of the neurobiology of intelligence.
6. Definition of intelligence: Neisser et al. 1996, p. 77; survey of experts: Snyderman and Rothman 1987.
7. Original study: Chorney et al. 1998, see also Plomin et al. 1994; failure to replicate: Hill, Chorney, Lubinski, Thompson, and Plomin 2002.
8. Gray and Thompson 2004, Thompson et al. 2001.
9. More technically, assuming simple additive genetic effects, heritability ($h^2$) can be quantified as double the difference between identical (monozygotic) and fraternal (dizygotic) twin correlations on a given trait ($h^2 = 2(r_{mz} - r_{dz})$); Bouchard 1998, Bouchard and McGue 1981, Bouchard, Lykken, Tellegen, and McGue 1996, Bouchard, Lykken, McGue, Segal, and Tellegan 1990, Devlin, Daniels, and Roeder 1997, Husén 1960, Loehlin and Nichols 1976, Scarr and Saltzman 1982, Vandenberg 1988.
10. Emergenesis: Lykken, McGue, Tellegen, and Bouchard 1992, p. 1565; emergenic nature of genius and creativity: Lykken et al. 1992, Waller, Bouchard, Lykken, Tellegen, and Blacker 1993; multiplicative models of talent: Simonton 2000, Walberg, Strykowski, Rovai, and Hung 1984.
11. Deacon 1997, Eriksson et al. 1998, Gage 2003, Kempermann and Gage 1999.
12. Deacon 1997, O'Leary and Stanfield 1989.
13. Darwinian nature of neural growth: Deacon 1997, Edelman 1987; neural pruning and learning: Deacon 1997, Gazzaniga and Heatherton 2003, Hebb 1949, Kandel and Hawkins 1995.
14. Bennett, Rosenzweig, and Diamond 1969, Deacon 1997, Globus, Rosenzweig, Bennett, and Diamond 1973, Parker and McKinney 1999, Rosenzweig, Bennett, and Diamond 1972, Rosenzweig, Krech, Bennett, and Diamond 1962, Skoyles 1999, Stiles 2000; neural growth after infancy: Kolb and Gibb 1999, Stiles 2000.
15. For example, Greenough and Chang 1989 found a 20–25 percent increase in number of dendrites in rats raised both socially and in enriched environments compared with those raised in either alone and in impoverished environments or together and in impoverished environments.
16. The longer the brain has to form: Deacon 1997, Finlay and Darlington 1995, Gibson 1991, Parker and McKinney 1999; brain size at birth: Portmann 1990, see also Parker and McKinney 1999, p. 332.
17. Clinical evidence for face recognition: Bower 2001, Sergent, Ohta, and McDonald 1992. Brain imaging evidence: Halgren, Raij, Marinkovic, Jousmaki, and Hari 2000, Kanwisher 2000, Kanwisher, McDermott, and Chun 1997, Klopp, Marinkovic, Chauvel,

Nenov, and Halgren 2000, Nakamura et al. 2000, O'Craven and Kanwisher 2000, Tarr and Gauthier 2000, Zeki 1999. There is some debate as to whether this region is dedicated solely to facial perception and should be named the "fusiform face area" (FFA) (Halgren et al. 2000, Kanwisher 2000, Kanwisher et al. 1997, cf. Bower 2001) or is more generalized and not intrinsically specific to faces (Tarr and Cheng 2003, Tarr and Gauthier 2000). Supporting the former position is the finding that perceptual processes involved in the face are distinct from processes involved in inanimate objects (Cooper and Wojan 2000). For a detailed brain map of each known brain region and papers supporting the function of each region, see http://hendrix.imm.dtu.dk/services/jerne/ninf/voi/index-alphabetic.html (accessed October 26, 2003) and P. Fox and Lancaster 1994; prosopagnosia: McNeil and Warrington 1993; "Exposure to faces during a sensitivity period . . .": Nelson 2001, p. 3; cf. Deacon 1997.

18. Adolphs, Tranel, and Damasio 2001, Baron-Cohen, Wheelwright, and Jolliffe 1997, Dolan 2000.
19. Chow and Cummings 1999, Gallagher and Frith 2003, B. Miller, Hou, Goldberg, and Mena 1999, Perry, Swerdlow, McDowell, and Braff 1999, Pincus 1999, 2001, Vogeley et al. 2001.
20. The general layout of the visual cortex: Carter 1998, Zeki 1999; spatial and navigational ability: Aguirre, Detre, Alsop, and D'Esposito 1996, O'Keefe and Nadel 1978, Maguire, Burgess, Jeffery, and O'Keefe 1999, Sergent et al. 1992; parahippocampal cortex or parahippocampal place area (PPA): R. Epstein and Kanwisher 1998, R. Epstein, DeYoe, Press, Rosen, and Kanwisher 2001.
21. McCarthy and Warrington 1988.
22. The general conclusion concerning number-math processing: Butterworth 1999, Cipolotti, Butterworth, and Warrington 1994, Dehaene, Spelke, Pinel, Stanescu, and Tsivkin 1999, Menon, Rivera, White, Eliez, Glover, and Reiss 2000; for instance, one patient: Caramazza and McCloskey 1987; neuroimaging research: Haier and Benbow 1995.
23. "Best connected of all cortical structures . . .": Fuster 2002, p. 377; frontal lobes as seat of "higher reaches": Banyas 1999, Chow and Cummings 1999, Deacon 1997, R. Dunbar 1993, Fuster 1999, 2002, Goel, Grafman, Sadato, and Hallet 1995, Jerison 1973, Krasnegor et al. 1997, B. Miller and Cummings 1999, Mithen 1996, Rumbaugh 1997, Semendeferi 1999, Stone, Baron-Cohen, and Knight 1998, Stuss, Gallup, and Alexander 2001.
24. Quote: Prum and Bush 2003, p. 86; see also Banyas 1999, Deacon 1997, Fuster 1999, 2002, Klein 1999. Many believe that the prefrontal cortex (PFC) is larger in humans than expected (that is, relatively larger), with Brodmann (1909, as cited in Fuster 2002), for instance, arguing that PFC constitutes 3.5 percent of the entire cortex in the cat, 11.5 percent in the macaque, 12.5 percent in the dog, 17 percent in the chimpanzee, and 29 percent in humans. Jerison (1997), however, has demonstrated that the enlargement of the PFC is more absolute than relative. The relative size of the PFC is what would be expected compared to other primates because biological scales are inherently nonlinear. They consist of doubling at each cycle of cell growth (1 becomes 2, 2 becomes 4, 4 becomes 8, etc.), and therefore the scales are logarithmic. When the PFC and overall brain

size are both placed on logarithmic scales, their ratios across many species of mammals, including primates and humans, are almost precisely what is predicted (see Jerison 1997, Uylings and Van Eden 1990).

25. Theory of mind: see Gallagher and Frith 2003 for a review of this literature; OFC and social knowledge: Chow and Cummings 1999, Stone et al. 1998, Stuss, Picton, and Alexander 2001; false belief in four-year-olds but not three-year-olds: Adolphs et al. 2001, Gopnik et al. 1999, Fuster 1999, Goel et al. 1995, Stone et al. 1998, Stuss, Gallup, and Alexander 2001.

26. Attentional disorders, schizophrenia, and sociopathic disorders: Perry et al. 1999, Pincus 2001, Raine, Buchsbaum, and LaCasse 1997, Raine, Lencz, Bihrle, LaCasse, and Colletti 2000, Raine et al. 1998, Richardson 2000; criminality and violence: Pincus 1999, 2001; volumetric assessments of gray matter: Raine et al. 2000, Raine et al. 1998.

27. In the 1950s the major theory was the psychoanalytic theory of Ernst Kris (1952), who argued for "regression in the service of the ego"; that is, highly creative people are able to revert to more primitive cognitive states but then do productive and creative things with these regressive experiences. J. P. Guilford (1959) developed a theory of creativity that argued for creativity to consist of fluency, flexibility, usefulness, and originality of ideas. In the 1960s, Mednick's (1962) theory underscores the associational richness of creative thinkers, who simply are able to have more remote associations to most ideas or objects. Frank Barron (1963) proposed that highly creative people have greater cognitive and behavioral latitude than less creative people; that is, "Thus the creative genius may be at once naïve and knowledgeable, being at home equally to primitive symbolism and to rigorous logic. He is both more primitive and more cultured, more destructive and more constructive, occasionally crazier and yet adamantly saner, than the average person" (p. 224). In the 1980s, Dean Simonton (1988a) began arguing that highly creative people simply produce more ideas than less creative people, some of which are likely to hit upon adaptive solutions to important problems, that is, be creative. The process of which ideas become the successful and adaptive ones is very much Darwinian and due to "chance configurations." And most recently in the 1990s, British psychologist Hans Eysenck (1995) argued that a major dimension of personality "psychoticism" (not psychosis) consists of disinhibited and creative thought, as well as aloofness, hostility, and eccentricity. Each of these theories has fluency of ideas, unusualness of associations, and disinhibited ideation as core components of creative thought, all of which we now know are primarily frontal lobe activities (see Feist 2004c).

28. Decreased latent inhibition: Carson, Peterson, and Higgins 2003, Lubow and Gewirtz 1995; consistent with frontal lobe functioning: Rieger, Gauggel, and Burmeister 2003, Schmajuk 2002; frontal activity and creativity: Chavez 2004, Chow and Cummings 1999, Mell, Howard, and Miller 2003; scientific reasoning: Kwon and Lawson 2000.

29. Beeman 1993, Beeman, Bowden, and Gernsbacher 2000, Bowden and Beeman 1998, Bradshaw 1989, Fiore and Schooler 1998, Galin 1974, Martindale, Hines, Mitchell, and Covello 1984, Weinsten and Graves 2001.

30. For dual processing models, see, for instance, S. Epstein 1994, Feist 1991b, Freud 1900/1953, Klaczynski and Robinson 2000, Stanovich 1999; parietal lobe and mathematical

functioning: Katz 1986; Pasini and Tessari 2001; right hemisphere and mathematical functioning: Benbow 1988, Langdon and Warrington 1997.

31. Feist 1991b, A. Miller 1996.

**CHAPTER 3. DEVELOPMENTAL PSYCHOLOGY OF SCIENCE**

1. Facial expressions: Gopnik et al. 1999, Montague and Walker-Andrews 2001; distinguish faces and things: Easterbrook, Kisilevsky, Muir, and Laplante 1999, Fantz 1963, Kagan, Hanker, Hen-Tov, and Lewis 1966, Valenza, Simion, Assia, and Umilta 1996; face as imprinted: M. Johnson and Morton 1991.
2. Physiologically disturbed: Kagan et al. 1966; newborn preference for attractive faces: Langlois, Ritter, Roggman, and Vaughn 1991, Rubenstein, Kalakanis, and Langlois 1999; consistent standards of attractiveness: Cunningham 1986, Langlois, Roggman, and Musselman 1994, Langlois et al. 2000.
3. Gopnik and Wellman 1994, Gopnik et al. 1999, Leslie 1987, 1994, Perner 1991.
4. Perner, Leekam, and Wimmer 1987.
5. Baron-Cohen 1989, 1991, 2001, Frith 1991, Karmiloff-Smith 1992, Rutter 1978. Autisim defined: American Psychiatric Association 1994.
6. See Rosengren, Johnson, and Harris (2000) for the view that magical and metaphysical thinking are not mutually exclusive with scientific thinking and development is not a simple matter of the former declining and the latter increasing. This point is made especially in the chapter by Nemeroff and Rozin 2000.
7. Piaget and Inhelder 1967.
8. Center-out and head-to-toe: Gesell and Thompson 1938, McGraw 1943; eye-preference technique and physical knowledge: Carey and Spelke 1994, Spelke 1990.
9. Baillargéon 1987, Baillargéon, Spelke, and Wasserman 1985, Hespos and Baillargéon 2001.
10. Possession of innards and self-propelled motion: Carey 1985, Karmiloff-Smith 1992; four or five other traits of living things: Keil 1994; research in the early 1980s: R. Gelman, Spelke, and Meck 1983; three- and four-year-old knowledge of brain function: C. Johnson and Wellman 1982.
11. Carey 1985.
12. R. Gelman 1990.
13. Piaget as pioneer: Piaget 1952; conservation of number: R. Gelman and Gallistel 1978, Piaget 1952; once they used smaller numbers of objects: Bever, Mehler, and J. Epstein 1968.
14. See, e.g., Antell and Keating 1983.
15. Subitizing numbers increase with age: Starkey and Cooper 1995. See Klahr 1973 for the argument that low-level perceptual ability precedes counting and R. Gelman and Gallistel 1978 and Mandler and Shebo 1982 for the argument that higher-level cognitive ability follows counting.
16. Wynn 1992, 1995, 1998. See also Simon, Hespos, and Rochat 1995.
17. Dehaene 1997, Devlin 2000, Gallistel 1990, Wynn 1998.
18. R. Gelman and Brenneman 1994.

19. See also chapter 7, note 22.
20. The concept of zero, fractions, and irrational numbers: R. Gelman and Brenneman 1994, Wynn 1998; the quote by Dehaene: 1997, pp. 245–46; analogues in formal mathematics: Pinker 1997.
21. Mild form of recapitulation: McCloskey 1983, Piaget and Garcia 1989, Wiser and Carey 1983; Piaget's general model of cognitive development: Inhelder and Piaget 1958, Piaget 1972.
22. Two general conclusions: Zimmerman 2000; metacognition: Flavell 1979, Sperber 1994, Sternberg 1985.
23. D. Kuhn and Pearsall 2000, p. 115.
24. Intelligence and metacognition: Chan 1996, Schwanenflugel, Stevens, and Carr 1997, Shore and Dover 1987; teaching metacognitive skills: Despete, Roeyers, and Buysse 2001, Georghiades 2000, Glynn and Muth 1994, Koch 2001, White and Frederiksen 1998.
25. Brewer and Samarapungavan 1991, Gopnik and Wellman 1994, Gopnik et al. 1999, Gopnik et al. 2001, Karmiloff-Smith 1992, Koslowski 1996; quote: Gopnik et al. 1999, p. 155.
26. Koslowski 1996.
27. R. Dunbar 1995, Mithen 1996.
28. Brewer and Samarapungavan 1991, Brewer, Chinn, and Samarapungavan 1998. More recent switch emphasizing differences: Chinn and Brewer 2000.
29. Fay, Klahr, and K. Dunbar 1990, Howe, Tolmie, and Sofroniou 1999, Klahr 2000, Klahr, Fay, and K. Dunbar 1993, D. Kuhn 1989, D. Kuhn, Amsel, and O'Loughlin 1988.
30. D. Kuhn and Pearsall 2000.
31. D. Kuhn 1989, 1993, D. Kuhn and Pearsall 2000, D. Kuhn, Amsel, and O'Loughlin 1988; see also Hogan and Maglienti 2001.
32. D. Kuhn 1993.
33. Klahr 2000, Klahr, Fay and K. Dunbar 1993.
34. D. Kuhn, Amsel, and O'Loughlin 1988, Schauble 1990.
35. Klaczynski and Gordon 1996, Klaczynski and Narasimham 1998, Klaczynski and Robinson 2000; see also Chinn and Brewer 2000.
36. Neuroimaging research: Fuster 2002, Head, Raz, Gunning-Dixon, Williamson, and Acker 2002, B. Miller and Cummings 1999, Miyake, Friedman, Emerson, Witzki, and Howerter 2000, Ragland et al. 2002; executive functions: Baddeley 1998, Fuster 2002, B. Miller and Cummings 1999, Miyake et al. 2000, Welsh, Pennington, and Groisser 1991; Fuster quote: Fuster 2002, p. 377; executive function and Piagetian reasoning: Kwon and Lawson 2000; executive functioning and mathematical reasoning: Bull and Scerif 2001; Menon et al. 2000; Rourke and Conway 1997.
37. Karmiloff-Smith 1992, pp. 17–18.
38. Compare Gopnik and Wellman 1994, Parker and McKinney 1999.
39. About 5.5 million out of 263 million people in 1995 made their livelihood doing science (commercial or academic), as reported by the National Science Foundation (1999).
40. Cognitive consistency: Greenwald et al. 2002; self-efficacy: Bandura 1982; increase in math and science self-efficacy: Luzzo, Hasper, Albert, Bibby, and Martinelli 1999; occupational interest: Kelly and Nelson 1999, Tobin, Tippins, and Hook 1995; fit of self-image: Nosek et al. 2002.

41. Feist 1998; compare, e.g., Bachtold and Werner 1972, Busse and Mansfield 1984, Chambers 1964, Eiduson 1962, Gough and Woodworth 1960, Rushton, Murray, and Paunonen 1987, Shaughnessy, Stockard, and Moore 1994; openness, conscientiousness, and scientific interest: Feist, Paletz, and Weitzer in preparation.
42. J. M. Cattell and Brimhall 1921, Clark and Rice 1982, Eiduson 1962, Galton 1874, Roe 1952a.
43. Chambers 1964, Datta 1967, Feist 1991a, Helson and Crutchfield 1970, Roe 1952a, Zuckerman 1996.
44. Protestant and Jewish: Chambers 1964, Datta 1967, Feist 1991a, Helson and Crutchfield 1970, Roe 1952a, Zuckerman 1996; 9 percent: Roe 1952a; 38 percent: Helson and Crutchfield 1970; most in the 20–30 percent range: Chambers 1964, Datta 1967, Feist 1991a, Zuckerman 1996; almost complete absence of religious affiliation in scientific elite: Chambers 1965, Feist 1991a, Roe 1952a, Terman 1954.
45. Eiduson 1962, Feist 1991a, Roe 1952a.
46. Berger 1994, Brockman 2004, Feist 2005, Helson and Crutchfield 1970, Simonton 1988a.
47. Meritocratic argument: Cole and Cole 1973, Merton 1973; work ethic argument: Tamir Druz as quoted in Berger 1994, p. 7; Simonton quote: 1988a, p. 126.
48. Stanley 1988, p. 206.
49. In high school: Benbow and Minor 1986, Benbow and Stanley 1982, Feist 2005; in college Lubinski and Benbow 1994; mathematicians and scientists average ninetieth percentile: Wise et al. 1979.
50. Farmer 1988, cf. Benbow 1988, Benbow and Lubinski 1993, Feist 2005, Subotnik and Steiner 1992. Poor predictive validity of IQ: Barron and Harrington 1981, Feist 2005, Gough 1976, Guilford 1959, Hudson 1958, MacKinnon 1960, Simonton 1988a, Sternberg 1988a, Taylor 1963. See chapter 9 for more discussion of the predictive validity of IQ tests.
51. Feist 2005.
52. Bayer and Dutton 1977, S. Cole 1979, Dennis 1956, Diamond 1986, Feist 2005, Horner, Rushton, and Vernon 1986, Lehman 1953, 1960, 1962, 1966, Over 1982, 1989, Simonton 1984, 1988a, 1988b, 1989, 1991, 1992a, Zuckerman 1996.
53. Criticisms of Lehman: S. Cole 1979; Dennis 1956, 1958, Horner et al. 1986; Over 1989, Zuckerman and Merton 1973; once controls taken into account: see Simonton 1988b; physics versus other fields: Lehman 1953, Moulin 1955, Simonton 1988b; physics contributions in one's twenties: Charness 1988, Lehman 1953, Simonton 1988b.
54. Quoted in Simonton 1988a, p. 67.
55. Age accounts for small percentage of variance: Bayer and Dutton 1977, S. Cole 1979, Horner et al. 1986; other individual difference and social factors: S. Cole 1979, Zuckerman 1996, Zuckerman and Merton 1973.
56. Simonton 1988b.
57. S. Cole 1979, Cole and Cole 1973, Hargens, McCann, and Reskin 1978, Merton 1973, Roe 1972, Simonton 1977, Zuckerman 1996.
58. Lu et al. 2004.
59. For example, Schaie 1984, 1994, Singer, Verhaeghen, Ghisletta, Lindenberger, and Baltes 2003, Willis, Jay, Diehl, and Marsiske 1992.

60. Beard 1874; Eiduson quote: 1974, p. 408.
61. Simonton 1984, 1988a, 1988b, 1989, 1991.
62. S. Cole 1979, Dennis 1954, 1966, Feist 2005, Helson and Crutchfield 1970, Horner et al. 1986, Lehman 1953, Over 1982, Reskin 1977, Roe 1965, Simonton 1988b, 1991, 1992a; for example, one study: Horner et al. 1986.
63. Matthew effect: S. Cole 1979, Merton 1973, Zuckerman and Merton 1973; one-tenth of the scientists produce half the work: Lotka 1926, Price 1963. Quantity matters more: Feist 1997.
64. Lehman 1953, 1960, 1966, Over 1989; longitudinal data: S. Cole 1979.
65. Planck, as quoted in Barber 1961, p. 597.
66. Darwin as quoted in Hull, Tessner, and Diamond 1978, p. 718.
67. Hull et al. 1978; more recently: Hull 2001.
68. Sulloway 1996.
69. Messeri 1988, Sulloway 1996; other researchers reanalyzed: Levin, Stephen, and Walker 1995; no researcher to date: Levin et al. 1995, p. 281.

## CHAPTER 4. COGNITIVE PSYCHOLOGY OF SCIENCE

1. Tweney 1989, p. 345, A. Miller 1996, pp. 227–28.
2. See Klahr and Simon 1999 for an alternative way of reviewing the literature around four methods: historical accounts, psychological experiments, direct observation of real world science, and computational modeling.
3. Gruber 1981, p. xxi.
4. As quoted in Gruber 1981, p. 173.
5. "The idea of evolution . . .": Gruber 1981, p. 134; "Darwin certainly began the notebooks . . .": Gruber 1981, p. 173 (emphasis in original).
6. As quoted in Gruber 1981, p. 123 (emphasis in original).
7. Gopnik and Glymour 2002, p. 124.
8. Not pie-in-the-sky: Darwin 1871, R. Dunbar 1995, G. Miller 2000, Mithen 1996, Papineau 2000, Shepard 1997; "A great stride in the development . . .": Darwin 1871, p. 633; evolved adaptation: D. Buss et al. 1998, Papineau 2000, Pinker 1997. See also chapter 1, note 22, above, and chapter 5, note 19, below.
9. K. Dunbar 1995, 2002, K. Dunbar and Blanchette 2001; quote: K. Dunbar 2002, p. 156.
10. Hanson 1958, Popper 1965, p. 46.
11. Gruber 1981, p. 174.
12. Ibid., p. 28.
13. Ibid., p. 161ff.
14. Brewer, Chinn, and Samarapungavan 1998, Carruthers 2002, K. Dunbar 2002, R. Dunbar 1995, Gopnik et al. 1999, Karmiloff-Smith 1992, Shepard 1997.
15. Polyani 1964, p. x, emphasis in original.
16. As quoted in A. Miller 1996, p. 351.
17. Tweney 1989, p. 347.
18. Fauconnier and Turner 2002, p. 14.

19. Clement 1989, De Mey 1989, K. Dunbar 1995, 2001, Fauconnier and Turner 2002, Gentner, Holyoak, and Kokinov 2001, Glynn, Britton, Semrud-Clikeman, and Muth 1989, Holyoak and Thagard 1995, John-Steiner 1985, Leary 1990, A. Miller 1996.
20. Gentner and Jeziorski 1989, Gorman 1995, Nersessian 1992, 2002.
21. Carey and Spelke 1994, p. 181.
22. Gentner, Bowdle, Wolff, and Boronat 2001.
23. Fauconnier and Turner 2002, Lakoff and Johnson 1980, Mithen 1996, Pinker 1997.
24. R. Dunbar 1995, Gruber 1981, John-Steiner 1985, T. Kuhn 1979, A. Miller 1996, Mithen 1996.
25. Cheng and Simon 1995, Finke, Ward, and Smith 1992, Gleick 1992, Gruber 1981, John-Steiner 1985, Larkin and Simon 1987, A. Miller 1989, 1996, Shepard 1978.
26. Einstein quoted in Wertheimer 1959, p. 228; Feynman quoted in Gleick 1992, p. 244.
27. Bohr quote: Gleick 1992, p. 242; Feynman quote: Gleick 1992, p. 244; String theory, see, e.g., Greene 1999.
28. Definition of creative thought: Amabile 1996, Feist 1998, Simonton 1988a, Sternberg 1988a; cognitive traits clustering around creative ability: Carson et al. 2003, Eysenck 1995, Finke et al. 1992, Guilford 1959, 1987, Martindale 1999, Mednick 1962, Mendelsohn 1976, Simonton 1988a, 1999, 2000, Taylor and Barron 1963.
29. Gruber 1981, p. 107ff.
30. See fig. 4 in Clement 1989 for a complex, dynamic, and interactive model of model formation, hypothesis testing, and empirical testing; Gruber quote, 1981, p. 108.
31. Recall the argument from chaps. 1 and 2 about the crucial role that language and written language in particular plays in externalizing thought and therefore facilitating metacognitive reflection.
32. See A. Miller 1996, pp. 8–10.
33. A. Miller 1996, Wertheimer 1959.
34. Wason 1960.
35. Wason selection task: Wason 1966; other cognitive researchers: Einhorn and Hogarth 1978, Kern, Mirels, and Hinshaw 1983, Mahoney and Kimper 1976.
36. Cosmides and Tooby 1992, Wason and Green 1984.
37. Klahr 2000, Klahr and K. Dunbar 1988, Klahr and Simon 1999, Klahr, K. Dunbar, and Fay 1990.
38. Gorman 1989, 1992.
39. Ibid.
40. Hanson 1962, Klayman and Ha 1987, Kruglanski 1994, Mahoney 1977, 1979, Mahoney and DeMonbreun 1977.
41. Gorman 1986, 1992, Gorman and Gorman 1984, Gorman, Stafford, and Gorman 1987.
42. Gruber 1981, pp. 143–44.
43. Tweney 1985, 1989, 1991.
44. Mynatt, Doherty, and Tweney 1977, 1978.
45. Chamberlain 1890/1965, p. 755; J. R. Platt (1964) built on Chamberlain's idea and further codified the strong inference methodology of explicitly testing two or more competing theories.
46. Freedman 1995, Gorman and Gorman 1984, Tweney, Doherty, and Mynatt 1981, Wharton, Cheng, and Wickens 1993.

47. See Pinker 1997 for a detailed and thoughtful review of the powers and limitations of artificial intelligence; BACON: Langley, Simon, Bradshaw, and Zykow 1987; KEKADA: Kulkarni and Simon 1988; HUYGENS: Cheng and Simon 1992, 1995.
48. See, e.g., Falkenhainer 1990, Gooding 1990, Gooding and Addis 1993, Grasshoff and May 1995, Pizzani 1990, Shen 1993, Thagard 1988, Thagard and Nowak 1990, and Valdes-Perez 1994.
49. Klahr and Simon 1999, p. 526.
50. See, e.g., A. Miller 1996 and Pinker 1997; Shrager and Langley quote: 1990, p. 15.
51. K. Dunbar 1995, 2002.
52. Definition of integrative complexity: Schroder, Driver, and Streufert 1967, Tetlock and Suedfeld 1988; APA presidents finding: Suedfeld 1985; eminent scientists' complexity: Feist 1994.
53. Anzai 1991, Chi, Feltovich, and Glaser 1981, Hogan and Maglienti 2001.
54. Anzai 1991, Carey 1992, Clement 1982, McCloskey 1983, Wiser and Carey 1983.
55. Larkin 1983, Larkin, McDermitt, Simon, and Simon 1980.
56. Clement 1989, 1991, Gentner and Genter 1983, K. Dunbar 1995, 2001, K. Dunbar and Blanchette 2001.

**CHAPTER 5. PERSONALITY PSYCHOLOGY OF SCIENCE**

1. See Digman 1990, John and Srivastava 1999, and McCrae and John 1992 for general reviews of the Big-Five approach.
2. D. Buss 1990, 1999, G. Miller 2000; Buss quote: 1990, p. 6.
3. See D. Buss 1990 and D. Buss and Greiling 1999 for discussion of the adaptive nature of personality traits. McCrae and Costa 1999 also situate the Big-Five dimensions in an evolutionary context.
4. See Allport 1937, Eysenck 1990a, Feist 1998, Feist and Barron 2003, Funder 1991, Rosenberg 1998.
5. Plomin and Caspi 1999, p. 262.
6. Benjamin et al. 1996, Ebstein et al. 1996, Hamer and Copeland 1998, Lesch et al. 1996, Plomin and Caspi 1999; for a critique of the QTL approach, see Wahlsten 1999.
7. Explains 40–50 percent of the variance: e,g., see Loehlin 1992, Loehlin and Nichols 1976, Loehlin, Willerman, and Horn 1987, Loehlin, McCrae, Costa, and John 1998, Plomin and Caspi 1999, Tellegen et al. 1988; unshared environmental influence: Bouchard and McGue 1990, Eysenck 1990b, Hamer and Copeland 1998, Loehlin et al. 1998, Plomin and Caspi 1999, Riemann, Angleitner, and Strelau 1997, Tellegen et al. 1988.
8. Definition of temperament: Gonzalez, Hynd, and Martin 1994, p. 238; Temperament as the foundation for personality: A. Buss and Plomin 1984, Rothbart, Ahadi, and Evans 2000; Rothbart quote: Rothbart et al. 2000, p. 122.
9. On the low end: A. Buss and Plomin 1984; on the high end: Thomas and Chess 1977.
10. Galton 1874.
11. J. M. Cattell 1910; J. M. Cattell and Brimhall 1921; Cox 1926; Terman 1925, 1954.
12. Feist 1998. Research on which the meta-analysis is based includes: Arvey and Dewhirst 1976, Bachtold 1976, Butcher 1969, R. Cattell and Drevdahl 1955, Eiduson 1962, Garwood 1964, Gough 1961, 1987, Ham and Shaughnessy 1992, Helson 1971, Kline and

Lapham 1992, Lacey and Erickson 1974, Mansfield and Busse 1981, McDermid 1965, Mossholder, Dewhirst, and Arvey 1981, Pearce 1968, Roe 1952a, Scott and Sedlacek 1975, Terman 1954, Wilson and Jackson 1994; also see Thomas, Benne, Marr, Thomas, and Hume 2000; for rules of thumb for meta-analytic effect sizes, see J. Cohen 1988.

13. In addition to Feist 1998, see Van Zelst and Kerr 1954; study of 116 female biologists and chemists: Bachtold and Werner 1972; male and female personality traits: Bachtold 1976, Barton and R. Cattell 1972; age and dominance: Scott and Sedlacek 1975.
14. Eysenck 1995, Storr 1988.
15. Gough 1987, p. 7.
16. Holland 1992, Lippa 1998, Prediger 1982.
17. Baron-Cohen, Wheelwright, Stott, et al. 1997, Baron-Cohen et al. 1998, Baron-Cohen et al. 1999, Baron-Cohen et al. 2001.
18. Most clinical psychologists in fact do no empirical research after graduate school (Mallinckrodt, Gelso, and Royalty 1990, Zachar and Leong 2000). Because so many clinical and counseling psychologists are not interested in research, professional schools of psychology and the doctorate in psychology (PsyD) have sprouted up that train people in the clinical side but not in the empirical side of psychology. Moreover, the rift between researchers and practitioners became so pronounced in the largest society of psychologists, American Psychological Association, that in the late 1980s a splinter group formed for the research and science side of psychology (American Psychological Society). Many of these studies in fact have used vocational interest and the vocational social-realistic-investigative dimensions found in the people-thing orientation as the measure of personality. Others have argued that vocational interests and personality are separate dimensions, even if vocational interests can be traits (e.g., Waller, Lykken, and Tellegen 1995).
19. Kahn and Scott 1997, Mallinckrodt, Gelso, and Royalty 1990, Royalty and Magoon 1985, Zachar and Leong 1992, 1997.
20. For a review of much of this literature, see Arthur 2001; see also Atwood and Tomkins 1976, Coan 1973, Conway 1988, Costa, McCrae, and Holland 1984, Hart 1982, J. Johnson, Germer, Efran, and Overton 1988, Simonton 2000, Tremblay, Herron, and Schultz 1986. A null result (no significant relation) has also been reported between theoretical orientation in counseling students and personality traits, Freeman 2003.
21. Sparkman 1994.
22. A. Buss and Plomin 1984, Thomas and Chess 1977, Rothbart et al. 2000.
23. Arvey, Dewhirst, and Brown 1978, S. Cole 1979, Diamond 1986, Eiduson 1974, Feist and Barron 2003, Hinrichs 1972, Horner et al. 1986, Kahn and Scott 1997, Roe 1965, Root-Bernstein, Bernstein, and Garnier 1995, Simonton 1991, 1992a, Subotnik, Duschl, and Selmon 1993, Terman 1954. The two that have looked at personality over time are Eiduson 1974 and Feist and Barron 2003.

### CHAPTER 6. SOCIAL PSYCHOLOGY OF SCIENCE

1. Allport's definition: 1985, p. 3; as others have noted: Shadish, Fuller, and Gorman 1994; *The Social Psychology of Science:* Shadish and Fuller 1994.
2. Barnes and Rosenthal 1985, Rosenthal 1976, 1994, Rosenthal and Fode 1963a, Stanton

and Baker 1942; experimenter expectancy with animals: Rosenthal and Fode 1963b; Pygmalion effect: Rosenthal and Jacobson 1992, compare also Rosenthal 1994.

3. Rosenthal and Rosnow 1991, p. 126.
4. Ibid., p. 129.
5. Rosenthal 1994, p. 220.
6. Kruglanski 1994, p. 211.
7. Simonton 1975, 1976a, 1976b, 1980. Chance configuration theory: Simonton 1988a, 1989, 1999.
8. Shadish 1989, Shadish et al. 1994, compare Feist 1997, Shadish quote: 1989, p. 407.
9. Shadish, Tolliver, Gray, and Gupta 1995, Sternberg and Gordeeva 1996.
10. Parents' education or career in science: Albert 1975, Berry 1981, Chambers 1964, Cox 1926, Eiduson 1962, Feist 1991a, Helson and Crutchfield 1970, Moulin 1955, Pheasant 1961, Roe 1952a, Simonton 1984, Terman 1954, Zuckerman 1996; achievement in science more likely if father is a scientist: Werts and Watley 1972; positive parental attitude: Byler 2000, Ferry, Fouad, and Smith 2000, Koutsoulis and Campbell 2001; science museum study: Crowley, Callanan, Tenenbaum, and Allen 2001.
11. Eiduson 1962, Feist 1991a, John-Steiner 1985, Subotnik et al. 1993, Subotnik and Steiner 1992.
12. John-Steiner 1985, Simonton 1992b, Zuckerman 1996.
13. Differences between individuals and groups: Gorman 1986, 1989, Gorman and Gorman 1984, Gorman, Gorman, Latta, and Cunningham 1984; small coacting groups more prone to confirmation bias: Gholson and Houts 1989.
14. Fox Keller 1985; cf. Nosek, Banaji, and Greenwald 2002.
15. Explicit attitudes: Eccles 1987, Fennema 1985, Hyde, Fennema, Ryan, Frost, and Hopp 1990; implicit attitudes: Nosek et al. 2002; performance on aptitude tests: Benbow and Stanley 1983, Benbow et al. 2000, Geary 1998, Halpern 2000; actual graduation and career data: J. Cole 1987, S. Cole and Zuckerman 1987, Farmer, Wardrop, and Rotella 1999, Jacobwitz 1983, Long 2001, National Science Foundation 1999, O'Brien, Martinez-Pons, and Kopala 1999, Reis and Park 2001, Stuessy 1988, Subotnik et al. 1993; attrition from science: Benbow et al. 2000, Feist 2005; Long 2001, Subotnik and Steiner 1994, Webb, Lubinski, and Benbow 2002; gender difference in mathematical domain: Benbow and Stanley 1983, Geary 1998, Halpern 2000, Kimura 1999.
16. Two percent of NAS is female: Long 2001, NSF 1999, Rosser 1988, Subotnik et al. 1993; the most extensive study: Long 2001.
17. Long 2001, NSF 1999, Webb et al. 2002.
18. Benbow et al. 2000.
19. Paletz study: see Feist, Paletz, and Weitzer in preparation; math skills as male domain: see Nosek et al. 2002, Kelly and Nelson 1999; Tobin et al. 1995; self-efficacy and training: Bandura 1982, Luzzo et al. 1999.
20. See, e.g., Davies, Spencer, Quinn, and Gerhardstein 2002, Nosek et al. 2002, Shih, Pittinsky, and Ambady 1999, Steele 1997, Steele and Aronson 1995; quote: Davies et al. 2002, p. 1616.
21. Gender effect: Achter, Lubinski, and Benbow 1996, Connellan, Baron-Cohen, Wheelwright, Batki, and Ahluwalia 2000, Lippa 1998, Lubinski 2000, Lubinski and Humph-

reys 1990, Prediger 1982, Schmidt, Lubinski, and Benbow 1998, Webb et al. 2002; people-thing orientation: Holland 1992, Lippa 1998, Prediger 1982; scientists score higher on autism and Aspergers than nonscientists: Baron-Cohen et al. 1999, Baron-Cohen, Wheelwright, Skinner, Martin, and Clubley 2001; autistic children more likely to have engineer for father or grandfather: Baron-Cohen, Wheelwright, Stott, 1997, Baron-Cohen et al. 1998; three-day-old preferences: Connellan et al. 2000.

22. Achter et al. 1996, Astin 1975, Backman 1972, Benbow 1988, Benbow and Stanley 1980, 1983, Benbow et al. 2000, Deaux 1985, Fischbein 1990, L. Fox 1976, Geary 1998, Holden 1987, Keating 1974, Leahey and Guo 2002, Lubinski 2000, Maccoby and Jacklin 1974, E. Moore and Smith 1987, Stanley 1988, Webb et al. 2002.
23. Benbow 1988 wrote the lead article for *Behavioral and Brain Sciences;* see all the commentaries in response (e.g., Bleier 1988, Eysenck 1988, Farmer 1988, Sternberg 1988b, Vandenberg 1988).
24. Prenatal exposure to testosterone: Geschwind and Behan 1982; Benbow quote: 1988, p. 182.
25. J. Cole 1979, 1987, Cole and Cole 1973, J. Cole and Zuckerman 1984, Guyer and Fidell 1973, Helmreich et al. 1980, Long 1992, 2001, Pasewark, Fitzgerald, and Sawyer 1975, Xie and Shauman 1998, Zuckerman and Cole 1975; increase over time: J. Cole 1987; decrease (but still persist) over time: Guyer and Fidell 1973, Long 1992, 2001, Xie and Shauman 1998.
26. Long 2001.
27. Married women outproduce single women: J. Cole 1979, 1987, S. Cole and Zuckerman 1987; likelihood of being in workforce: Long 2001; quote: Long 2001, p. 91.
28. Gender differences cannot be explained by institution: Cole and Cole 1973, Long 2001; women less likely to work full time: Long 2001; hours per week: Benbow et al. 2000.
29. Rudwick 1985, Shadish et al. 1994; cf. Gorman 1992.
30. Asch 1956, Moscovici and Nemeth 1974.
31. Gorman and Rosenwein 1995, Jacobs and Campbell 1961, Mynatt et al. 1978, Rosenwein 1994.
32. See Shadish and Fuller, eds. 1994 for a more complete discussion of these topics.
33. International Conference at the National Institute for Science, Technology and Development Studies (NISTADS) on "Women in Science: Is the Glass Ceiling Disappearing," New Delhi, India, March 8 to 10, 2004.
34. Kelley 1967; see Eflin and Kite 1996 for a study of college students on attribution theory and scientific reasoning.
35. Feist 1991a.

## CHAPTER 7. THE APPLICATIONS AND FUTURE OF PSYCHOLOGY OF SCIENCE

1. The question of assessing and selecting scientific talent in students can be and is a topic for educational psychologists as well, but since I am most interested in scientific talent and achievement in careers, I am putting this topic mostly under the rubric of "I/O psychology."

2. Feist under review. See also Cole and Cole 1973, Lehman 1953, Simonton 1988a.
3. IQ predicts: Cox 1926, Eiduson 1962, Eysenck 1995, Simonton 1988a, Terman 1954; quantitative reasoning: Achter et al. 1999, Benbow and Stanley 1983, Benbow et al. 2000, Gustin and Corazza 1994, Schoon 2001; spational reasoning: Baker 1985, Cooper 2000, Gardner 1983, Piburn 1980, Reuhkala 2001, Shea, Lubinski, and Benbow 2001.
4. SAT predicts GPA: Weitzman 1982; SAT-verbal better than SAT-quantitative: Lawler, Richman, and Richman 1997, Mauger and Kolmodin 1975; SAT distinguishes graduation rates: H. Stumpf and Stanley 2002; SAT costs not justified: Crouse and Trusheim 1991.
5. Narrow meta-analytic reviews: Goldberg and Alliger 1992, Morrison and Morrison 1995, Schneider and Briel 1990; broad reviews: Kuncel, Hezlett, and Ones 2001, Sternberg and Williams 1997.
6. IQ and creativity: Barron 1963, Barron and Harrington 1981, Cox 1926, Eysenck 1995, Jensen 1996, MacKinnon 1978 (chap. 13), Roe 1952a, Runco 1999, Simonton 1988a, Sternberg and O'Hara 1999, Wallach 1970, Wallach and Kogan 1972; broaden and spread out associations: Getzels 1987, Guilford 1959, 1987, MacKinnon 1978, Simonton 1988a, 1989, Sternberg 1988a, 1988c, Wallach 1970, Wallach and Kogan 1972.
7. Sternberg 1988c.
8. Expanded view of intelligence: Gardner 1983, 1999, Sternberg 1988c; see Feist 2004a and Sternberg, Grigorenko, and Singer 2004 for a general discussion and review by different scholars with a range of views on the question of domain specific talent; emotional intelligence: Goleman 1995, Kihlstrom and Cantor 2000, Mayer, Salovey, and Caruso 2000, Salovey and Mayer 1990.
9. See MacKinnon 1978, chap. 13, for similar discussion of augmenting intelligence and scholastic criteria in selecting students with creative potential. See also Feist and Barron 2003.
10. Feist and Barron 2003.
11. Berger 1994, p. 18.
12. Feist 2005, Subotnik and Steiner 1994, Webb et al. 2002.
13. Webb et al. 2002.
14. Long 2004. For institutional forces behind gender differences, see Long 2001. For summaries and reviews of some of the intervention programs in gender and science, see M. Fox 1998 and Matyas and Dix 1992.
15. Shadish, Fuller, and Gorman (with Amabile, Kruglanski, Rosenthal, and Rosenwein) 1994.
16. Gholson, Shadish, Neimeyer, and Houts 1989, Simonton 1988a.
17. Matarazzo 1987.

#### CHAPTER 8. EVOLUTION OF THE HUMAN MIND

1. For similar arguments about evolution of scientific thinking, see Atran 1990, Carruthers 2002, R. Dunbar 1995, Mithen 1996, Shepard 1997.
2. Indeed, I am somewhat consistently asked when I lecture on an evolutionary perspective

in psychology whether such a view is valid or whether it has been discredited. Although I acknowledge that not all in the social sciences like or are sympathetic to an evolutionary perspective, to say that it is invalid or discredited is simply factually incorrect. I could garner all kinds of evidence from basic psychology textbooks, from experts, from standard psychological conferences, from new journals, from course syllabi throughout the country, etc., etc., to show that evolution is moving front and center in thought for a large portion of current psychologists. For instance, a survey of twenty introductory psychology textbooks now on the market indicates that every one of them lists evolutionary psychology and evolutionary theory as one of the major perspectives in psychology. Evolution may not quite be as central in psychology as it is in biology, but it is looking more and more as though twenty to thirty years from now almost all questions in psychology will be framed in an evolutionary perspective. In other words, the field has passed the point of no return with regards to conceptualizing human behavior, like all animal traits, in the framework of evolutionary theory.

3. Donald 1991, G. Miller 2000, Mithen 1996, Pinker 1997, 2002, Tooby and Cosmides 1992.
4. R. Gelman and Brenneman 1994, p. 371; Domains are not synonymous with modules: cf. Fodor 1983, Karmiloff-Smith 1992, Sperber 1994; for language, art, and music, see, e.g., Aiken 1998, Dissanayake 1992, Feist 2001, Gardner 1983, 1999, Pinker 1996, 1997, Wallin, Merker, and Brown 2001.
5. For comparison of criteria for domains of mind, see Gardner 1983; for the sake of space, developmental evidence was discussed in more detail in chapter 3 and genetic evidence in more detail in chapters 2 and 5.
6. Theory of mind: Gardner 1983, 1999, Goleman 1995, Premack and Woodruff 1978, Salovey and Mayer 1990, Thorndike 1920. Granted there are distinctions to be made between social, personal, and emotional intelligence, they each capture the abilities involving social interaction and personal insight. Also, I intentionally avoid using the word "intelligence" because as Gardner pointed out, and others after him have criticized, "intelligence" runs the risk of being reified and spread too thinly if it simply means ability. He addresses this issue directly when he writes: "that nothing much hangs on the particular use of this term [intelligence], and I would be satisfied to substitute such phrases as 'intellectual competences,' 'thought processes,' 'cognitive capacities,' 'cognitive skills,' 'forms of knowledge,' or other cognate mentalistic terminology. What is crucial is not the label, but rather the conception: that individuals have a number of domains of potential intellectual competence, which they are in a position to develop" (1983, p. 284). Indeed, many traditional intelligence researchers who defend the notion of generalized intelligence (g) are critical of broadening the definition of intelligence to include talents or competencies, arguing that these are more appropriately classified as talents or personality traits (Herrnstein and Murray 1994, Scarr 1985, 1989). I completely agree with the thrust of Gardner's argument, that the concept of developing intellectual competencies is the key idea, and that there are more of them than traditional intelligence researchers are willing to acknowledge, but I think applying the word "intelligences" to these has distracted from the main argument. One stands little to gain and much to lose by calling them intelligences, and therefore I use the phrase "domains of mind."

7. Byrne 2001, Byrne and Whiten 1988, Cheney and Seyfarth 1990, de Waal 1998, Parker and McKinney 1999, Povinelli 1993, Premack 1998, Premack and Woodward 1978. See Heyes 1998 for a critical review of this literature.
8. Ekman 1994, Ekman and Friesen 1971, Ekman, Sorenson, and Friesen 1969, Ekman et al. 1987, cf. Eibl-Eibesfeldt 1970 for an even stronger argument for universality; more accurate at in-group emotion-recognition: Elfenbein and Ambady 2002.
9. Universality of theory of mind: P. Harris 1990, Scholl and Leslie 1999, Segal 1996; what vs. that: Lillard 1998.
10. Elias, Arnold, and Hussey 2003, Goleman 1995, Hatch 1997, Schmitt and Grammar 1997.
11. R. Gelman, Spelke, and Meck 1983, Mithen 1996, Pinker 2002, Spelke 1990; some archeologists: e.g., Byrne 2001, Mithen 1996.
12. Donald 1991, Leaky and Lewin 1977, Mithen 1996, Nobel and Davidson 1996.
13. Observations of chimps in the wild: Goodall 1986; complex tool use rare among primates: Byrne 2001, Matsuzawa 1996, McGrew 1992, Parker and McKinney 1999, Visalberghi, Fragaszy, and Savage-Rumbaugh 1995.
14. Baron-Cohen, Wheelwright, Stone, and Rutherford 1999, Baron-Cohen et al. 2001, Baron-Cohen, Wheelwright, Stott, et al. 1997, Baron-Cohen et al. 1998, Rimland and Fein 1988, Treffert and Wallace 2002.
15. Boesch and Boesch 1989, Goodall 1986, McGrew 2001, Parker and McKinney 1999, Stanford 2001, Wrangham 1977.
16. Atran 1990, Berlin 1992, quotes by Lévi-Strauss 1966, pp. 39 and 43, respectively.
17. Atran 1990, 1994.
18. Gardner 1999, pp. 50–51, Howe 1999.
19. Butterworth 1999, Devlin 2000, Wynn 1992, 1995, 1998.
20. Crump 1990, quote from p. 5.
21. Bell 1937, Gleick 1992, Kanigel 1991, Wiener 1953.
22. Language, see Bloom 1998, Gardner 1999, Mithen 1996, Pinker 1996, 1997; music, quote by Tramo 2001, p. 54. Also see, for instance, R. Gelman and Brenneman 1994, Wallin, Merker, and Brown 2001; art and aesthetics, see Aiken 1998, Barrow 1995, Bradshaw 2001, Dissanayake 1992, Feist 2001, Gardner 1983, Karmiloff-Smith 1992, G. Miller 2000, Orians 2001, Orians and Heerwagan 1992.
23. The question of what is an adaptation versus what is a co-opted by-product of an adaptation ("exaptation") is a big and rather controversial issue in evolutionary biology and psychology and is beyond the scope of this book (see Buller and Hardcastle 2000, D. Buss, Haselton, Shackelford, Bleske, and Wakefield 1998, Gould 1991, Gould and Lewontin 1979, Feist 2001, G. Miller 2000, Pinker 1997, Thornhill 1997, Tooby and Cosmides 1992, Williams 1966). The debate is particularly contentious when it comes to intelligence and creativity. Darwin himself implied that most mental abilities were evolved by-products of the brain's capacity for language: "A great stride in the development of the intellect will have followed, as soon as the half-art and half-instinct of language came into use. . . . The higher intellectual powers of man, such as those of ratiocination, abstraction, self-consciousness, etc., probably follow from the continued improvement and exercise of the other mental faculties" (Darwin 1871, p. 633).

24. Feist 2001.
25. Donald 1991, Klein 1999, Mithen 1996, Parker and McKinney 1999.
26. Parker and McKinney 1999.
27. Haile-Selassie 2001, Klein 1999, Mithen 1996.
28. Mithen 1996.
29. Klein 1999, Mithen 1996.
30. Klein 1999, Mithen 1996.
31. Aiello and R. Dunbar 1993, R. Dunbar 1992, Falk 1983, Klein 1999, Mithen 1996, Tobias 1987.
32. Donald 1991 labeled this phase "episodic"; Mithen 1996 labeled it "generalized intelligence"; and Parker and McKinney 1999 labeled it the "apprenticeship stage"; implicit cognition: Bargh and Chartrand 1999, Kandel and Hawkins 1995, Kihlstrom 1987, Reber 1967, Schacter 1987, Underwood 1996; 95 percent of mental processing without awareness: see Bargh and Chartrand 1999, Baumeister, Bratslavsky, Muraven, and Tice 1998.
33. R. Dunbar 1995.
34. Donald 1991, Klein 1999, Mithen 1996, Parker and McKinney 1999.
35. Klein 1999, Mithen 1996.
36. Ibid.; see Goren-Inbar et al. 2004 for evidence of fire making in hominins (either *erectus, ergaster,* or *archaic sapiens*) around 790 kya.
37. Donald 1991 uses the term "mimetic" to describe this phase, Mithen 1996 "domain specific intelligence," and Parker and McKinney 1999 the "joint attention model"; first order and second order: Sperber 1994; explicit representation: Karmiloff-Smith 1992.
38. Mithen 1996.
39. Donald 1991, Klein 1999, Mithen 1996, Tattersall 1997.
40. Donald 1991, Klein 1999, Mithen 1996, Tattersall 1997.
41. Enard et al. 2002; compare Nobel and Davidson (1996) who define language more narrowly (syntactical) and therefore also reach the conclusion that it is but about 100,000 years old; Falk 2000, Lieberman 1993, Lieberman and Crelin 1971, Mithen 1996, Nobel and Davidson 1996.
42. Klein 1999, Mithen 1996, Tattersall 1997
43. Enlarged frontal lobe: Deacon 1997, R. Dunbar 1993, Jerison 1973, Krasnegor, Lyon, and Goldman-Rakic 1997, Rumbaugh 1997; creative and cognitively fluid mind: Krasnegor et al. 1997, Mithen 1996, Rumbaugh 1997, Semendeferi 1999; think before behaving: Stenhouse 1974.
44. Neural change 50 kya: Klein 1999; more recent evidence by Henshilwood, d'Errico, Vanhaeren, van Niekerk, and Jacobs 2004 casts doubt on this theory because of their finding that symbolic thought (in the form of bead making) was expressed as early as 77 kya; greater neural connectivity: Buxhoeveden, Switala, Roy, Litaker, and Casanova 2001, Gibson 1991, Purves 1988.
45. Byrne 1998, Donald 1991, Mithen 1996, 1998, Parker and McKinney 1999.
46. Neural connectivity among creative individuals: Ramachandran and Hubbard 2003; intuition and creativity: Chandrasekar 1987, Curtin 1980, Poincare 1952, Simonton 1988a.
47. See Freud 1900/1953; also see Solms and Turnbull 2002 for the view that Freud and neuroscience are converging and that many of Freud's insights concerning the unconscious,

dreams, and defense mechanisms are being supported by current findings in cognitive neuroscience; vibes, see S. Epstein 1994; hemispheric functioning and creativity: Beeman and Bowden 2000, Beeman, Bowden, and Gernsbacher 2000, Fiore and Schooler 1998, Solms and Turnbull 2002; highly creative people facile: see, e.g., Barron 1963, Feist 1991b, Kris 1952, Simonton 1988a, 1989.

## CHAPTER 9. ORIGINS OF THE SCIENTIFIC THINKING

1. Frazer 1922/1996, pp. 853–54.
2. Parker and MacKinney 1999.
3. For a different evolutionary view of the development of science in hominid history, see Mithen 2002.
4. This is not meant to be a formal definition of animal, but rather a folk one. We now know that some animals (sponges, for example) do not move and that plants reproduce. But neither of these traits is obvious and not part of our evolved or perceptual understanding of the world.
5. Einstein 1950, p. 60.
6. See Givon and Maller 2002 for various perspectives on the evolution of language out of prelanguage.
7. Popper 1965, p. 47, emphasis in the original.
8. Mithen 1996.
9. Indeed, this is a major argument of Karl Popper against induction in *Logic of Scientific Discovery* and again in *Conjectures and Refutations.* Observation cannot be divorced from theory. One cannot observe the world without some sort of guiding theory of how the world works. So we do not passively go from observation to theory (induction), but rather the two are inherently interconnected.
10. Genetic change in language 120,000 years ago: Enard et al. 2002; language components coming together 50,000 or 60,000 years ago: MacWhinney 2002; from phonology to morphology to syntax, see, e.g., Brown 1973, Tomasello 2002; language development in children, see Gazzaniga and Heatherton 2003, Pinker 1996, 2002, Tomasello 2002.
11. Klein 1999, Mithen 1996, Pfeiffer 1982, Tattersall 1997.
12. Atran 1990, Cajete 2000, Donald 1991, R. Dunbar 1995, Lévi-Strauss 1966.
13. Donald 1991, p. 214.
14. Ibid., p. 267.
15. Lévi-Strauss 1966, p. 13.
16. Frazer 1922/1996.
17. See Hellmans and Bunch 1991.
18. The following time line is reconstructed mainly from Hellmans and Bunch's *Timetables of Science* and George Sarton's *Ancient Science through the Golden Age of Greece* and other sources cited in the text.
19. d'Errico and Cacho 1994, Eccles 1989, Marshack 1972, 1985, Maynard and Edwards 1971.
20. Not only in the Middle East was there a flourishing archeoastronomy but in North America and central Africa as well. The Mayans, for instance, made astronomical-

astrological inscriptions and built structures to mark astronomical events approximately 10 to 11 kya. And in modern day Congo (formerly Zaire), in central Africa, bones dating to around 8.5 kya have been found that seem to have been marked to record the months and lunar phases.

21. Macrone 1992.
22. Burnham 1983. As an interesting side note, the term "dog days of summer" come from the first rising of the "dog star" and the summer solstice.
23. See Gregory 1981 and Kramer 1998.
24. Donald 1991.
25. As way of introduction to the development of writing systems, let me begin with a caveat: the origins and classifications of early writing systems is a terribly complex field of study, and it cannot begin to be awarded anything other than a very superficial summary here. The interested reader can find more detailed discussion of these ideas in the books that were used in gathering the information summarized here, namely, Coulmas 1989, Daniels and Bright 1996, DeFrancis 1989, Diringer 1962, Gelb 1963, R. Harris 1986, and Robinson 1995.
26. Daniels and Bright 1996, DeFrancis 1989.
27. Klein 1999.
28. Klein 1999; see Soffer 1985 for similar structures built in Russia around 15 to 20 kya.
29. Hellmans and Bunch 1991, Mithen 1996, 2002, A. Moore 1978, Simmons, Kohler-Rollefson, Rollefson, Mandel, and Kafafi 1988.
30. Sarton 1952a; see also "The great pyramid of Cheops (Khufu)" at http://www.touregypt.net/cheops.htm (accessed March 1, 2003).
31. Hellemans and Bunch 1991, Sarton 1952a.
32. Hellemans and Bunch 1991; Pare 2000, Tylecote 1992.
33. Hellemans and Bunch 1991.
34. Much of what I summarize in the remainder of the chapter is from Butterfield 1960, Durant 1926/1961, and Sarton 1952a.
35. Dean K. Simonton, personal communication, March 11, 2002.
36. A noteworthy aspect of Socrates was his ridicule of astronomy and cosmology, for he considered these topics unsolvable and full of folly and should only be attempted once knowledge of human affairs was complete (see Sarton 1952a, p. 259ff.).
37. Atran 1990, Durant 1926/1961.
38. Durant 1926/1961, p. 90. I have focused only on some of the major thinkers of ancient Greece and emphasized astronomy, physics, and mathematics, but I would be remiss if I did not at least point out that the ancient Greeks also made great advances in geography, zoology, architecture, biology, botany, mineralogy, and medicine (see Sarton 1952a). For sake of space and time, I do not elaborate on these developments.
39. Crump 2002.
40. Roger Bacon 1268/1907, p. 369; Locke quote: S. Stumpf 1975, p. 204.
41. Butterfield 1960, I. Cohen 1985, Crump 2002.
42. I. Cohen 1985, p. 142.
43. Skinner 1953.

44. Sagan 1987, p. 42.
45. T. Kuhn 1977.
46. Taubes 1993.
47. They are not evolved adaptations because they do not directly increase survivability or reproductive success (see Pinker 2002). But they are co-opted adaptations (or exaptations), as they do indirectly increase our survivability. They are built upon more basic domains of mind and the first principles therein. At some point the knowledge and ideas of science do move beyond folk knowledge and sometimes support "common sense" but often do not. To the extent that they are at odds with "common sense," they will be difficult for many to learn, except those with special domain-specific talents.

### CHAPTER 10. SCIENCE, PSEUDOSCIENCE, AND ANTISCIENCE

Epigraph. Written in a letter to Hans Muesham, July 9, 1951 (Calaprice 2005, p. 245). As an interesting and cautious side note, this quote is also readily attributed to Mark Twain on many "Favorite Quotes" Internet sites. It makes one leery of the ubiquitous and poorly referenced quotes attributed to everyone's favorite sources of quotes, Twain and Einstein. These two figures are quoted on most every topic, and indeed the other king of quotes, baseball Hall-of-Famer Yogi Berra, is reported to have said, "I didn't really say all the things I said." This is, in fact, one of those things Twain never said. The phenomenon of how quotes or chain letters proliferate and evolve on the Internet has itself become an interesting topic of study (see Bennett, Li, and Ma 2003).

1. Shermer 1997 p. 26; originally reported by Gallup and Newport 1991 in *Skeptical Inquirer* 15: 137–47.
2. Popper 1965, p. 37.
3. "About CSICOP," http://www.csicop.org/about/ (accessed April 16, 2004).
4. Derry 1999. Others have discussed similar criteria. For instance, in an article originally published in the skeptic magazine *Rational Enquirer*, Lee Moller (L. Moller, "BCS Debates a Qi Gong Master," *Rational Enquirer* 6, no. 4 [1994], published by the British Columbia Skeptics Society and reproduced at http://physics.syr.edu/courses/modules/PSEUDO/moller.html [accessed April 15, 2004]) argued that one should ask oneself at least sixteen questions when attempting to answer this question. Many are quite similar to the standard criteria I discuss concerning pseudoscience, so I will not go through all sixteen, but some not already discussed include: (1) does the discipline use technical words (such as "energy") without defining them or that are meaningless (such as "vibrational energy"); (2) when criticized do the defenders attack the critic rather than the criticism; (3) does the proponent make appeals to history (i.e., it has been around a long time, so it must be true); (4) is the evidence offered mostly if not exclusively anecdotal? And finally, (5) is the subject taught only at noncredited institutions? If the answer to most of these questions is yes, we are probably dealing with a topic that is at best fringe and probably fake science.
5. Velikovsky 1950; see also Derry 1999.
6. Barrow 1998, p. 3.

7. Careful and controlled manipulation of an "independent variable" and random assignment of "participants" to control or experimental conditions are the hallmarks of experimental research and are often not possible, for ethical and practical reasons, in much research with humans (as well as some of the physical sciences, such as astronomy). Instead of experimental research, correlational or observational research is done that observes rather than manipulations relationships.
8. "Atlantis in Human Imagination" http://www.ddg.com/LIS/InfoDesignF97/car/Atlantis1.htm (accessed April 16, 2004).
9. Latour 2004, p. 228.
10. http://www.nap.edu/readingroom/books/obas/contents/values.html (accessed April 15, 2004).
11. Original article published in "April Fool's" version of *The Onion* 35, no. 12 (April 1, 1999), repr. in Gleick 2000; quotes are from pp. 145–46 of Gleick.
12. As quoted in Shermer 1997.
13. "Theodore Kaczynski," http://en.wikipedia.org/wiki/Theodore Kaczynski (accessed May 7, 2004).
14. As quoted in Plotkin 1997, p. 88.
15. Feist 1991a.
16. Gross and Levitt 1998, pp. 2–3.
17. Ibid., p. 13.
18. "Scientific questions are decided . . .": ibid., p. 46; "such propositions have . . .": ibid., pp. 46–47.
19. Latour 1985, Latour and Woolgar 1986.
20. "It is a reminder . . .": Latour and Woolgar 1986, p. 284; "the concluding chapter . . .": ibid., p. 284.
21. S. Cole 1996; T. Kuhn as quoted in S. Cole 1996, p. 276.
22. As quoted in Gross and Levitt 1998, p. 96.
23. Ibid., p. 91.
24. Sokal 1996a, p. 217.
25. "Deep analyses one step further . . .": ibid., p. 217; Sokal 1996b.
26. Sokal 1996b, pp. 62–63.
27. Fox Keller 1996.
28. Sagan 1996, p. 263.
29. R. Dunbar quote: 1995, p. 96, see also Atran 1990, Shepard 1997.
30. Also see Gleick 2000 for other examples of lucid and yet "popular" science writing. In fact, *The Best American Science Writing* is an annual series edited each year by a different well-known popularizer of science, e.g., Matt Ridley (2002) or Oliver Sacks (2003).

# Bibliography


Achter, J. A., Lubinski, D., and Benbow, C. P. 1996. Multipotentiality among the intellectually gifted: "It was never there and already it's vanishing." *Journal of Counseling Psychology* 43: 65–76.

Achter, J. A., Lubinski, D., Benbow, C. P., and Eftekhari-Sanjani, H. 1999. Assessing vocational preferences among intellectually gifted adolescents adds incremental validity to abilities: A discriminant analysis of educational outcomes over a 10-year interval. *Journal of Educational Psychology* 91: 777–86.

Adolphs, R., Tranel, D., and Damasio, A. R. 2001. The human amygdala in social judgment. *Nature* 393: 470–74.

Aguirre, G. K., Detre, J. A., Alsop, D. C., and D'Esposito, M. 1996. The parahippocampus subserves topographical learning in man. *Cerebral Cortex* 6: 823–29.

Aiello, L., and Dunbar, R. I. M. 1993. Neocortex size, group size and the evolution of language. *Current Anthropology* 34: 184–93.

Aiken, N. E. 1998. *The biological origins of art.* Westport, Conn.: Praeger.

Albert, R. S. 1975. Toward a behavioral definition of genius. *American Psychologist* 30: 140–51.

Allport, G. W. 1937. *Personality: A psychological interpretation.* New York: Holt, Rinehart, and Winston.

———. 1985. The historical background of social psychology. In *Handbook of social psychology,* ed. G. Lindzey and E. Aronson. Hillsdale, N.J.: Erlbaum.

Amabile, T. 1996. *Creativity in context.* New York: Westview.

American Psychiatric Association. 1994. *Diagnostic and statistical manual for mental disorders,* 4th ed. Washington, D.C.: APA Publishing.

Antell, S. E., and Keating, D. P. 1983. Perception of numerical invariance in neonates. *Child Development* 54: 695–701.

Anzai, Y. 1991. Learning and use of representations for physics expertise. In *Toward a general theory of expertise,* ed. K. A. Ericsson and J. Smith. Cambridge: Cambridge University Press.

Arthur, A. R. 2001. Personality, epistemology and psychotherapists' choice of theoretical model: A review and analysis. *European Journal of Psychotherapy, Counseling and Health* 4: 45–64.

Arvey, R. D., and Dewhirst, H. D. 1976. Goal-setting attributes, personality variables, and job satisfaction. *Journal of Vocational Behavior* 9: 179–89.

Arvey, R. D., Dewhirst, H. D., and Brown, E. M. 1978. A longitudinal study of the impact of changes in goal setting on employee satisfaction. *Personnel Psychology* 31: 595–608.

Asch, S. E. 1956. Studies of independence and conformity: A minority of one against a unanimous majority. *Psychological Monographs* 709: Whole No. 16.

Astin, H. S. 1975. Sex differences in mathematical and scientific precocity. *Journal of Special Education* 9: 79–91.

Atran, S. 1990. *Cognitive foundations of natural history: Towards an anthropology of science.* Cambridge: Cambridge University Press.

———. 1994. Core domains versus scientific theories: Evidence from systematics and Itza-Maya folkbiology. In *Mapping the mind: Domain specificity in cognition and culture,* ed. L. A. Hirschfeld and S. A. Gelman. Cambridge: Cambridge University Press.

Atwood, G. E., and Tomkins, S. S. 1976. On subjectivity of personality theory. *Journal of the History of the Behavioral Sciences* 12: 166–77.

Bachtold, L. M. 1976. Personality characteristics of women of distinction. *Psychology of Women Quarterly* 1: 70–78.

Bachtold, L. M., and Werner, E. E. 1972. Personality characteristics of women scientists. *Psychological Reports* 31: 391–96.

Backman, M. E. 1972. Patterns of mental abilities: Ethnic, socioeconomic, and sex differences. *American Educational Research Journal* 9: 1–12.

Bacon, R. 1268/1907. On experimental science. In *The Library of Original Sources,* vol. 4: *The Early Medieval World,* ed. O. J. Thatcher. Milwaukee, Wis.: University Research Extension.

Baddeley, A. D. 1998. The central executive: A concept and some misconceptions. *Journal of International Neuropsychological Society* 4: 523–26.

Baillargéon, R. 1987. Object permanence in 3.5- and 4.5-month-old infants. *Developmental Psychology* 23: 655–64.

Baillargéon, R., Spelke, E., and Wasserman, S. 1985. Object permanence in five-month-old infants. *Cognition* 20: 191–208.

Baker, D. R. 1985. Predictive value of attitude, cognitive ability, and personality to science achievement in middle school. *Journal of Research in Science Teaching* 22: 103–13.

Bandura, A. 1982. Self-efficacy mechanism in human agency. *American Psychologist* 37: 122–47.

Banyas, C. A. 1999. Evolution and phylogenetic history of the frontal lobes. In *The human frontal lobes: Functions and disorders,* ed. B. L. Miller and J. L. Cummings. New York: Guilford Press.

Barber, B. 1952. *Science and the social order.* New York: Free Press.

———. 1961. Resistance by scientists to scientific discovery. *Science* 134: 596–602.

Bargh, J. A., and Chartrand, T. L. 1999. The unbearable automaticity of being. *American Psychologist* 54: 462–79.

Barkow, J. H., Cosmides, L., and Tooby, J., eds. 1992. *The adapted mind: Evolutionary psychology and the generation of culture.* New York: Oxford.

Barnes, M. L., and Rosenthal, R. 1985. Interpersonal effects of experimenter attractiveness, attire, and gender. *Journal of Personality and Social Psychology* 48: 435–46.

Baron-Cohen, S. 1989. Perceptual role-taking and proto-declarative pointing in autism. *British Journal of Developmental Psychology* 7: 113–27.

———. 1991. Precursors to a theory of mind: Understanding attention in others. In *Natural theories of mind: Evolution, development and simulation of everyday mindreading,* ed. A. Whiten. Cambridge: Basil Blackwell.

———. 2001. Theory of mind and autism: A review. In *International review of research in mental retardation: Autism,* vol. 23, ed. L. M. Glidden. San Diego, Calif.: Academic Press.

Baron-Cohen, S., Bolton, P., Wheelwright, S., Short, L., Mead, G., Smith, A., and Scahill, V. 1998. Autism occurs more often in families of physicists, engineers, and mathematicians. *Autism* 2: 296–301.

Baron-Cohen, S., Wheelwright, S., and Jolliffe, T. 1997. Is there a "language of the eyes"? Evidence from normal adults and adults with autism or Asperger syndrome. *Visual Cognition* 4: 311–32.

Baron-Cohen, S., Wheelwright, S., Skinner, R., Martin, J., and Clubley, E. 2001. The Autism-Spectrum Quotient (AQ): Evidence from Asperger syndrome/high-functioning autism, males and females, scientists and mathematicians. *Journal of Autism and Developmental Disorders* 31: 5–17.

Baron-Cohen, S., Wheelwright, S., Stone, V. E., and Rutherford, M. 1999. A mathematician, a physicist, and a computer scientist with Asperger syndrome: Performance on folk psychology and folk physics tests. *Neurocase* 5: 475–83.

Baron-Cohen, S., Wheelwright, S., Stott, C., Bolton, P., and Goodyer, I. 1997. Is there a link between engineering and autism? *Autism* 1: 101–9.

Barron, F. 1963. *Creativity and psychology health.* New York: Van Nostrand.

Barron, F., and Harrington, D. 1981. Creativity, intelligence, and personality. *Annual Review of Psychology* 32: 439–76.

Barrow, J. D. 1995. *The artful universe.* Boston: Little, Brown.

———. 1998. *Impossibility: The limits of science and the science of limits.* Oxford: Oxford University Press.

Barton, K., and Cattell, H. 1972. Personality characteristics of female psychology, science and art majors. *Psychological Reports* 31: 807–13.

Baumeister, R. F., Bratslavsky, E., Muraven, M., and Tice, D. M. 1998. Ego depletion: Is the active self a limited resource? *Journal of Personality and Social Psychology* 74: 1252–65.

Bayer, A. E., and Dutton, J. E. 1977. Career age and research—professional activities of academic scientists: Tests of alternative non-linear models and some implications for higher education faculty policies. *Journal of Higher Education* 48: 259–82.

Beard, G. M. 1874. *Legal responsibilities in old age.* New York: Russell Sage.

Beeman, M. J. 1993. Semantic processing in the right hemisphere may contribute to drawing inferences during comprehension. *Brain and Language* 44: 80–120.

Beeman, M. J., and Bowden, E. 2000. The right hemisphere maintains solution-related activation for yet-to-be solved insight problems. *Memory and Cognition* 28: 1231–41.

Beeman, M. J., Bowden, E. M., and Gernsbacher, M. A. 2000. Right and left hemisphere cooperation for drawing predictive and coherence inferences during normal story comprehension. *Brain and Language* 71: 310–36.

Bell, E. T. 1937. *Men of mathematics.* New York: Simon and Schuster.

Benbow, C. P. 1988. Sex differences in mathematical reasoning ability in intellectually talented preadolescents: Their nature, effects, and possible causes. *Behavioral and Brain Sciences* 11: 169–83.

Benbow, C. P., and Lubinski, D. 1993. Psychological profiles of the mathematically talented: Some sex differences and evidence supporting their biological basis. In *The origins and development of high ability,* ed. G. R. Bock and K. Ackrill. Chichester, Eng.: John Wiley and Sons.

Benbow, C. P., Lubinski, D., Shea, D. L., and Eftekhari- Sanjani, H. E. 2000. Sex differences in mathematical reasoning ability at age 13: Their status 20 years later. *Psychological Science* 11: 474–80.

Benbow, C. P., and Minor, L. L. 1986. Mathematically talented students and achievement in the high school sciences. *American Educational Research Journal* 23: 425–36.

Benbow, C. P., and Stanley, J. C. 1980. Sex differences in mathematical ability: Fact or artifact? *Science* 210: 1262–64.

———. 1982. Consequences in high school and college of sex differences in mathematical reasoning ability: A longitudinal perspective. *American Educational Research Journal* 19: 598–622.

———. 1983. Sex differences in mathematical ability: More facts. *Science* 222: 1029–31.

Benjamin, J., Li, L., Patterson, C., Greenburg, B. D., Murphy, D. L., and Hamer, D. H. 1996. Population and familial association between the D4 dopamine receptor gene and measures of novelty seeking. *Nature Genetics* 12: 81–84.

Bennett, C. H., Li, M., and Ma, B. 2003. Chain letters and evolutionary histories. *Scientific American* 288: 76–81.

Bennett, E. L., Rosenzweig, M. R., and Diamond, M. C. 1969. Rat brain: Effects of environmental enrichment on wet and dry weights. *Science* 163: 825–26.

Berger, J. 1994. *The young scientists: America's future and the winning of the Westinghouse.* Reading, Mass.: Addison-Wesley.

Berlin, B. 1992. *Ethnobiological classification: Principles of categorization of plants and animals in traditional societies.* Princeton: Princeton University Press.

Bernal, B. 1939. *The social function of science.* New York: MacMillan.

Berry, C. 1981. The Nobel scientists and the origins of scientific achievement. *British Journal of Sociology* 32: 381–91.

Bever, T. G., Mehler, J., and Epstein, J. 1968. What children do in spite of what they know. *Science* 162: 921–24.

Bleier, R. 1988. The plasticity of the human brain and human potential. *Behavioral and Brain Sciences* 11: 184–85.

Bloom, P. 1998. Some issues in the evolution of language and thought. In *The evolution of mind,* ed. D. D. Cummins and C. Allen. New York: Oxford University Press.

Boesch, C., and Boesch, H. 1989. Hunting behavior of wild chimpanzees in the Taï National Park. *American Journal of Physical Anthropology* 78: 547–73.

Bolton, P., and Rutter, M. 1990. Genetic influences in autism. *International Review of Psychiatry* 2: 67–80.

Bouchard, T. 1998. Genetic and environmental influences on adult intelligence and special mental ability. *Human Biology* 70: 257–59.

Bouchard, T. J., Jr., Lykken, D. T., McGue, M., Segal, N. L., and Tellegan, A. 1990. Sources of human psychological differences: The Minnesota study of twins reared apart. *Science* 250: 223–28.

Bouchard, T. J., Jr., Lykken, D. T., Tellegan, A., and McGue, M. 1996. Genes, drives, environment, and experience: EPD theory revised. In *Intellectual talent,* ed. C. P. Benbow and D. Lubinski. Baltimore: Johns Hopkins University Press.

Bouchard, T. J., Jr., and McGue, M. 1981. Familial studies of intelligence: A review. *Science* 212: 1055–59.

———. 1990. Genetic and rearing environmental influences on adult personality: An analysis of adopted twins reared apart. *Journal of Personality* 58: 263–92.

Bouchard, T. J., Jr., and Segal, N. L. 1990. Advanced mathematical reasoning ability: A behavioral perspective. *Behavioral and Brain Sciences* 13: 191–92.

Bowden, E. M., and Beeman, M. J. 1998. Getting the right idea: Semantic activation in the right hemisphere may help solve insight problems. *Psychological Science* 6: 435–40.

Bower, B. 2001. Faces of perception. *Science News* 160: 10.

Boyd, R. 1991a. Confirmation, semantics, and the interpretation of scientific theories. In *The philosophy of science,* ed. R. Boyd, P. Gasper, and J. Trout. Cambridge: MIT Press.

———. 1991b. On the current status of scientific realism. In *The philosophy of science,* ed. R. Boyd, P. Gasper, and J. D. Trout. Cambridge: MIT Press.

Bradshaw, G. F., Langley, P. W., and Simon, H. A. 1983. Studying scientific discovery by computer simulation. *Science* 222: 971–75.

Bradshaw, J. L. 1989. *Hemispheric specialization and psychological function.* Chichester, Eng.: John Wiley and Sons.

———. 2001. Ars brevis, Vita longa: The possible evolutionary antecedents of art and aesthetics. *Bulletin of Psychology and the Arts* 2: 7–11.

Brewer, W. F., and Chinn, C. A. 1992. Entrenched beliefs, inconsistent information, and knowledge change. In *Proceedings of 1991 international conferences on the learning sciences,* ed. L. Birnbaum. Charlottesville, Va.: Association for the Advancement of Computing in Education.

Brewer, W. F., Chinn, C. A., and Samarapungavan, A. 1998. Explanation in scientists and children. *Mind and Machines* 8: 119–36.

Brewer, W. F., and Samarapungavan, A. 1991. Child theories vs. scientific theories: Differences in reasoning or differences in knowledge? In *Cognition and the symbolic processes: Applied and ecological perspectives,* ed. R. R. Hoffman and D. S. Palermo. Hillsdale, N.J.: Erlbaum.

Brockman, J., ed. 2004. *Curious minds: How a child becomes a scientist.* New York: Pantheon Books.

Brown, R. 1973. *A first language: The early stages.* Cambridge: Harvard University Press.

Bull, R., and Scerif, G. 2001. Executive functioning as a predictor of children's mathematics ability: Inhibition, switching, and working memory. *Developmental Neuropsychology* 19: 273–93.

Buller, D. J., and Hardcastle, V. G. 2000. Evolutionary psychology, meet developmental neurobiology: Against promiscuous modularity. *Brain and Mind* 1: 307–25.

Burgess, N., Jeffery, K. J., and O'Keefe, J., eds. 1999. *The hippocampal and parietal foundations of spatial cognition.* New York: Oxford University Press.

Burnham, R. 1983. *Burnham's celestial handbook,* vol. 1. New York: Dover Publications.

Buss, A. H., and Plomin, R. 1984. *Temperament: Early personality traits.* Hillsdale, N.J.: Erlbaum.

Buss, D. M. 1990. Toward a biologically informed psychology of personality. *Journal of Personality* 58: 1–16.

———. 1999. *Evolutionary psychology: The new science of the mind.* Needham Heights, Mass.: Allyn and Bacon.

Buss, D. M., and Greiling, H. 1999. Adaptive individual differences. *Journal of Personality* 67: 209–43.

Buss, D. M., Haselton, M. G., Shackelford, T. K., Bleske, A. L., and Wakefield, J. C. 1998. Adaptations, exaptations, and spandrels. *American Psychologist* 53: 533–48.

Busse, T. V., and Mansfield, R. S. 1984. Selected personality traits and achievement in male scientists. *Journal of Psychology* 116: 117–31.

Butcher, H. J. 1969. The structure of abilities, interests and personality in 1,000 Scottish school children. *British Journal of Educational Psychology* 39: 154–65.

Butterfield, H. 1960. *The origins of modern science, 1300–1800.* New York: Macmillan.

Butterworth, B. 1999. *What counts: How every brain is hardwired for math.* New York: Free Press.

Buxhoeveden, D. P., Switala, A. E., Roy, E., Litaker, M., and Casanova, M. F. 2001. Morphological differences between minicolumns in human and nonhuman primate cortex. *American Journal of Physical Anthropology* 115: 361–71.

Byler, P. L. 2000. Middle school girls' attitudes toward math and science: Does the setting make a difference? *Dissertation Abstracts International Section A: Humanities and Social Sciences* 61 (6_A): 2181.

Byrne, R. W. 1998. The early evolution of creative thinking. In *Creativity in human evolution and prehistory,* ed. S. Mithen. London: Routledge.

———. 2001. Social and technical forms of primate intelligence. In *Tree of origin: What primate behavior can tell us about human social evolution,* ed. F. B. M. deWaal. Cambridge: Harvard University Press.

Byrne, R. W., and Whiten, A. 1988. *Machiavellian intelligence: Social expertise and the evolution of intellect in monkeys, apes, and humans.* Oxford: Clarendon Press.

Cajete, G. 2000. *Native science: Natural laws of interdependence.* Santa Fe, N.M.: Clear Light.

Calaprice, A. 2005. *The new quotable Einstein.* Princeton, N.J.: Princeton University Press.

Campbell, D. T. 1960. Blind variation and selection retention in creative thought as in other knowledge processes. *Psychological Review* 67: 380–400.

Caramazza, A., and McCloskey, M. 1987. Dissociations of calculation processes. In *Mathematical disabilities: A cognitive neuropsychological perspective,* ed. G. Deloche and X. Seron. Hillsdale, N.J.: Erlbaum.

Carey, S. 1985. *Conceptual change in childhood.* Cambridge: MIT Press.

———. 1992. The origin and evolution of everyday concepts. In *Cognitive models of science,* ed. R. N. Giere. Minneapolis: University of Minnesota Press.

Carey, S., and Spelke, E. 1994. Domain specific knowledge and conceptual change. In *Mapping the mind: Domain specificity in cognition and culture,* ed. L. A Hirschfeld and S. A. Gelman. Cambridge: Cambridge University Press.

Carruthers, P. 2002. The roots of scientific reasoning. In *The cognitive basis of science,* ed. P. Carruthers, S. Stich, and M. Siegal. Cambridge: Cambridge University Press.

Carson, S. H., Peterson, J. B., and Higgins, D. M. 2003. Decreased latent inhibition is associated with increased creative achievement in high-functioning individuals. *Journal of Personality and Social Psychology* 85: 499–506.

Carter, R. 1998. *Mapping the mind.* Berkeley, Calif.: University of California Press.

Cattell, J. M. 1910. A further statistical study of American men of science. *Science* 32: 633–48.

Cattell, J. M., and Brimhall, D. R. 1921. *American Men of Science,* 3rd ed. Garrison, N.Y.: Science Press.

Cattell, R. B., and Drevdahl, J. E. 1955. A comparison of the personality profile 16PF of eminent researchers with that of eminent teachers and administrators, and the general population. *British Journal of Psychology* 46: 248–61.

Chamberlain, T. C. 1890/1965. The method of multiple working hypotheses. *Science* 148: 754–59.

Chambers, J. A. 1964. Relating personality and biographical factors to scientific creativity. *Psychological Monographs: General and Applied* 78: 1–20.

———. 1965. Comments. *Science* 147: 67.

Chan, L. K. S. 1996. Motivational orientations and metacognitive abilities of intellectually gifted students. *Gifted Child Quarterly* 40: 184–93.

Chandrasekhar, S. 1987. *Truth and beauty: Aesthetics and motivations in science.* Chicago, Ill.: University of Chicago Press.

Charness, N. 1988. Expertise in chess, music, and physics: A cognitive perspective. In *The exceptional brain: Neuropsychology of talent and special abilities,* ed. L. K. Obler and D. Fein. New York: Guilford Press.

Chavez, R. A. 2004. On the neurobiology of the creative process. *Bulletin of Psychology and the Arts* 5: 29–35.

Cheney, D., and Seyfarth, R. 1990. *How monkeys see the world.* Chicago: University of Chicago Press.

Cheng, P. C. H., and Simon, H. A. 1992. The right representation for discovery: Finding the conservation of momentum. In *Machine Learning: Proceedings of the Ninth International Conference,* ed. D. Sleeman and P. Edwards. San Mateo, Calif.: Morgan Kaufmann.

———. 1995. Scientific discovery and creative reasoning with diagrams. In *The creative cognition approach,* ed. S. M. Smith, T. B. Ward, and R. A. Finke. Cambridge: MIT Press.

Chi, M. T. H., Feltovich, P. J., and Glaser, R. 1981. Categorization and representation of physics problems by experts and novices. *Cognitive Science* 5: 121–52.

Chinn, C. A., and Brewer, W. F. 2000. Knowledge change in science, religion, and magic. In *Imagining the impossible: Magical, scientific, and religious thinking in children,* ed. K. S. Rosegren, C. N. Johnson, and P. L. Harris. Cambridge: Cambridge University.

Chorney, M. J., Chorney, K., Seese, N., Owen, M. J., Daniels, J., McGuffin, P., Thompson, L. A., Detterman, D. K., Benbow, C. P., Lubinski, D., Eley, T., and Plomin, R. 1998. A quantitative trait locus associated with cognitive ability in children. *Psychological Science* 9: 159–66.

Chow, T. W., and Cummings, J. L. 1999. Frontal-subcortical circuits. In *The human frontal lobes: Functions and disorders,* ed. B. L. Miller and J. L. Cummings. New York: Guilford Press.

Cipolotti, L., Butterworth, B., and Warrington, E. K. 1994. From "one thousand nine hundred and forty-five" to 1000,945. *Neuropsychologia* 32: 503–9.

Clark, R. D., and Rice, G. A. 1982. Family constellations and eminence: The birth-orders of Nobel Prize winners. *Journal of Psychology* 110: 281–87.

Clark, W. R., and Grunstein, M. 2000. *Are we hardwired? The role of genes in human behavior.* Oxford: Oxford University Press.

Clement, J. 1982. Students preconceptions in introductory mechanics. *American Journal of Physics* 50: 66–71.

———. 1989. Learning via model construction and criticism: Protocol evidence on sources of creativity in science. In *Handbook of creativity: Perspectives on individual differences,* ed. J. A. Glover, R. R. Ronning, and C. R. Reynolds. New York: Plenum.

———. 1991. Experts and science students: The use of analogies, extreme cases, and physical intuition. In *Informal reasoning and education,* ed. J. F. Voss, D. N. Perkins, and J. W. Segal. Hillsdale, N.J.: Erlbaum.

Coan, R. W. 1973. Toward a psychological interpretation of psychology. *Journal of the History of the Behavioral Sciences* 9: 313–27.

Cohen, I. B. 1985. *Revolution in science.* Cambridge: Belknap Press of Harvard University Press.

Cohen, J. 1988. *Statistical power analysis for behavioral sciences,* 2nd ed. Hillsdale, N.J.: Erlbaum.

Cole, J. R. 1979. *Fair science: Women in the scientific community.* New York: Free Press.

———. 1987. Women in science. In *Scientific excellence,* ed. D. Jackson and P. Rushton. Beverly Hills, Calif.: Sage.

Cole, J. R., and Cole, S. 1973. *Social stratification in science.* Chicago: University of Chicago Press.

Cole, J. R., and Zuckerman, H. 1984. The productivity puzzle: Persistence and change in patterns of publication in men and women scientists. In *Advances in motivation and achievement,* ed. P. Maeher and M. W. Steinkamp. Greenwich, Conn.: JAI Press.

Cole, S. 1979. Age and scientific performance. *American Journal of Sociology* 84: 958–77.

———. 1996. Voodoo sociology: Recent developments in the sociology of science. In *The flight from science and reason,* ed. P. R. Gross, M. Levitt, and M. W. Lewis. New York: New York Academy of Sciences.

Cole, S., and Zuckerman, H. 1987. Marriage, motherhood, and research performance in science. *Scientific American* 256: 119–25.

Collins, H. M. 1981. Stages in the empirical program of relativism. *Social Studies of Science* 12: 3–10.

Connellan, J., Baron-Cohen, S., Wheelwright, S., Batki, A., and Ahluwalia, J. 2000. Sex differences in human neonatal social perception. *Infant Behavior and Development* 23: 113–18.

Conway, J. B. 1988. Differences among clinical psychologists: Scientists, practitioners, and science-practitioners. *Professional Psychology: Research and Practice* 19: 642–55.

Cooper, E. E. 2000. Spatial-temporal intelligence: Original thinking processes of gifted inventors. *Journal for the Education of the Gifted* 24: 170–93.

Cooper, E. E., and Wojan, T. J. 2000. Differences in the coding of spatial relations in face identification and basic-level object recognition. *Journal of Experimental Psychology: Learning, Memory, and Cognition* 26: 470–88.

Cosmides, L., and Tooby, J. 1992. Cognitive adaptations for social exchange. In *The Adapted mind: Evolutionary psychology and the generation of culture,* ed. J. H. Barkow, L. Cosmides, and J. Tooby. Oxford: Oxford University Press.

Costa, P. T., McCrae, R. R., and Holland, J. L. 1984. Personality and vocational interests in an adult sample. *Journal of Applied Psychology* 69: 390–400.

Coulmas, F. 1989. *The writing systems of the world.* New York: Basil Blackwell.

Cox, C. 1926. *Genetic studies of genius,* vol. 2: *The early mental traits of three hundred geniuses.* Stanford: Stanford University Press.

Crouse, J., and Trusheim, D. 1991. How colleges can correctly determine selection benefits from the SAT. *Harvard Educational Review* 61: 125–47.

Crowley, K., Callanan, M. A., Tenenbaum, H. R., and Allen, E. 2001. Parents explain more often to boys than to girls during shared scientific thinking. *Psychological Science* 12: 258–61.

Crump, T. 1990. *The anthropology of numbers.* Cambridge: Cambridge University Press.

———. 2002. *A brief history of science.* New York: Carroll and Graf.

Cunningham, M. R. 1986. Measuring the physical in physical attractiveness: Quasiexperiments on the sociobiology of female facial beauty. *Journal of Personality and Social Psychology* 50: 925–35.

Curtin, D. W., ed. 1980. *The aesthetic dimension of science.* New York: Philosophical Library.

Daniels, P. T., and Bright, W., eds. 1996. *The world's writing systems.* New York: Oxford University Press.

Darwin, C. 1859. *The origin of species.* London: Murray (repr. 1964, Harvard University Press).

———. 1871. *The descent of man.* London: Murray (repr. 1998, Prometheus Books).

Datta, L. E. 1967. Family religious background and early scientific creativity. *American Sociological Review* 32: 626–35.

Davies, P. G., Spencer, S. J., Quinn, D. M., and Gerhardstein, R. 2002. All consuming images: How television commercials that elicit stereotype threat can restrain women academically and professionally. *Personality and Social Psychology* 28: 615–28.

Deacon, T. 1997. *The symbolic species: The co-evolution of language and the brain.* New York: W. W. Norton.

Deaux, K. 1985. Sex and gender. *Annual Review of Psychology* 36: 49–81.

DeFrancis, J. 1989. *Visible speech: The diverse oneness of writing systems.* Honolulu: University of Hawaii Press.

Dehaene, S. 1997. *The number sense: How the mind creates mathematics.* New York: Oxford University Press.

Dehaene, S., Dehaene-Lambertz, G., and Cohen, L. 1998. Abstract representations of numbers in the animal and human brain. *Trends in Neurosciences* 21: 355–61.

Dehaene, S., Spelke, E., Pinel, P., Stanescu, R., and Tsivkin, S. 1999. Sources of mathematical thinking: Behavioral and brain-imaging evidence. *Science* 284: 970–74.

De Mey, M. 1989. Cognitive paradigms and the psychology of science. In *Psychology of science: Contributions to metascience,* ed. B. Gholson, W. R. Shadish, R. A. Neimeyer, and A. C. Houts. Cambridge: Cambridge University Press.

Dennis, W. 1954. Predicting scientific productivity in later decades from records of earlier decades. *Journal of Gerontology* 9: 465–67.

———. 1956. Age and productivity among scientists. *Science* 123: 724–25.

———. 1958. The age decrement in outstanding scientific contributions: Fact or artifact? *American Psychologist* 13: 457–60.

———. 1966. Creative productivity between the ages of 20 and 80 years. *Journal of Gerontology* 21: 1–8.

d'Errico, F., and Cacho, C. 1994. Notation versus decoration in the Upper Paleolithic: A case study from Tossal de la Roca, Alicante, Spain. *Journal of Archeological Science* 21: 185–200.

Derry, G. 1999. *What is science and how it works.* Princeton: Princeton University Press.

Despete, A., Roeyers, H., and Buysse, A. 2001. Metacognition and mathematical problem solving in Grade 3. *Journal of Learning Disabilities* 34: 435–49.

Devlin, B., Daniels, M., and Roeder, K. 1997. The heritability of IQ. *Nature* 388: 368–471.

Devlin, K. 2000. *The math gene: How mathematical thinking evolved and why numbers are like gossip.* New York: Basic Books.

Diamond, A. M. 1986. The life-cycle research productivity of mathematicians and scientists. *Journal of Gerontology* 41: 520–25.

Digman, J. M. 1990. Personality structure: Emergence of the Five-Factor Model. *Annual Review in Psychology* 41: 417–40.

Diringer, D. 1962. *Writing.* New York: Frederick A. Praeger.

Dissanayake, E. 1992. *Homo aestheticus: Where art comes from and why.* New York: Free Press.

Dolan, R. J. 2000. Emotion processing in the human brain revealed though functional neuroimaging. In *The new cognitive neurosciences,* ed. M. S. Gaszzaniga. Cambridge: MIT Press.

Donald, M. 1991. *Origins of the modern mind: Three stages in the evolution of culture and cognition.* Cambridge: Harvard University Press.

Dunbar, K. 1995. How scientists really reason: Scientific reasoning in real-world laboratories. In *The nature of insight,* ed. R. J. Sternberg and J. Davidson. Cambridge: MIT Press.

———. 2001. The analogical paradox: Why analogy is so easy in naturalistic settings, yet so difficult in the psychological laboratory. In *The analogical mind: Perspectives from cognitive science,* ed. D. Gentner, K. Holyoak, and B. N. Kokinov. Cambridge: MIT Press.

———. 2002. Understanding the role of cognition in science: The *Science as Category* framework. In *The cognitive basis of science,* ed. P. Carruthers, S. Stich, and M. Siegal. Cambridge: Cambridge University Press.

Dunbar, K., and Blanchette, I. 2001. The *in vivo/in vitro* approach to cognition: The case of anology. *Trends in Cognitive Sciences* 5: 334–39.

Dunbar, R. I. M. 1992. Neocortex size as a constraint on group size in primates. *Journal of Human Evolution* 20: 469–93.

———. 1993. Coevolution of neocortical size, group size and language in humans. *Behavioral and Brain Sciences* 16: 681–735.

———. 1995. *The trouble with science.* Cambridge: Harvard University Press.

———. 2001. Brains on two legs: Group size and the evolution of intelligence. In *Tree of origin: What primate behavior can tell us about human social evolution,* ed. F. B. M. deWaal. Cambridge: Harvard University Press.

Durant, W. 1926/1961. *The story of philosophy.* New York: Washington Square Press.

Easterbrook, M. A., Kisilevsky, B. S., Muir, D. W., and Laplante, D. P. 1999. Newborns discriminate schematic faces from scrambled faces. *Canadian Journal of Experimental Psychology* 53: 231–41.

Ebstein, R. P., Novick, O., Umansky, R., Priel, B., Osher, Y., Blaine, D., Bennet, E. R., Nemanov, L., Katz, M., and Belmaker, R. H. 1996. Dopamine D4 receptor D4DR exon III polymorphism associated with the human personality trait Novelty Seeking. *Nature Genetics* 12: 78–80.

Eccles, J. 1987. Gender roles and women's achievement-related decisions. *Psychology of Women Quarterly* 11: 135–72.

———. 1989. *Evolution of the brain.* London: Routledge.

Edelman, G. 1987. *Neural darwinism.* New York: Basic Books.

Eflin, J. T., and Kite, M. E. 1996. Teaching scientific reasoning through attribution theory. *Teaching of Psychology* 23: 87–91.

Eibl-Eibesfeldt, I. 1970. *Ethology, the biology of behavior.* New York: Holt, Rinehart and Winston.

Eiduson, B. T. 1962. *Scientists: Their psychological world.* New York: Basic Books.

———. 1974. 10 year longitudinal Rorschachs on research scientists. *Journal of Personality Assessment* 38: 405–10.

Einhorn, H. J., and Hogarth, R. M. 1978. Confidence in judgment: Persistence of the illusion of validity. *Psychological Review* 85: 395–416.

Einstein, A. 1950. *Out of my later years.* Totowa, N.J.: Littlefield, Adams.

Ekman, P. 1994. Strong evidence for universals in facial expressions: A reply to Russell's mistaken critique. *Psychological Bulletin* 115: 268–87.

Ekman, P., and Friesen, W. V. 1971. Constants across cultures in the face and emotion. *Journal of Personality and Social Psychology* 17: 124–29.

Ekman, P., Friesen, W. V., O'Sullivan, M., Chan, A., Diacoyanni-Tarlatzis, I., Heider, K., Krause, R., LeCompte, W. A., Pitcairn, T., Ricci Bitti, P. E., Scherer, K. R., Tomita, M., and Tzavaras, A. 1987. Universals and cultural differences in the judgments of facial expressions of emotion. *Journal of Personality and Social Psychology* 53: 712–17.

Ekman, P., Sorenson, E. R., and Friesen, W. V. 1969. Pan-cultural elements in facial displays of emotion. *Science* 164: 86–88.

Elfenbein, H. A., and Ambady, N. 2002. On the universality of cultural specificity of emotion recognition: A meta-analysis. *Psychological Bulletin* 128: 203–35.

Elias, M. J., Arnold, H., and Hussey, C. S., eds. 2003. *EQ + IQ = best leadership practices for caring and successful schools.* Thousand Oaks, Calif.: Corwin Press.

Elms. A. C. 1994. *Uncovering lives: The uneasy alliance between biography and psychology.* Oxford: Oxford University Press.

Enard, W., Przeworski, M., Fisher, S. E., Lai, C. S. L., Wiebe, V., Kitano, T., Monoco, A. P., and Pääbo, S. 2002. Molecular evolution of FOXP2, a gene involved in speech and language. *Nature* 418: 869–72.

Epstein, R., DeYoe, E. A., Press, D. Z., Rosen, A. C., and Kanwisher, N. 2001. Neuro-psychological evidence for a topographical learning mechanism in parahippocampal cortex. *Cognitive Neuropsychology* 18: 481–508.

Epstein, R., and Kanwisher, N. 1998. A cortical representation of the local visual environment. *Nature* 392: 598–601.

Epstein, S. 1994. Integration of the cognitive and psychodynamic unconscious. *American Psychologist* 49: 709–24.

Eriksson, P. S., Perfilieva, E., Bjork-Eriksson, T., Alborn, A. M., Nordborg, C., Peterson, D. A., and Gage, F. H. 1998. Neurogenesis in the adult human hippocampus. *Nature Medicine* 4: 1313–17.

Eysenck, H. J. 1988. O tempora, o mores! *Behavioral and Brain Sciences* 11: 189–90.

———. 1990a. Biological dimensions of personality. In *Handbook of personality: Theory and research,* ed. L. A. Pervin. New York: Guilford Press.

———. 1990b. Genetic and environmental contributions to individual differences: The three major dimensions of personality. *Journal of Personality* 58: 245–61.

———. 1995. *Genius: The natural history of creativity.* Cambridge: Cambridge University Press.

Falk, D. 1983. Cerebral cortices of East African early hominids. *Science* 221: 1072–74.

———. 2000. Hominid brain evolution and the origins of music. In *The origins of music,* ed. N. L. Wallin, B. Merker, and S. Brown. Cambridge: MIT Press.

Falkenhainer, B. 1990. A unified approach to explanation and theory formation. In *Computational models of scientific discovery and theory formation,* ed. J. Shrager and P. Langley. San Mateo, Calif.: Morgan Kaufman.

Fantz, R. L. 1963. Pattern vision in newborn infants. *Science* 140: 296–97.

Farmer, H. S. 1988. Predicting who our future scientists and mathematicians will be. *Behavioral and Brain Sciences* 11: 190–91.

Farmer, H. S., Wardrop, J. L., and Rotella, S. C. 1999. Antecedent factors differentiating women and men in science/nonscience careers. *Psychology of Women Quarterly* 23: 763–80.

Fauconnier, G., and Turner, M. 2002. *The way we think: Conceptual blending and the mind's hidden complexities.* New York: Basic Books.

Fay, A. L., Klahr, D., and Dunbar, K. 1990. Are there developmental milestones in scientific reasoning? In *Proceedings of the twelfth annual conference of cognitive science society,* Conference Proceedings. Hillsdale, N.J.: Erlbaum.

Feist, G. J. 1991a. *The psychology of science: Personality, cognitive, motivational and working styles of eminent and less eminent scientists.* PhD diss., University of California, Berkeley.

———. 1991b. Synthetic and analytic thought: Similarities and differences among art and science students. *Creativity Research Journal* 4: 144–55.

———. 1993. A structural model of scientific eminence. *Psychological Science* 4: 366–71.

———. 1994. Personality and working style predictors of integrative complexity: A study of scientists' thinking about research and teaching. *Journal of Personality and Social Psychology* 67: 474–84.

———. 1995. Psychology of science and history of psychology: Putting behavioral generalizations to the test. *Psychological Inquiry* 6: 119–23.

———. 1997. Quantity, impact, and depth of research as influences on scientific eminence: Is quantity most important? *Creativity Research Journal* 10: 325–35.

———. 1998. A meta-analysis of the impact of personality on scientific and artistic creativity. *Personality and Social Psychological Review* 2: 290–309.

———. 2001. Natural and sexual selection in the evolution of creativity. *Bulletin of Psychology and the Arts* 2: 11–16.

———. 2004a. The evolved fluid specificity of human creative talent. In *Creativity: From potential to realization,* ed. R. J. Sternberg, E. L. Grigorenko, and J. L. Singer. Washington, D.C.: American Psychological Association.

———. 2004b. Domain-specific creativity in the physical sciences. In *Creativity across domains: Faces of the muse,* ed. J. C. Kaufman and J. Baer. Mahwah, N.J.: Erlbaum.

———. 2004c. Creativity and the frontal lobes. *Bulletin of Psychology and the Arts* 5: 21–28.

———. 2005. The development of scientific talent in Westinghouse finalists and members of the National Academy of Sciences. *Journal of Adult Development.*

Feist, G. J., and Barron, F. X. 2003. Predicting creativity from early to late adulthood: Intellect, potential and personality. *Journal of Research in Personality* 37: 62–88.

Feist, G. J., and Gorman, M. E. 1998. Psychology of science: Review and integration of a nascent discipline. *Review of General Psychology* 2: 3–47.

Feist, G. J., Paletz, S., and Weitzer, W. In preparation. Predicting scientific interest in college students: The influence of quantitative skills, gender, self-image, and personality.

Fennema, E. 1985. Attribution theory and achievement in mathematics. In *The development of reflection,* ed. S. R. Yussen. New York: Academic Press.

Ferry, T. R., Fouad, N. A., and Smith, P. L. 2000. The role of family context in a social cognitive model for career related choice behavior: A math and science perspective. *Journal of Vocational Behavior* 57: 348–64.

Finke, R. A., Ward, T. B., and Smith, S. M. 1992. *Creative cognition: Theory, research, and applications.* Cambridge: The MIT Press.

Finlay, B. L., and Darlington, R. B. 1995. Linked regularities in the development and evolution of mammalian brains. *Science* 286: 1578–84.

Fiore, S. M., and Schooler, J. W. 1998. Right hemisphere contributions to creative problem solving: Converging evidence for divergent thinking. In *Right hemisphere language comprehension: Perspectives from cognitive neuroscience,* ed. M. Beeman and C. Chiarello. Mahwah, N.J.: Erlbaum.

Fisch, R. 1977. Psychology of science. In *Science, technology, and society: A cross disciplinary perspective,* ed. I. Spiegel-Rösing and D. de Solla Price. London: Sage.

Fischbein, S. 1990. Biosocial influences on sex differences for ability and achievement test results as well as marks at school. *Intelligence* 14: 127–39.

Flavell, J. H. 1979. Metacognition and cognitive monitoring: A new area of cognitive-developmental inquiry. *American Psychologist* 34: 906–11.

Fodor, J. A. 1983. *The modularity of mind: An essay on faculty psychology.* Cambridge: MIT Press.

Fox, L. H. 1976. Sex differences in mathematical precocity: Bridging the gap. In *Intellectual talent: Research and development,* ed. D. P. Keating. Baltimore: Johns Hopkins University Press.

Fox, M. F. 1998. Women in science and engineering: Theory, practice and policy in programs. *Signs* 24: 201–23.

Fox, P. T., and Lancaster, J. L. 1994. Neuroscience on the Net. *Science* 266: 994–96.

Fox Keller, E. 1985. *Reflections on gender and science.* New Haven, Conn.: Yale University Press.

———. 1996. Letter to editor. *Lingua Franca,* July/August.

Frazer, J. 1922/1996. *The golden bough: A study in magic and religion.* New York: Penguin Books.

Freedman, E. 1995. October. *Working memory and testing multiple hypotheses.* Paper presented at Annual Convention of the Society for Social Studies of Science, Charlottesville, Va.

Freeman, M. S. 2003. Personality traits as predictors of a preferred theoretical orientation in beginning counselor education students. *Dissertation Abstracts International Section A: Humanities and Social Sciences* 2-A: 407.

Freud, S. 1900/1953. Interpretation of dreams. In *The standard edition of the complete psychological works of Sigmund Freud,* vol. 3, ed. and trans. by J. Strachey. London: Hogarth Press.

Frith, U. 1991. *Autism and Asperger's syndrome.* Cambridge: Cambridge University Press.

Fuller, S. 1988. *Social epistemology.* Bloomington: Indiana University Press.

Funder, D. C. 1991. Global traits: A neo-Allportian approach to personality. *Psychological Science* 2: 31–39.

Fuster, J. M. 1999. Cognitive functions of the frontal lobes. In *The human frontal lobes: Functions and disorders,* ed. B. L. Miller and J. L. Cummings. New York: Guilford Press.

———. 2002. Frontal lobe and cognitive development. *Journal of Neurocytology* 31: 373–85.

Gage, F. H. 2003. Brain, repair yourself. *Scientific American* 289: 47–53.

Galin, D. 1974. Implications for psychiatry of left and right cerebral specialization: A neurophysiological context for unconscious processes. *Archives of General Psychiatry* 31: 572–83.

Gallagher, H. L., and Frith, C. D. 2003. Functional imaging of "theory of mind." *Trends in Cognitive Sciences* 7: 77–83.

Gallistel, C. R. 1990. *The organization of learning.* Cambridge: MIT Press.

Galton, F. 1874. *English men of science.* London: Macmillan.

Gantz, B. S., Erickson, C. O., and Stephenson, R. W. 1972. Some determinants of promotion in a research and development population. *Proceedings of 80th Convention, American Psychological Association,* 451–52.

Gardner, H. 1983. *Frames of mind: The theory of multiple intelligences.* New York: Basic Books.

———. 1999. *Intelligence reframed: Multiple intelligences for the 21st century.* New York: Basic Books.

Garwood, D. S. 1964. Personality factors related to creativity in young scientists. *Journal of Abnormal and Social Psychology* 68: 413–19.

Gazzaniga, M., and Heatherton, T. 2003. *Psychological science: Mind, brain, and behavior.* New York: W. W. Norton.

Geary, D. C. 1998. *Male, female: The evolution of human sex differences.* Washington, D.C.: American Psychological Association.

Geary, D. C., and Huffman, K. J. 2002. Brain and cognitive evolution: Forms of modularity and functions of mind. *Psychological Bulletin* 128: 667–98.

Gelb, I. J. 1963. *A study of writing,* 2nd ed. Chicago: University of Chicago Press.

Gelman, R. 1990. First principles organize attention to and learning about relevant data: Number and the animate-inanimate distinction as examples. *Cognitive Science* 14: 79–106.

Gelman, R., and Brenneman, L. 1994. First principles can support both universal and culture-specific learning about number and music. In *Mapping the mind: Domain specificity in cognition and culture,* ed. L. A. Hirschfeld and S. A. Gelman. New York: Cambridge University Press.

Gelman, R., and Gallistel, C. R. 1978. *The child's understanding of number.* Cambridge: Harvard University Press.

Gelman, R., Spelke, E., and Meck, E. 1983. What preschoolers know about animate and inanimate objects. In *The development of symbolic skills,* ed. D. Rogers and J. A. Sloboda. London: Plenum.

Gentner, D., Bowdle, B., Wolff, P., and Boronat, C. 2001. Metaphor is like analogy. In *The analogical mind: Perspective from cognitive science,* ed. D. Gentner, K. Holyoak, and B. Kokinov. Cambridge: MIT Press.

Gentner, D., and Gentner, G. R. 1983. Flowing waters or teeming crowds: Mental models of electricity. In *Mental models,* ed. D. Gentner and A. L. Stevens. Hillsdale, N.J.: Erlbaum.

Gentner, D., Holyoak, K., and Kokinov, B., eds. 2001. *The analogical mind: Perspective from cognitive science.* Cambridge: MIT Press.

Gentner, D., and Jeziorski, M. 1989. Historical shifts in the use of analogy in science. In *Psychology of science: Contributions to metascience,* ed. B. Gholson, W. R. Shadish, R. A. Neimeyer, and A. C. Houts. Cambridge: Cambridge University Press.

Georghiades, P. 2000. Beyond conceptual change learning in science education: Focusing on transfer, durability and metacognition. *Educational Research* 42: 119–39.

Geschwind, N., and Behan, P. 1982. Left-handedness: Association with immune disease, migraine, and developmental learning disorder. *Proceedings of the National Academy of Sciences* 79: 5097–100.

Gesell, A., and Thompson, H. 1938. *The psychology of early growth including norms of infant behavior and a method of genetic analysis.* New York: Macmillan.

Getzels, J. W. 1987. Creativity, intelligence, and problem finding: Retrospect and prospect. In *Frontiers of creativity research,* ed. S. G. Isaksen. Buffalo, N.Y.: Bearly.

Gholson, B., and Houts, A. C. 1989. Toward a cognitive psychology of science. *Social Epistemology* 3: 107–27.

Gholson, B., Shadish, W. R., Neimeyer, R. A., and Houts, A. C., eds. 1989. *The psychology of science: Contributions to metascience.* Cambridge: Cambridge University Press.

Gibson, K. R. 1991. Myelination and behavioral development: A comparative perspective on questions of neoteny, altriciality, and intelligence. In *Brain maturation and cognitive development: Comparative and cross-cultural perspectives,* ed. K. R. Gibson and A. C. Peterson. New York: Adeline Gruyter.

Givon, T., and Maller, B. F. 2002. *The evolution of language out of pre-language.* Amsterdam: John Benjamins Publishing.

Gleick, J. 1992. *Genius: Richard Feynman and modern physics.* New York: Pantheon.

Gleick, J., ed. 2000. *The best American science writing 2000.* New York: Harper Collins.

Globus, A., Rosenzweig, M. R., Bennett, E. L., and Diamond, M. C. 1973. Effects of differential experience on dendritic spine counts in rat cerebral cortex. *Journal of Comparative and Physiological Psychology* 82: 175–81.

Glynn, S. M., Britton, B. K., Semrud-Clikeman, M., and Muth, K. D. 1989. Analogical reasoning and problem solving in science textbooks. In *Handbook of creativity: Perspectives on individual differences,* ed. J. A. Glover, R. R. Ronning, and C. R. Reynolds. New York: Plenum.

Glynn, S. M., and Muth, K. D. 1994. Reading and writing to learn science: Achieving scientific literacy. *Journal of Research in Science Teaching* 31: 1057–73.

Goel, V., Grafman, J., Sadato, N., and Hallett, M. 1995. Modeling other minds. *Neuroreport* 6: 1741–46.

Goldberg, E. L., and Alliger, G. M. 1992. Assessing the validity of the GRE for students in psychology: A validity generalization approach. *Educational and Psychological Measurement* 52: 1019–27.

Goleman, D. 1995. *Emotional intelligence.* New York: Bantam Books.

Gonzalez, J. J., Hynd, G. W., and Martin, R. P. 1994. Neuropsychology of temperament. In *The neuropsychology of individual differences,* ed. P. A. Vernon. San Diego, Calif.: Academic Press.

Goodall, J. 1986. *Chimpanzees of the Gombie.* Cambridge: Harvard University Press.

Gooding, D. C. 1990. *Experiment and the making of meaning: Human agency in scientific observation and experiment.* Dordrecht: Kluwer Academic Publishers.

Gooding, D. C., and Addis, T. R. 1993. *Modeling Faraday's experiments with visual functional programming 1: Models, methods and examples.* Working Paper: Joint Research Councils' Initiative on Cognitive Science and Human Computer Interaction Special Project Grant #9107137.

Gopnik, A., and Glymour, C. 2002. Causal maps and Bayes nets: A cognitive and computational account of theory formation. In *The cognitive basis of science,* ed. P. Carruthers, S. Stich, and M. Siegal. Cambridge: Cambridge University Press.

Gopnik, A., Meltzoff, A. N., and Kuhl, P. K. 1999. *The scientist in the crib: Minds, brains, and how children learn.* New York: William Morrow.

Gopnik, A., Sobel, D. M., Schulz, L. E., and Glymour, C. 2001. Causal learning mechanisms in very young children: Two-, three-, and four-year-olds infer causal relations from patterns of variation and covariation. *Developmental Psychology* 37: 620–29.

Gopnik, A., and Wellman, H. M. 1994. The theory theory. In *Mapping the mind: Domain specificity in cognition and culture,* ed. L. A. Hirschfeld and S. A. Gelman. Cambridge: Cambridge University Press.

Goren-Inbar, N., Alperson, N., Kislev, M. E., Simchoni, O., Melamed, Y., Ben-Nun, A., and Werker, E. 2004. Evidence of hominin control of fire at Gesher Benot Ya'qov Israel. *Science* 304: 725–27.

Gorman, M. E. 1986. How the possibility of error affects falsification on a task that models scientific problem-solving. *British Journal of Psychology* 77: 85–96.

———. 1989. Error, falsification and scientific inference: An experimental investigation. *Quarterly Journal of Experimental Psychology* 41: 385–412.

———. 1992. *Simulating science: Heuristics, mental models and technoscientific thinking.* Bloomington: Indiana University Press.

———. 1995. Confirmation, disconfirmation and invention: The case of Alexander Graham Bell and the telephone. *Thinking and Reasoning* 11: 31–53.

Gorman, M. E., ed. 2004. *New directions for the cognitive study of scientific and technological thinking.* Mahwah, N.J.: Erlbaum.

Gorman, M. E., and Gorman, Margaret E. 1984. A Comparison of disconfirmatory, confirmatory and a control strategy on Wason's 2–4–6 task. *Quarterly Journal of Experimental Psychology* 36: 629–48.

Gorman, M. E., Gorman, Margaret E., Latta, R. M., and Cunningham, G. 1984. How disconfirmatory, confirmatory and combined strategies affect group problem-solving. *British Journal of Psychology* 75: 65–79.

Gorman, M. E., and Rosenwein, R. 1995. Simulating social epistemology. *Social Epistemology* 91: 71–9.

Gorman, M. E., Stafford, A., and Gorman, Margaret E. 1987. Disconfirmation and dual hypotheses on a more difficult version of Wason's 2–4–6 task. *Quarterly Journal of Experimental Psychology* 39: 1–28.

Gough, H. G. 1961, February. *A personality sketch of the creative research scientist.* Paper presented at 5th Annual Conference on Personnel and Industrial Relations Research, UCLA, Los Angeles, Calif.

———. 1976. What happens to creative medical students. *Journal of Medical Education* 51: 461–67.

———. 1987. *California Psychological Inventory manual.* Palo Alto: Calif.: Consulting Psychologists Press.

Gough, H. G., and Woodworth, D. G. 1960. Stylistic variations among professional research scientists. *Journal of Psychology* 49: 87–98.

Gould, S. J. 1991. Exaptation: A crucial tool for evolutionary psychology. *Journal of Social Issues* 47: 43–65.

Gould, S. J., and Lewontin, R. C. 1979. The spandrals of San Marco and the Panglossian program: A critique of the adaptationist programme. *Proceedings of the Royal Society of London* 205: 281–88.

Grasshoff, G., and May, M. 1995. From historical case studies to systematic methods of discovery. *Working notes: AAAI spring symposium on systematic methods of scientific discovery.* Stanford, Calif.: AAAI.

Gray, J. R., and Thompson, P. M. 2004. Neurobiology of intelligence: Science and ethics. *Nature Reviews: Neuroscience* 5: 471–82.

Green, A. J. K., and Gilhooly, K. J. 1992. Empirical advances in expertise research. In *Ad-*

*vances in the psychology of thinking,* ed. M. T. Keane and K. J. Gilhooly. London: Harvester and Wheatchief.

Greene, B. 1999. *Elegant universe: Superstrings, hidden dimensions, and the quest for the ultimate theory.* New York: W. W. Norton.

Greenough, W. T., and Chang, F. F. 1989. Plasticity of synapse structure and pattern in the cerebral cortex. *Cerebral Cortex* 7: 391–440.

Greenwald, A. G., Banaji, M. R., Rudman, L. A., Farnham, S. D., and Nosek, B. A. 2002. A unified theory of implicit attitudes, stereotypes, self-esteem, and self-concept. *Psychological Review* 109: 3–25.

Gregory, R. L. 1981. *Mind in science: A history of explanations in psychology and physics.* Cambridge: Cambridge University Press.

Gross, P. R., and Levitt, N. 1998. *Higher superstition: The academic left and its quarrels with science.* Baltimore: Johns Hopkins University Press.

Grover, S. C. 1981. *Toward a psychology of the scientist: Implications of psychological research for contemporary philosophy of science.* Washington, D.C.: University of Press America.

Gruber, H. E. 1981. *Darwin on man: A psychological study of scientific creativity,* 2nd ed. Chicago: University of Chicago Press.

Guilford, J. P. 1959. Traits of creativity. In *Creativity and its cultivation,* ed. H. H. Anderson. New York: Harper.

———. 1987. A review of a quarter century progress. In *Frontiers of creativity research,* ed. S. G. Isaksen. Buffalo, N.Y.: Bearly.

Gustin, W. C., and Corazza, L. 1994. Mathematical and verbal reasoning as predictors of science achievement. *Roeper Review* 16: 160–62.

Guyer, L., and Fidell, L. 1973. Publications of men and women psychologists: Do women publish less? *American Psychologist* 28: 157–60.

Haier, R. J., and Benbow, C. P. 1995. Sex differences and lateralization in temporal lobe glucose metabolism during mathematical reasoning. *Developmental Neuropsychology* 11: 405–14.

Haile-Selassie, Y. 2001. Late Miocene hominids from the Middle Awash, Ethiopia. *Nature* 412: 178–81.

Halgren, E., Raij, T., Marinkovic, K., Jousmaki, V., and Hari, R. 2000. Cognitive response profile of the human fusiform face area as determined by MEG. *Cerebral Cortex* 10: 69–81.

Halpern, D. F. 2000. *Sex differences in cognitive abilities,* 3rd ed. Mahwah, N.J.: Erlbaum.

Ham, S., and Shaughnessy, M. F. 1992. Personality and scientific promise. *Psychological Reports* 70: 971–75.

Hamer, D., and Copeland, P. 1998. *Living with our genes.* New York: Doubleday.

Hanson, N. R. 1958. *Patterns of discovery.* Cambridge: Cambridge University Press.

———. 1962. Scientists and logicians: A confrontation. *Science* 138: 1311–13.

Hargens, L. L., McCann, J. C., and Reskin, B. F. 1978. Productivity and reproductivity: Fertility and professional achievement among research scientists. *Social Forces* 57: 154–63.

Harris, P. 1990. The child's theory of mind and its cultural context. In *Causes of development,* ed. G. Butterworth and P. Bryant. New York: Harvester Wheatsheaf.

Harris, R. 1986. *The origin of writing.* London: Duckworth.

Hart, J. J. 1982. Psychology of the scientists: XLVI: Correlation between theoretical orientation in psychology and personality type. *Psychological Reports* 50: 795–801.

Hatch, T. 1997. Friends, diplomats, and leaders in kindergarten: Interpersonal intelligence in play. In *Emotional development and emotional intelligence: Educational implications,* ed. P. Salovey and D. J. Sluyter. New York: Basic Books.

Head, D., Raz, R., Gunning-Dixon, F., Williamson, A., and Acker, J. D. 2002. Age-related differences in the course of cognitive skill acquisition: The role of regional cortical shrinkage and cognitive resources. *Psychology and Aging* 17: 72–84.

Hebb, D. O. 1949. *Organization of behavior: A neuropsychological theory.* New York: John Wiley.

Hellmans, A., and Bunch, B. 1991. *The timetables of science,* 2nd ed. New York: Simon and Schuster.

Helmreich, R. L., Spence, J. T., Beane, W. E., Lucker, G. W., and Matthews, K. A. 1980. Making it in academic psychology: Demographic and personality correlates of attainment. *Journal of Personality and Social Psychology* 39: 896–908.

Helson, R. 1971. Women mathematicians and the creative personality. *Journal of Consulting and Clinical Psychology* 36: 210–20.

Helson, R., and Crutchfield, R. S. 1970. Mathematicians: The creative researcher and the average PhD. *Journal of Consulting and Clinical Psychology* 34: 250–57.

Henshilwood, C., d'Errico, F., Vanhaeren, M., van Niekerk, K., and Jacobs, Z. 2004. Middle stone age shell beads from South Africa. *Science* 304: 404.

Herrnstein, R. J., and Murray, C. 1994. *The bell-curve: Intelligence and class structure in American life.* New York: Free Press.

Hespos, S. J., and Baillargéon, R. 2001. Infants' knowledge about occlusion and containment events: A surprising discrepancy. *Psychological Science* 121: 141–47.

Heyes, C. M. 1989. Uneasy chapters in the relationship between psychology and epistemology. In *Psychology of science: Contributions to metascience,* ed. B. Gholson, W. R. Shadish, R. A. Neimeyer, and A. C. Houts. Cambridge: Cambridge University Press.

———. 1998. Theory of mind in nonhuman primates. *Behavioral and Brain Sciences* 21: 101–34.

Hill, L., Chorney, M. J., Lubinski, D., Thompson, L. A., and Plomin, R. 2002. A quantitative trait locus not associated with cognitive ability in children: A failure to replicate. *Psychological Science* 13: 561.

Hinrichs, J. R. 1972. Value adaptation of new PhD's to academic and industrial environments: A comparative longitudinal study. *Personnel Psychology* 25: 545–65.

Hogan, K., and Maglienti, M. 2001. Comparing the epistemological underpinnings of students' and scientists' reasoning about conclusions. *Journal of Research in Science Teaching* 38: 663–87.

Holden, C. 1987. Female math anxiety on the wane. *Science* 236: 600–601.

Holland, J. L. 1992. *Making vocational choices,* 2nd ed. Odessa, Fl.: Psychological Assessment Resources.

Holton, G. 1973. *Thematic origins of scientific thought: Kepler to Einstein.* Cambridge: Harvard University Press.

Holyoak, K. J., and Thagard, P. 1995. *Mental leaps.* Cambridge: MIT Press.

Horner, K. L., Rushton, J. P., and Vernon, P. A. 1986. Relation between aging and research productivity of academic psychologists. *Psychology and Aging* 4: 319–24.

Houts, A. 1989. Contributions of the psychology of science to metascience: A call for explorers. In *Psychology of science: Contributions to metascience,* ed. B. Gholson, W. R. Shadish, R. A. Neimeyer, and A. C. Houts. Cambridge: University of Cambridge Press.

Howe, M. J. A. 1999. Prodigies and creativity. In *Handbook of creativity,* ed. R. J. Sternberg. New York: Cambridge University Press.

Howe, C., Tolmie, A., and Sofroniou, N. 1999. Experimental appraisal of personal beliefs in science: Constraints on performance in the 9 to 14 age group. *British Journal of Educational Psychology* 69: 243–74.

Hudson, L. 1958. Undergraduate academic record of Fellows of the Royal Society. *Nature* 182: 1326.

Hull, D. L. 1988. *Science as a process: An evolutionary account of the social and conceptual development of science.* Chicago: University of Chicago Press.

———. 2001. *Science and selection: Essays on biological evolution and the philosophy of science.* Cambridge: Cambridge University Press.

Hull, D. L., Tessner, P. D., and Diamond, A. M. 1978. Planck's principle: Do younger scientists accept new scientific ideas with greater alacrity than older scientists? *Social Studies of Science* 202: 717–23.

Husén, T. 1960. Abilities of twins. *Scandinavian Journal of Psychology* 1: 25–35.

Hyde, J. S., Fennema, E., Ryan, M., Frost, L. A., and Hopp, C. 1990. Gender comparisons of mathematics attitudes and affect: A meta-analysis. *Psychology of Women Quarterly* 14: 299–324.

Inhelder, B., and Piaget, J. 1958. *The growth of logical thinking from childhood to adolescence,* trans. A. Parsons and S. Milgram. New York: Basic Books.

Jacobs, R. C., and Campbell, D. T. 1961. The perpetuation of an arbitrary tradition through several generations of a laboratory microculture. *Journal of Abnormal and Social Psychology* 83: 649–58.

Jacobwitz, T. 1983. Relationship of sex, achievement, and science self-concept to the science career preferences of Black students. *Journal of Research in Science Teaching* 20: 621–28.

Jensen, A. R. 1996. Giftedness and genius: Crucial differences. In *Intellectual talent,* ed. C. P. Benbow and D. Lubinski. Baltimore: Johns Hopkins University Press.

Jerison, H. J. 1973. *The evolution of the brain and intelligence.* New York: Academic Press.

———. 1997. Evolution of prefrontal cortex. In *Development of the prefrontal cortex,* ed. N. A. Krasnegor, G. R. Lyon, and P. S. Goldman-Rakic. Baltimore, Md.: P. H. Brookes.

Jevons, W. 1874. *The principles of science: A treatise on logic and scientific method.* London: MacMillan.

John, O. P., and Srivastava, S. 1999. The Big Five trait taxonomy: History, measurement, and theoretical perspectives. In *Handbook of personality theory and research,* 2nd ed., ed. L. A. Pervin and O. P. John. New York: Guilford.

Johnson, C. N., and Wellman, H. M. 1982. Children's developing conceptions of the mind and the brain. *Child Development* 53: 222–34.

Johnson, J. A., Germer, C. K., Efran, J. S., and Overton, W. F. 1988. Personality as the basis for theoretical predilections. *Journal of Personality and Social Psychology* 55: 824–35.

Johnson, M. H., and Morton, J. 1991. *Biology and cognitive development: The case of face recognition.* New York: Blackwell.

John-Steiner, V. 1985. *Notebooks of the mind.* Albuquerque: University of New Mexico Press.

Kagan, J., Hanker, B. A., Hen-Tov, A., and Lewis, M. 1966. Infants' differential reactions to familiar and distorted faces. *Child Development* 37: 519–32.

Kahn, J. H., and Scott, N. A. 1997. Predictors of research productivity and science-related career goals among counseling psychology doctoral students. *Counseling Psychologist* 25: 38–67.

Kandel, E. R., and Hawkins, R. D. 1995. Neuronal plasticity and learning. In *Neuroscience, memory, and language.* vol. 1: *Decade of the brain,* ed. R. D. Broadwell. Washington, D.C.: Library of Congress.

Kanigel, R. 1991. *The man who knew infinity: A life of the genius Ramanujan.* New York: Scribner's and Sons.

Kantorovich, A. 1993. *Scientific discovery: Logic and tinkering.* Albany, N.Y.: SUNY Press.

Kanwisher, N. 2000. Domain specificity in face perception. *Nature Neuroscience* 3: 759.

Kanwisher, N., McDermott, J., and Chun, M. M. 1997. The fusiform face area: A module in human extrastriate cortex specialized for face perception. *Journal of Neuroscience* 17: 4302–11.

Karlson, J. L. 1991. *Genetics of human mentality.* New York: Praeger.

Karmiloff-Smith, A. 1992. *Beyond modularity: A developmental perspective on cognitive science.* Cambridge: MIT Press.

Katz, A. N. 1986. The relationship between creativity and cerebral hemisphericity for creative architects, scientists, and mathematicians. *Empirical Studies of the Arts* 4: 97–108.

Katzko, M. W., and Monks, F. J., eds. 1995. *Nurturing talent: Individual needs and social ability.* Assen, Neth.: Van Gorcum.

Kaufer, D. I., and Lewis, D. A. 1999. Frontal lobe anatomy and cortical connectivity. In *The human frontal lobes: Functions and disorders,* ed. B L. Miller and J. L. Cummings. New York: Guilford Press.

Keating, D. P. 1974. The study of mathematically precocious youth. In *Mathematical talent: Discovery, description, and development,* ed. J. C. Stanley, D. P. Keating, and L. H. Fox. Baltimore: Johns Hopkins University Press.

Keil, F. C. 1994. The birth and nurturance of concepts by domains: The origins of concepts of living things. In *Mapping the mind: Domain specificity in cognition and culture,* ed. L. A. Hirschfeld and S. A. Gelman. Cambridge: Cambridge University Press.

Kelley, H. H. 1967. Attribution theory in social psychology. In *Nebraska symposium on motivation,* ed. D. L. Vine. Lincoln: University of Nebraska Press.

Kelly, K. R., and Nelson, R. C. 1999. Task-Specific Occupational Self-Efficacy Scale: A predictive validity study. *Journal of Career Assessment* 7: 381–92.

Kempermann, G., and Gage, F. H. 1999. New nerve cells for the adult brain. *Scientific American* 280: 48–53.

Kern, L. J., Mirels, H. L., and Hinshaw, V. G. 1983. Scientists' understanding of propositional logic: An experimental investigation. *Social Studies of Science* 13: 131–46.

Kihlstrom, J. F. 1987. The cognitive unconscious. *Science* 237: 1445–52.

Kihlstrom, J. F., and Cantor, N. 2000. Social intelligence. In *Handbook of intelligence,* ed. R. J. Sternberg. Cambridge: Cambridge University Press.

Kimura, D. 1999. *Sex and cognition.* Cambridge: MIT Press.

Klaczynski, P., and Gordon, D. H. 1996. Everyday statistical reasoning during adolescence and young adulthood: Motivational, general ability, and developmental influences. *Child Development* 67: 2873–92.

Klaczynski, P., and Narasimham, G. 1998. Development of scientific reasoning biases: Cognitive versus ego-protective explanations. *Developmental Psychology* 34: 175–87.

Klaczynski, P., and Robinson, B. 2000. Personal theories, intellectual ability, and epistemological beliefs: Adult age differences in everyday reasoning biases. *Psychology and Aging* 15: 400–16.

Klahr, D. 1973. Quantification processes. In *Visual information processing,* ed. W. G. Chase. New York: Academic Press.

———. 2000. *Exploring science: The cognition and development of discovery processes.* Cambridge: MIT Press.

Klahr, D., and Dunbar, K. 1988. Dual space search during scientific reasoning. *Cognitive Science* 12: 1–48.

Klahr, D., Dunbar, K., and Fay, A. L. 1990. Designing good experiments to test bad hypotheses. In *Computational models of discovery and theory formation,* ed. J. Shrager and P. Langley. San Mateo, Calif.: Morgan Kaufmann Publishers.

Klahr, D., Fay, A. L., and Dunbar, K. 1993. Heuristics for scientific experimentation: A developmental study. *Cognitive Psychology* 25: 111–46.

Klahr, D., and Robinson, M. 1981. Formal assessment of problem-solving and planning processes in preschool children. *Cognitive Psychology* 13: 113–48.

Klahr, D., and Simon, H. 1999. Studies of scientific discovery: Complementary approaches and convergent findings. *Psychological Bulletin* 125: 524–43.

Klayman, J., and Ha, Y. W. 1987. Confirmation, disconfirmation and information in hypothesis testing. *Psychological Review* 94: 211–28.

Klein, R. G. 1999. *The human career: Human biological and cultural origins,* 2nd ed. Chicago: University of Chicago Press.

Kline, P., and Lapham, S. L. 1992. Personality and faculty in British universities. *Personality and Individual Differences* 13: 855–57.

Klopp, J., Marinkovic, K., Chauvel, P., Nenov, V., and Halgren, E. 2000. Early widespread cortical distribution of coherent fusiform face selective activity. *Human Brain Mapping* 11: 286–93.

Kluckholm, C., and Murray, H. 1953. *Personality in nature, culture, and society.* New York: Knopf.

Knorr-Cetina, K. D. 1981. *The manufacture of knowledge: An essay on the constructivist and contextual nature of science.* Oxford: Pergammon.

Koch, A. 2001. Training in metacognition and comprehension of physics texts. *Science Education* 85: 758–68.

Kolb, B., and Gibb, R. 1999. Neuroplasticity and recovery of function after brain injury. In *Cognitive neurorehabilitation,* ed. D. T. Stuss, G. Winocur, and I. H. Robertson. New York: Cambridge University Press.

Kolb, B., and Whishaw, I. Q. 1998. Brain plasticity and behavior. *Annual Review of Psychology* 29: 43–64.

Koslowski, B. 1996. *Theory and evidence: The development of scientific reasoning.* Cambridge: MIT Press.

Koutsoulis, M. K., and Campbell, J. R. 2001. Family processes affect students' motivation, and science and math achievement in Cypriot high schools. *Structural Equation Modeling* 8: 108–27.

Kramer, S. N. 1998. *History begins at Sumer: Thirty-nine firsts in recorded history,* 3rd ed. Philadelphia: University of Pennsylvania Press.

Krasnegor, N. A., Lyon, G. R., and Goldman-Rakic, S., eds. 1997. *Development of the prefrontal cortex: Evolution, neurobiology, and behavior.* Baltimore, Md.: P. H. Brookes.

Kris, E. 1952. *Psychoanalytic explorations in art.* New York: Schoken Books.

Kruglanski, A. W. 1994. The social-cognitive bases of scientific knowledge. In *The social psychology of science,* ed. W. R. Shadish and S. Fuller. New York: Guilford Press.

Kuhn, D. 1989. Children and adults as intuitive scientists. *Psychological Review* 964: 674–89.

———. 1993. Connecting scientific and informal reasoning. *Merrill-Palmer Quarterly* 39: 74–103.

Kuhn, D., Amsel, E., and O'Loughlin, M. 1988. *The development of scientific thinking skills.* Orlando Fl.: Academic.

Kuhn, D., and Pearsall, S. 2000. Developmental origins of scientific thinking. *Journal of Cognition and Development* 1: 113–29.

Kuhn, D., Schauble, L., and Garcia-Mila, M. 1992. Cross-domain development of scientific reasoning. *Cognition and Instruction* 9: 285–327.

Kuhn, T. S. 1962. *The structure of scientific revolutions,* 2nd ed. Chicago: University of Chicago Press.

———. 1970. Logic of discovery or psychology of research? In *Criticism and the growth of knowledge,* ed. I. Lakatos and A. Musgrave. Cambridge: Cambridge University Press.

———. 1977. *The essential tension.* Chicago: University of Chicago Press.

———. 1979. Metaphor in science. In *Metaphor and thought,* ed. A. Ortony. Cambridge: Cambridge University Press.

Kulkarni, D., and Simon, H. A. 1988. The processes of scientific discovery: The strategies of experimentation. *Cognitive Science* 12: 139–75.

Kuncel, N. R., Hezlett, S. A., and Ones, D. S. 2001. A comprehensive meta-analysis of the predictive validity of the Graduate Record Examinations: Implications for graduate student selection and performance. *Psychological Bulletin* 127: 162–81.

Kwon, Y. J., and Lawson, A. E. 2000. Linking brain growth with the development of scientific reasoning ability and conceptual change during adolescence. *Journal of Research in Science Teaching* 37: 44–62.

Lacey, L. A., and Erickson, C. E. 1974. Psychology of the scientist: XXXI: Discriminability of a creativity scale for the Adjective Check List among scientists and engineers. *Psychological Reports* 34: 755–58.

Lake, M. 1998. "Homo": The creative genus? In *Creativity in human evolution and prehistory,* ed. S. Mithen. London: Routledge.

Lakoff, G., and Johnson, M. 1980. *Metaphors we live by.* Chicago: University of Chicago Press.

Langdon, D. W., and Warrington, E. K. 1997. The abstraction of numerical relations: A role for the right hemisphere in arithmetic? *Journal of the International Neuropsychological Society* 3: 260–68.

Langley, P., Simon, H. A., Bradshaw, G. L., and Zykow, J. M. 1987. *Scientific discovery: Computational explorations of the creative processes.* Cambridge: MIT Press.

Langlois, J. H., Kalakanis, L., Rubenstein, A. J., Larson, A., Hallam, M., and Smoot, M. 2000. Maxims or myths of beauty? A meta-analytic and theoretical review. *Psychological Bulletin* 126: 390–423.

Langlois, J. H., Ritter, J. M., Roggman, L. A., and Vaughn, L. S. 1991. Facial diversity and infant preferences for attractive faces. *Developmental Psychology* 27: 79–84.

Langlois, J. H., and Roggman, L. A. 1990. Attractive faces are only average. *Psychological Science* 1: 115–21.

Langlois, J. H., Roggman, L. A., and Musselman, L. 1994. What is average and what is not average about attractive faces? *Psychological Science* 5: 214–20.

Larkin, J. H. 1983. The role of problem representation in physics. In *Mental models,* ed. D. Gentner and A. L. Stevens. Hillsdale, N.J.: Erlbaum.

Larkin, J. H., McDermitt, J., Simon, D. P., and Simon, H. A. 1980. Expert and novice performance in solving physics problems. *Science* 208: 1335–42.

Larkin, J. H., and Simon, H. A. 1987. Why a diagram is (sometimes) worth ten thousand words. *Cognitive Science* 11: 65–99.

Latour, B. 1985. *Science in action: How to follow scientists and engineers through society.* Cambridge: Harvard University Press.

———. 2004. Why has critique run out of steam? From matters of fact to matters of concern. *Critical Inquiry* 30: 225–48.

Latour, B., and Woolgar, S. 1986. *Laboratory life: The construction of scientific facts.* Princeton: Princeton University Press.

Lawler, S., Richman, S., and Richman, C. L. 1997. The validity of using the SAT as a criterion for Black and White students' admission to college. *College Student Journal,* 507–13.

Leahey, E., and Guo, G. 2002. Gender differences in mathematical trajectories. *Social Forces* 80: 713–32.

Leaky, R. E., and Lewin, R. 1977. *Origins.* New York: E. P. Dutton.

Leary, D. E., ed. 1990. *Metaphors in the history of psychology.* New York: Cambridge University Press.

Lederberg, J. 2001. The meaning of epigenetics. *Scientist* 15: 6.

Lehman, H. C. 1953. *Age and achievement.* Princeton: Princeton University Press.

———. 1960. The age decrement in outstanding scientific creativity. *American Psychologist* 15: 128–34.

———. 1962. The creative production rates of present versus past generations of scientists. *Journal of Gerontology* 17: 409–17.

———. 1966. The psychologist's most creative years. *American Psychologist* 21: 363–69.

Lesch, K. P., Bengel, D., Heils, A., Sabol, S. Z., Greenburg, B. D., Petri, S., Benjamin, J., Müller, C. R., Hamer, D. H., and Murphy, D. L. 1996. Association of anxiety-related traits with a polymorphism in the serotonin transporter gene regulatory region. *Science* 274: 1527–31.

Leslie, A. M. 1987. Pretense and representation: The origins of "theory of mind." *Psychological Review* 94: 412–26.

———. 1994. ToMM, ToBy, and agency: Core architecture and domain specificity. In *Mapping the mind: Domain specificity in cognition and culture,* ed. L. A. Hirschfeld and S. A. Gelman. Cambridge: Cambridge University Press.

Levin, S. G., Stephen, P. E., and Walker, M. B. 1995. Planck's principle revisited: A special note. *Social Studies of Science* 25: 275–83.

Lévi-Strauss, C. 1966. *The savage mind.* Chicago: Chicago University Press.

Lieberman, P. 1993. On the Kebara KMH 2 hyoid and Neanderthal speech. *Current Anthropology* 34: 172–75.

Lieberman, P., and Crelin, E. S. 1971. On the speech of Neanderthal man. *Linguistics Enquiry* 2: 203–22.

Lillard, A. 1998. Ethnopsychologies: Cultural variations in theories of mind. *Psychological Bulletin* 123: 3–32.

Lippa, R. 1998. Gender related individual differences and structure of vocational interests: The importance of the people-things dimension. *Journal of Personality and Social Psychology* 74: 996–1009.

Liu, T., Pan, Y., Kao, S-Y., Li, C., Kohane, I., Chan, J., and Yankner, B. A. 2004. Gene regulation and DNA damage in the aging human brain. *Nature* (June 9, 2004, online issue; http://dx.doi.org/10.1038/nature02662; accessed June 16, 2004).

Loehlin, J. C. 1992. *Genes and environment in personality development.* Newbury Park, Calif.: Sage.

Loehlin, J. C., McCrae, R. R., Costa, P. T., and John, O. P. 1998. Heritabilities of common and measure specific components of the Big Five personality factors. *Journal of Research in Personality* 32: 431–53.

Loehlin, J. C., and Nichols, R. C. 1976. *Heredity, environment, and personality: A study of 850 sets of twins.* Austin: University of Texas Press.

Loehlin, J. C., Willerman, L., and Horn, J. M. 1987. Personality resemblance in adoptive families: A 10-year follow-up. *Journal of Personality and Social Psychology* 53: 961–69.

Long, J. S. 1992. Measures of sex differences in scientific productivity. *Social Forces* 71: 159–78.

———. 2004. "Demographic inertia and the glass ceiling in science" talk given at the NISTADS International Conference on "Women in Science: Is the Glass Ceiling Disappearing?" New Delhi, India, March 9, 2004.

Long, J. S., ed. 2001. *From scarcity to visibility: Gender differences in the careers of doctoral scientists and engineers.* Washington, D.C.: National Academy Press.

Lotka, A. J. 1926. The frequency distribution of scientific productivity. *Journal of the Washington Academy of Sciences* 16: 317–23.

Lu, T., Pan, Y., Kao, S. Y., Li., C., Kohane, I., Chan, J., and Yankner, B. A. 2004. Gene regulation and DNA damage in the ageing human brain. *Nature* 429: 883–91.

Lubinski, D. 2000. Scientific and social significance of assessing individual differences: "Sinking shafts at a few critical points." *Annual Review of Psychology* 51: 405–44.

Lubinski, D., and Benbow, C. P. 1994. The study of mathematically precocious youth: The first three decades of a planned 50-year study of intellectual talent. In *Beyond Terman:*

*Longitudinal studies of giftedness and talent,* ed. R. F. Subotnik and K. D. Arnold. Norwood, N.J.: Ablex.

Lubinski, D., and Humphreys, L. G. 1990. A broadly based analysis of mathematical giftedness. *Intelligence* 14: 327–55.

Lubow, R. E., and Gewirtz, J. C. 1995. Latent inhibition in humans: Data, theory, and implications. *Psychological Bulletin* 117: 87–103.

Luzzo, D. A., Hasper, P., Albert, K. A., Bibby, M. A., and Martinelli, E. A. 1999. Effects of self-efficacy-enhancing interventions on the math/science self-efficacy and career interests, goals, and actions of career undecided college students. *Journal of Counseling Psychology* 46: 233–43.

Lykken, D. T., McGue, M., Tellegen, A., and Bouchard, T. J., Jr. 1992. Emergenesis: Genetic traits that may not run in families. *American Psychologist* 47: 1565–77.

Maccoby, E. E., and Jacklin, C. N. 1974. *The psychology of sex differences.* Stanford: Stanford University Press.

Mach, E. 1883. *Die Mechanik in iherer Entwicklung: Historisch-kritisch dargestellt.* Leipzig, Ger.: F. A. Brockhaus.

MacKinnon, D. W. 1960. The highly effective individual. *Teachers College Record* 61: 367–78.

———. 1978. *In search of human effectiveness.* Buffalo, N.Y.: Bearly.

Macrone, M. 1992. *By Jove!* New York: Harper-Collins.

MacWhinney, B. 2002. The gradual emergence of language. In *The evolution of language out of pre-language,* ed. T. Givon and B. F. Maller. Amsterdam: John Benjamins Publishing.

Maguire, E. A., Burgess, N., and O'Keefe, J. 1999. Human spatial navigation: Cognitive maps, sexual dimorphism, and neural substrates. *Current Opinion in Neurobiology* 9: 171–77.

Mahoney, M. J. 1977. Publication prejudices: An experimental study of confirmatory bias in the peer review system. *Cognitive Therapy and Research* 1: 161–75.

———. 1979. Psychology of the scientist: An evaluative review. *Social Studies of Science* 9: 349–75.

Mahoney, M. J., and DeMonbreun, B. J. 1977. Psychology of the scientist: An analysis of problem solving bias. *Cognitive Therapy and Research* 1: 229–38.

Mahoney, M. J., and Kimper, T. P. 1976. From ethics to logic: A survey of scientists. In *Scientist as subject: The psychological imperative,* ed. M. J. Mahoney. Cambridge: Ballinger.

Mallinckrodt, B., Gelso, C. J., and Royalty, G. M. 1990. Impact of the research training environment and counseling psychology students' Holland personality type on interest in research. *Professional Psychology: Research and Practice* 21: 26–32.

Mandler, G., and Shebo, B. J. 1982. Subitizing: An analysis of its component processes. *Journal of Experimental Psychology, General* 111: 1–22.

Mansfield, R. S., and Busse, T. V. 1981. *The psychology of creativity and discovery: Scientists and their work.* Chicago: Nelson-Hall.

Marshack, A. 1972. Cognitive aspects of Upper Paleolithic engraving. *Current Anthropology* 13: 445–77.

———. 1985. *Hierarchical evolution of the human capacity,* New York: American Museum of Natural History.

Martindale, C. 1999. Biological bases of creativity. In *Handbook of creativity,* ed. R. J. Sternberg. Cambridge: Cambridge University Press.

Martindale, C., Hines, D., Mitchell, L., and Covello, E. 1984. EEG alpha asymmetry and creativity. *Personality and Individual Differences* 5: 77–86.

Maslow, A. 1966. *The psychology of science.* New York: Harper and Row.

Matarazzo, J. 1987. Relationships of health psychology to other segments of psychology. In *Health psychology: A discipline and a profession,* ed. G. S. Stone et al. Chicago: University of Chicago Press.

Matsuzawa, T. 1996. Chimpanzee intelligence in nature and in captivity: Isomorphism of symbol use and tool use. In *Great ape societies,* ed. W. McGrew, L. Marchant, and T. Nishida. Cambridge: Cambridge University Press.

Matyas, M. L., and Dix, L. S., eds. 1992. *Science and engineering programs: On target for women?* Washington, D.C.: National Academy Press.

Mauger, P. A., and Kolmodin, C. A. 1975. Long-term predictive validity of the Scholastic Aptitude Test. *Journal of Educational Psychology* 67: 847–51.

Mayer, J. D., Salovey, P., and Caruso, D. 2000. Models of emotional intelligence. In *Handbook of intelligence,* ed. R. J. Sternberg. Cambridge: Cambridge University Press.

Maynard, L., and Edwards, R. 1971. Wall markings. In *Archaeology of the Gallus Site, Koonalda Cave,* ed. R. V. S. Wright. Canberra: Australian Institute of Aboriginal Studies.

McCarthy, R. A., and Warrington, E. K. 1988. Evidence for modality-specific meaning systems in the brain. *Nature* 334: 428–30.

McCloskey, M. 1983. Naive theories of motion. In *Mental models,* ed. D. Gentner and A. L. Stevens. Hillsdale, N.J.: Erlbaum.

McCrae, R. R., and Costa, P. T. 1999. A Five-Factor theory of personality. In *Handbook of personality theory and research,* ed. L. A. Pervin and O. P. John. New York: Guilford Press.

McCrae, R. R., and John, O. P. 1992. An introduction to the Five-Factor Model and its applications. *Journal of Personality* 60: 175–215.

McDermid, C. D. 1965. Some correlates of creativity in engineering personnel. *Journal of Applied Psychology* 49: 14–19.

McGraw, M. 1943. *The neuromuscular maturation of the human infant.* New York: Columbia University Press.

McGrew, W. 1992. *Chimpanzee material culture.* New York: Cambridge University Press.

———. 2001. The nature of culture: Prospects and pitfalls of cultural primatology. In *Tree of origin: What primate behavior can tell us about human social evolution,* ed. F. B. M. de-Waal. Cambridge: Harvard University Press.

McMullin, E. 1970. The history and philosophy of science: A taxonomy. In *Minnesota studies in the philosophy of science,* vol. 5, ed. R. Stuewer. Minneapolis: University of Minnesota Press.

McNeil, J. E., and Warrington, E. K. 1993. Prosopagnosia: A face-specific disorder. *Quarterly Journal of Experimental Psychology* 46: 1–10.

Mednick, S. A. 1962. The associative basis of the creative process. *Psychological Review* 69: 220–32.

Mell, J. C., Howard, S. M., and Miller, B. L. 2003. Art and the brain: The influence of frontotemporal dementia on an accomplished artist. *Neurology* 60: 1707–10.

Mendelsohn, G. A. 1976. Associative and attentional processes in creative performance. *Journal of Personality* 44: 341–69.

Menon, V., Rivera, S. M., White, C. D., Eliez, S., Glover, G. H., and Reiss, A. L. 2000. Functional optimization of arithmetic processing in perfect performers. *Cognitive Brain Research* 9: 343–45.

Merton, R. K. 1945. Sociology of knowledge. In *Twentieth century sociology,* ed. G. Gurvitch and W. E. Moore. New York: Philosophical Library.

———. 1973. *The sociology of science: Theoretical and empirical investigations.* Chicago: University of Chicago Press.

Messeri, P. A. 1988. Age differences in the reception of new scientific theory: The case of plate tectonics theory. *Social Studies of Science* 18: 91–112.

Miller, A. I. 1989. Imagery, metaphor, and physical reality. In *Psychology of science: Contributions to metascience,* ed. B. Gholson, W. R. Shadish, R. A. Neimeyer, and A. C. Houts. Cambridge: Cambridge University Press.

———. 1996. *Insights of genius: Imagery and creativity in science and art.* New York: Springer Verlag.

Miller, B. L., and Cummings, J. L., eds. 1999. *The human frontal lobes: Functions and disorders.* New York: Guilford Press.

Miller, B. L., Hou, C., Goldberg, M., and Mena, I. 1999. Anterior temporal lobes. In *The human frontal lobes: Functions and disorders,* ed. B. L. Miller and J. L. Cummings. New York: Guilford Press.

Miller, G. F. 2000. *The mating mind: How sexual choice shaped the evolution of human nature.* New York: Doubleday.

Mithen, S. 1996. *The prehistory of the mind: The cognitive origins of art and science.* London: Thames and Hudson.

———. 1998. Introduction to Part II. In *Creativity in human evolution and prehistory,* ed. S. Mithen. London: Routledge.

———. 2002. Human evolution and the cognitive basis of science. In *The cognitive basis of science,* ed. P. Carruthers, S. Stich, and M. Siegal. Cambridge: Cambridge University Press.

Miyake, A., Friedman, N. P., Emerson, M. J., Witzki, A. H., and Howerter, A. 2000. The unity and diversity of executive functions and their contributions to complex "frontal lobe" tasks: A latent variable approach. *Cognitive Psychology* 41: 49–100.

Montague, D. P. F., and Walker-Andrews, A. S. 2001. Peekaboo: A new look at infants' perception of emotion expressions. *Developmental Psychology* 37: 826–38.

Moore, A. M. T. 1978. *The neolithic of the Levant.* Oxford: Oxford University Press.

Moore, E. G. J., and Smith, A. W. 1987. Sex and ethnic group differences in mathematics achievement: Results from the National Longitudinal Study. *Journal for Research in Mathematics Education* 18: 25–36.

Morrison, T., and Morrison, M. 1995. A meta-analytic assessment of the predictive validity of the Quantitative and Verbal components of the Graduate Record Examination with graduate grade point average representing the criterion of graduate success. *Educational and Psychological Measurement* 55: 309–16.

Moscovici, S., and Nemeth, C. 1974. Social influence II: Minority influence. In *Social psychology: Classic and contemporary integrations,* ed. C. Nemeth. Chicago: Rand McNally.

Mossholder, K. W., Dewhirst, H. D., and Arvey, R. D. 1981. Vocational interest and personality differences between development and research personnel: A field study. *Journal of Vocational Behavior* 19: 233–43.

Moulin, L. 1955. The Nobel Prizes for the sciences from 1901 to 1950: An essay in sociological analysis. *British Journal of Sociology* 6: 246–63.

Mullins, N. 1973. *Theories and theory groups in contemporary American sociology.* New York: Harper and Row.

Mynatt, C. R., Doherty, M. E., and Tweney, R. D. 1977. Confirmation bias in simulated research environment: An experimental study of scientific inference. *Quarterly Journal of Experimental Psychology* 29: 85–95.

———. 1978. Consequences of confirmation and disconfirmation in a simulated research environment. *Quarterly Journal of Experimental Psychology* 30: 395–406.

Nakamura, K., Kawashima, R., Sato, N., Nakamura, A., Sugiura, M., Kato, T. Hatano, K., Ito, K., Fukuda, H., Schormann, T., and Zilles, K. 2000. Functional delineation of the human occipito-temporal areas related to face and scene processing: A PET study. *Brain* 123: 1903–12.

National Science Foundation. 1999. *Women, minorities, and persons with disabilities in science and engineering: 1998* (NSF 99–87). Arlington, Va.: National Science Foundation.

Neisser, U., Boodoo, G., Bouchard, T. J., Boykin, A. W., Brody, N., Ceci, S. J., Halpern, D., Loehlin, J. C., Perloff, R., Sternberg, R. J., and Urbina, S. 1996. Intelligence: Knowns and unknowns. *American Psychologist* 51: 77–101.

Nelson, C. A. 2001. The development and neural bases of face recognition. *Infant and Child Development* 10: 3–18.

Nemeroff, C., and Rozin, P. 2000. The makings of the magical mind. In *Imagining the impossible: Magical, scientific, and religious thinking in children,* ed. K. S. Rosegren, C. N. Johnson, and P. L. Harris. Cambridge: Cambridge University Press.

Nersessian, N. J. 1992. How do scientists think? Capturing the dynamics of conceptual change in science. In *Cognitive models of science,* ed. R. N. Giere. Minneapolis: University of Minnesota Press.

———. 2002. The cognitive basis of model-based reasoning in science. In *The cognitive basis of science,* ed. P. Carruthers, S. Stich, and M. Siegal. Cambridge: Cambridge University Press.

Nobel, W., and Davidson, I. 1996. *Human evolution, language, and mind: A psychological and archeological inquiry.* Cambridge: Cambridge University Press.

Nosek, B. A., Banaji, M. R., and Greenwald, A. G. 2002. Math = male, me = female, therefore math ≠ me. *Journal of Personality and Social Psychology* 83: 44–59.

O'Brien, V., Martinez-Pons, M., and Kopala, M. 1999. Mathematics self-efficacy, ethnic identity, gender, and career interests related to mathematics and science. *Journal of Educational Research* 92: 231–35.

O'Craven, K. M., and Kanwisher, N. 2000. Mental imagery of faces and places activates corresponding stimulus specific brain-regions. *Journal of Cognitive Neuroscience* 12: 1013–23.

O'Keefe, J., and Nadel, L. 1978. *The hippocampus as a cognitive map.* New York: Oxford University Press.

O'Leary, D. D. M., and Stanfield, B. 1989. Selective elimination of axons extended by devel-

oping cortical neurons is dependent on regional locale experiments utilizing fetal cortical transplants. *Journal of Neuroscience* 9: 2230–46.
Orians, G. H. 2001. An evolutionary perspective on aesthetics. *Bulletin of Psychology and the Arts* 2: 25–29.
Orians, G. H., and Heerwagen, J. H. 1992. Evolved responses to landscapes. In *The adapted mind,* ed. J. H. Barkow, L. Cosmides, and J. Tooby. New York: Oxford University Press.
Over, R. 1982. Is age a good predictor of research productivity? *Australian Psychologist* 17: 129–39.
———. 1989. Age and scholarly impact. *Psychology and Aging* 4: 222–25.
Papineau, D. 2000. The evolution of knowledge. In *Evolution and the human mind: Modularity, language and meta-cognition,* ed. P. Carruthers and A. Chamberlain. Cambridge: Cambridge University Press.
Pare, C. F. E., ed. 2000. *Metals make the world go round: The supply and circulation of metals in Bronze Age Europe.* Oxford, Eng.: Oxbow Publishing.
Parker, S. T., and McKinney, M. L. 1999. *Origins of intelligence.* Baltimore: Johns Hopkins University Press.
Pasewark, R. A., Fitzgerald, B. J., and Sawyer, R. N. 1975. Psychology of the scientist: XXXII. God at the synapse: Research activities of clinical, experimental, and physiological psychologists. *Psychological Reports* 36: 671–74.
Pasini, M., and Tessari, A. 2001. Hemispheric specialization in quantification processes. *Psychological Research* 65: 57–63.
Pearce, C. 1968. Creativity in young science students. *Exceptional Children* 35: 121–26.
Pearson, K. 1892. *Grammar of science.* New York: Schribner's Sons.
Perner, J. 1991. *Understanding the representational mind.* Cambridge: MIT Press.
Perner, J., Leekam, S., and Wimmer, H. 1987. Three year olds' difficulty with false belief: The case for a conceptual deficit. *British Journal of Developmental Psychology* 5: 125–37.
Perry, W., Swerdlow, N. R., McDowell, J. E., and Braff, D. L. 1999. Schizophrenia and frontal lobe functioning. In *The human frontal lobes: Functions and disorders,* ed. B. L. Miller and J. L. Cummings. New York: Guilford Press.
Pfeiffer, J. 1982. *The creative explosion.* New York: Harper and Row.
Pheasant, J. H. 1961. The influence of the school on the choice of science careers. *British Journal of Educational Psychology* 31: 38–42.
Piaget, J. 1952. *The child's concept of number.* New York: W. W. Norton.
———. 1972. Intellectual evolution from adolescence to adulthood. *Human Development* 15: 1–12.
Piaget, J., and Garcia, R. 1989. *Psychogenesis and the history of science.* New York: Columbia University Press.
Piaget, J., and Inhelder, B. 1967. *The child's conception of space.* New York: Norton.
Piburn, M. 1980. Spatial reasoning as a correlate of formal thought and science achievement for New Zealand students. *Journal for Research in Science Teaching* 17: 577–82.
Pincus, J. H. 1999. Aggression, criminality, and the frontal lobes. In *The human frontal lobes: Functions and disorders,* ed. B. L. Miller and J. L. Cummings. New York: Guilford Press.
———. 2001. *Base instincts: What makes killers kill?* New York: W. W. Norton.

Pinker, S. 1996. *Language learnability and language development,* rev. ed. Cambridge: Harvard University Press.

———. 1997. *How the mind works.* New York: W. W. Norton.

———. 2002. *The blank slate: The modern denial of human nature.* New York: Viking.

Pizzani, M. J. 1990. *Creating a memory of causal relationships.* Hillsdale, N.J.: Erlbaum.

Platt, J. R. 1964. Strong inference. *Science* 146: 347–53.

Plomin, R., and Caspi, A. 1999. Behavioral genetics and personality. In *Handbook of personality theory and research,* ed. L. A. Pervin and O. P. John. New York: Guilford Press.

Plomin, R., McClearn, G. E., Smith, D. L., Vignetti, S., Chorney, M. J., Chorney, K., Venditti, C. P., Kasarda, L., Thompson, L. A., Detterman, D. K., Daniels, J., Owen, M., and McGuffin, P. 1994. DNA markers associated with high versus low IQ: The Quantitiative Trait Loci QTL project. *Behavior genetics* 24: 107–18.

Plotkin, H. 1997. *Evolution in mind: An introduction to evolutionary psychology.* Cambridge: Harvard University Press.

Poincare, H. 1952. Mathematical creation. In *The creative process,* ed. B. Ghiselin. New York: Plume and Meridan Books.

Polanyi, M. 1964. *Personal knowledge.* New York: Harper and Row.

Popper, K. 1959. *The logic of scientific discovery.* New York: Science Editions.

———. 1965. *Conjectures and refutations: The growth of scientific knowledge.* New York: Harper Torchbooks.

———. 1970. Normal science and its dangers. In *Criticism and the growth of knowledge,* ed. I. Lakatos and A. Musgrave. Cambridge: Cambridge University Press.

———. 1974. *Unended quest: An intellectual autobiography.* LaSalle, Ill.: Open Court.

Portmann, A. 1990. *A zoologist looks at humankind,* trans. J. Schaefer. New York: Columbia University Press.

Povinelli, D. J. 1993. Reconstructing the evolution of mind. *American Psychologist* 48: 493–509.

Prediger, D. J. 1982. Dimensions underlying Holland's hexagon: Missing link between interests and occupations? *Journal of Vocational Behavior* 21: 259–87.

Premack, D. 1988. "Does the chimpanzee have a theory of mind?" revisited. In *Machivellian intelligence,* ed. R. Byrne and A. Whiten. New York: Oxford University Press.

Premack, D., and Woodruff, G. 1978. Does the chimpanzee have a theory of mind? *Behavioral and Brain Sciences* 1: 515–26.

Price, D. 1963. *Little science, big science.* New York: Columbia University Press.

Prum, R. O., and Bush, A. H. 2003. Which came first, the feather or the bird? *Scientific American* 288: 84–93.

Purves, D. 1988. *Body and brain: A trophic theory of neural connections.* Cambridge: Harvard University Press.

Ragland, J. D., Turetsky, B. I., Gur, R. C., Gunning-Dixon, G., Turner, T., Schroeder, L., Chan, R., and Gur, R. E. 2002. Working memory for complex figures: An fMRI comparison of letter and fractal n-back tasks. *Neuropsychology* 16: 370–79.

Raine, A., Buchsbaum, M., and LaCasse, L. 1997. Brain abnormalities in murderers indicated by positron emission tomography. *Biological Psychiatry* 42: 495–508.

Raine, A., Lencz, T., Bihrle, S., LaCasse, L., and Colletti, P. 2000. Reduced prefrontal gray matter volume and reduced autonomic activity in antisocial personality disorder. *Archives of General Psychiatry* 57: 119–27.

Raine, A., Meloy, J. R., Bihrle, S., Stoddard, J., LaCasse, L., and Buchsbaum, M. S. 1998. Reduced prefrontal and increased subcortical brain functioning assessed using positron emission tomography in predatory and affective murderers. *Behavioral Sciences and the Law* 16: 319–32.

Ramachandran, V. S., and Hubbard, E. M. 2003. Hearing colors, tasting shapes. *Scientific American* 288: 53–59.

Reber, A. S. 1967. Implicit learning of artificial grammars. *Journal of Verbal Learning and Verbal Behavior* 6: 855–63.

Reis, S. M., and Park, S. 2001. Gender differences in high-achieving students in math and science. *Journal for the Education of the Gifted* 25: 52–73.

Reskin, B. F. 1977. Scientific productivity and the reward structure of science. *American Sociological Review* 42: 491–504.

Reuhkala, M. 2001. Mathematical skills in ninth-graders: Relationship with visuo-spatial abilities and working memory. *Educational Psychology* 21: 387–99.

Reznikoff, M., Domino, G., Bridges, C., and Honeyman, M. 1973. Creative abilities in identical and fraternal twins. *Behavior Genetics* 3: 365–77.

Richardson, W. 2000. Criminal behavior fueled by attention deficit hyperactivity disorder and addiction. In *The science, treatment, and prevention of antisocial behaviors: Application to the criminal justice system,* ed. D. H. Fishbein. Kingston, N.J.: Civic Research Institute.

Rieger, M., Gauggel, S., and Burmeister, K. 2003. Inhibition of ongoing responses following frontal, nonfrontal, and basal ganglia lesions. *Neuropsychology* 17: 272–82.

Riemann, R., Angleitner, A., and Strelau, J. 1997. Genetic and environmental influences on personality: A study of twins reared together using self- and peer-report NEO-FFI scales. *Journal of Personality* 65: 449–76.

Rimland, B., and Fein, D. 1988. Special talents of autistic savants. In *The exceptional brain: Neuropsychology of talent and special abilities,* ed. L. K. Obler and D. Fein. New York: Guilford Press.

Robinson, A. 1995. *The story of writing: Alphabets, hieroglyphs, and pictograms.* London: Thames and Hudson.

Roe, A. 1952a. *The making of a scientist.* New York: Dodd, Mead.

———. 1952b. A psychologist examines 64 eminent scientists. *Scientific American* 187: 21–25.

———. 1953. A psychological study of eminent psychologists and anthropologists, and a comparison with biological and physical scientists. *Psychological Monographs: General and Applied* 67: 1–55.

———. 1965. Changes in scientific activities with age. *Science* 150: 313–18.

———. 1972. Maintenance of creative output through the years. In *Climate for creativity,* ed. C. W. Taylor. New York: Pergamon.

Root-Bernstein, R. S., Bernstein, M., and Garnier, H. 1995. Correlations between avocations, scientific style, work habits, and professional impact of scientists. *Creativity Research Journal* 8: 115–37.

Rosenberg, E. L. 1998. Levels of analysis and the organization of affect. *Review of General Psychology* 2: 247–70.

Rosengren, K. S., Johnson, C. N., and Harris, P. L., eds. 2000. *Imagining the impossible: Magical, scientific, and religious thinking in children.* Cambridge: Cambridge University Press.

Rosenthal, R. 1976. *Experimenter effects in behavioral research enlarged edition.* New York: Irvington Publishers.

———. 1994. On being one's own case study: Experimenter effects in behavioral research—30 years later. In *The social psychology of science,* ed. W. Shadish and S. Fuller. New York: Guilford Press.

Rosenthal, R., and Fode, K. L. 1963a. Psychology of the scientist: V: Three experiments in experimenter bias. *Psychological Reports* 12: 491–511.

———. 1963b. The effect of experimenter bias on the performance of the albino rat. *Behavioral Science* 8: 183–89.

Rosenthal, R., and Jacobson, L. 1992. *Pygmalion in the classroom,* 2nd ed. New York: Irvington.

Rosenthal, R., and Rosnow, R. L. 1991. *Essentials of behavioral research: Methods and data analysis,* 2nd ed. New York: McGraw-Hill.

Rosenwein, R. 1994. Social influence in science: Agreement and dissent in achieving scientific consensus. In *The social psychology of science,* ed. W. R. Shadish and S. Fuller. New York: Guilford Press.

Rosenzweig, M. R., Bennett, E. L., and Diamond, M. C. 1972. Cerebral effects of differential experience in hypophysectomized rats. *Journal of Comparative and Physiological Psychology* 79: 56–66.

Rosenzweig, M. R., Krech, D., Bennett, E. L., and Diamond, M. C. 1962. Effects of environmental complexity and training on brain chemistry and anatomy: A replication and extension. *Journal of Comparative and Physiological Psychology* 55: 429–37.

Rosser, S., ed. 1988. *Feminism within the science and healthcare professions: Overcoming resistance.* Exeter, Eng.: A. Wheaton.

Rothbart, M. K., Ahadi, S. A., and Evans, D. E. 2000. Temperament and personality: Origins and outcomes. *Journal of Personality and Social Psychology* 78: 122–35.

Rourke, B. P., and Conway, J. A. 1997. Disabilities of arithmetic and mathematical reasoning: Perspectives from neurology and neuropsychology. *Journal of Learning Disabilities* 30: 34–46.

Royalty, G. M., and Magoon, T. M. 1985. Correlates of scholarly productivity among counseling psychologists. *Journal of Counseling Psychology* 32: 458–61.

Rubenstein, A. J., Kalakanis, L., and Langlois, J. H. 1999. Infant preference for attractive faces: A cognitive explanation. *Developmental Psychology* 35: 848–55.

Rudwick, M. J. S. 1985. *The Devonian controversy: The shaping of scientific knowledge among gentlemanly specialists.* Chicago: University of Chicago Press.

Rumbaugh, D. M. 1997. Competence, cortex, and primate models: A comparative primate perspective. In *Development of the prefrontal cortex,* ed. N. A. Krasnegor, G. R. Lyon, and P. S. Goldman-Rakic. Baltimore, Md.: P. H. Brookes.

Runco, M. A. 1999. Divergent thinking. In *Encyclopedia of creativity,* vol. 1, ed. M. A. Runco and S. R. Pritzker. San Diego, Calif.: Academic Press.

Runyan, W. M. 1982. *Life histories and psychobiography.* New York: Oxford University Press.

Rushton, J. P., Murray, H. G., and Paunonen, S. V. 1987. Personality characteristics associated with high research productivity. In *Scientific excellence,* ed. D. Jackson and J. P. Rushton. Beverly Hills, Calif.: Sage.

Russell, B. 1929. *Mysticism and logic.* New York: W. W. Norton.

Rutter, M. 1978. Diagnosis and definition. In *Autism: A reappraisal of concepts and treatment,* ed. M. Rutter and E. Schopler. New York: Plenum.

Sagan, C. 1980. *Cosmos.* New York: Random House.

———. 1987. The burden of skepticism. *The Skeptical Inquirer* 12: 38–46.

———. 1996. *A demon-haunted world: Science as a candle in the dark.* New York: Random House.

Saklofske, D. H., and Zeidner, M., eds. 1995. *International handbook of personality and intelligence.* New York: Plenum.

Salovey, P., and Mayer, J. D. 1990. Emotional intelligence. *Imagination, Cognition and Personality* 9: 185–211.

Sarton, G. 1952a. *Ancient science through the golden age of Greece.* New York: Dover Publications.

———. 1952b. *A guide to the history of science.* Waltham, Mass.: Chronica Botanica.

Scarr, S. 1985. An author's frame of mind [Review of *Frames of mind* by H. Gardner]. *New Ideas in Psychology* 3: 95–100.

———. 1989. Protecting general intelligence: Constructs and consequences for interventions. In *Intelligence: Measurement, theory, and public policy,* ed. R. L. Linn. Urbana: University of Illinois Press.

Scarr, S., and Saltzman, L. C. 1982. Genetics and intelligence. In *Handbook of human intelligence,* ed. R. J. Sternberg. Cambridge: Cambridge University Press.

Schacter, D. L. 1987. Implicit memory: History and current status. *Journal of Experimental Psychology: Learning, Memory, and Cognition* 13: 501–18.

Schaie, K. W. 1984. Midlife influences upon intellectual functioning in old age. *International Journal of Behavior Development* 7: 463–78.

———. 1994. The course of adult intellectual development. *American Psychologist* 49: 304–13.

Schauble, L. 1990. Belief revision in children: The role of prior knowledge and strategies for generating evidence. *Journal of Experimental Child Psychology* 49: 31–57.

Schlick, M. 1932–33/1991. Positivism and realism. In *The philosophy of science,* ed. R. Boyd, P. Gasper, and J. D. Trout. Cambridge: MIT Press.

Schmajuk, N. A. 2002. *Latent inhibition and its neural substrates.* Dordrecht, Neth.: Kluwer Academic Publishers.

Schmidt, D. B., Lubinski, D., and Benbow, C. P. 1998. Validity of assessing educational-vocational preference dimensions among intellectually talented 13-year-olds. *Journal of Counseling Psychology* 45: 436–53.

Schmitt, A., and Grammar, K. 1997. Social intelligence and success: Don't be too clever in order to be smart. In *Machiavellian intelligence II: Extensions and evaluations,* ed. A. Whiten and R. W. Byrne. Cambridge: Cambridge University Press.

Schneider, L. M., and Briel, J. B. 1990. *Validity of the GRE: 1988–89 summary report.* Princeton, N.J.: Educational Testing Service.

Scholl, B. J., and Leslie, A. M. 1999. Modularity, development and "theory of mind." *Mind and Language* 14: 131–53.

Schoon, I. 2001. Teenage job aspirations and career attainment in adulthood: A 17-year follow-up study of teenagers who aspired to become scientists, health professionals, or engineers. *International Journal of Behavioral Development* 25: 124–32.

Schroder, H. M., Driver, M. J., and Streufert, S. 1967. *Human information processing.* New York: Holt, Rinehart, and Winston.

Schwanenflugel, P. J., Stevens, T. P. M., and Carr, M. 1997. Metacognitive knowledge of gifted children and nonidentified children in early elementary school. *Gifted Child Quarterly* 41: 25–35.

Scott, N. A., and Sedlacek, W. E. 1975. Personality differentiation and prediction of persistence in physical science and engineering. *Journal of Vocational Behavior* 6: 205–16.

Segal, G. 1996. The modularity of theory of mind. In *Theories of theory of mind,* ed. P. Carruthers and P. Smith. Cambridge: Cambridge University Press.

Semendeferi, K. 1999. The frontal lobes of the great apes with a focus on the gorilla and orangutan. In *The mentality of gorillas and orangutans,* ed. S. T. Parker, R. W. Mitchell, and H. L. Miles. New York: Cambridge University Press.

Sergent, J., Ohta, S., and MacDonald, B. 1992. Functional neuroanatomy of face and object processing. *Brain* 115: 15–36.

Shadish, W. R. 1989. The perception and evaluation of quality in science. In *Psychology of science: Contributions to metascience,* ed. B. Gholson, W. R. Shadish, R. A. Neimeyer, and A. C. Houts. Cambridge: Cambridge University Press.

Shadish, W. R., and Fuller, S., eds. 1994. *Social psychology of science.* New York: Guilford Press.

Shadish, W. R., Fuller, S., and Gorman, M. E. (with Amabile, T., Kruglanski, A., Rosenthal, R., and Rosenwein, R. E.). 1994. Social psychology of science: A conceptual and empirical research program. In *Social psychology of science,* ed. W. R. Shadish and S. Fuller. New York: Guilford Press.

Shadish, W. R., Houts, A. C., Gholson, B., and Neimeyer, R. A. 1989. The psychology of science: An introduction. In *Psychology of science: Contributions to metascience,* ed. B. Gholson, W. R. Shadish, R. A. Neimeyer, and A. C. Houts. Cambridge: Cambridge University Press.

Shadish, W. R., Tolliver, D., Gray, M., and Gupta, S. K. 1995. Author judgments about works they cite: Three studies from psychology journals. *Social Studies of Science* 253: 477–98.

Shaughnessy, M. F., Stockard, J., and Moore, J. 1994. Scores on the 16 Personality Factor Questionnaire and success in college calculus. *Psychological Reports* 75: 348–50.

Shea, D. L., Lubinski, D., and Benbow, C. P. 2001. Importance of assessing spatial ability in intellectually talented young adolescents: A 20-year longitudinal study. *Journal of Educational Psychology* 93: 604–14.

Shen, W. M. 1993. Discovery as autonomous learning from the environment. *Machine Learning* 12: 143–65.

Shepard, R. N. 1978. The mental image. *American Psychologist* 33: 125–37.

———. 1997. The genetic basis of human scientific knowledge. In *Characterizing human psychological adaptations,* ed. G. R. Bock and G. Cardew. New York: Wiley.

Shermer, M. 1997. *Why people believe weird things: Pseudoscience, superstition, and other confusions of our time.* New York: W. H. Freeman.

Shih, M., Pittinsky, T. L., and Ambady, N. 1999. Stereotype susceptibility: Identity salience and shifts in quantitative performance. *Psychological Science* 10: 80–83.

Shore, B. M., and Dover, A. C. 1987. Metacognition, intelligence, and giftedness. *Gifted Child Quarterly* 31: 37–39.

Shrager, J., and Langley, P. 1990. *Computational models of scientific discovery and theory formation.* San Mateo, Calif.: Morgan Kaufmann Publishers.

Simmons, A. H., Kohler-Rollefson, I., Rollefson, G. O., Mandel, R., and Kafafi, Z. 1988. 'Ain Ghazal: A major neolithic settlement in central Jordon. *Science* 240: 35–39.

Simon, H. A. 1966. Scientific discovery and the psychology of problem solving. In *Mind and cosmos,* ed. R. Colodny. Pittsburgh: University of Pittsburgh Press.

Simon, H. A., Langley, P. W., and Bradshaw, G. 1981. Scientific discovery as problem solving. *Syntheses* 47: 1–27.

Simon, T. J., Hespos, S. J., and Rochat, P. 1995. Do infants understand simple arithmetic? A replication of Wynn (1992). *Cognitive Development* 10: 253–69.

Simonton, D. K. 1975. Invention and discovery among the sciences: A p-technique factor analysis. *Journal of Vocational Behavior* 7: 275–81.

———. 1976a. The causal relation between war and scientific discovery. *Journal of Cross-Cultural Psychology* 7: 133–44.

———. 1976b. Interdisciplinary and military determinants of scientific productivity: A cross-lagged correlation analysis. *Journal of Vocational Behavior* 9: 53–62.

———. 1977. Creative productivity, age, and stress: A biographical time-series analysis of 10 classical composes. *Journal of Personality and Social Psychology* 35: 791–804.

———. 1980. Techno-scientific activity and war: A yearly time-series analysis, 1500–1903 A.D. *Scientometrics* 2: 251–55.

———. 1984. Creative productivity and age: A mathematical model based on a two-step cognitive process. *Developmental Review* 4: 77–111.

———. 1988a. *Scientific genius: A psychology of science.* Cambridge: Cambridge University Press.

———. 1988b. Age and outstanding achievement: What do we know after a century of research? *Psychological Bulletin* 104: 251–67.

———. 1989. Chance-configuration theory of scientific creativity. In *Psychology of science: Contributions to metascience,* ed. B. Gholson, W. R. Shadish, R. A. Neimeyer, and A. C. Houts. Cambridge: University of Cambridge Press.

———. 1991. Career landmarks in science: Individual differences and interdisciplinary contrasts. *Developmental Psychology* 27: 119–30.

———. 1992a. The social context of career success and course for 2,026 scientists and inventors. *Personality and Social Psychology Bulletin* 18: 452–63.

———. 1992b. Leaders in American psychology, 1879–1967: Career development, creative output and professional achievement. *Journal of Personality and Social Psychology* 62: 5–17.

———. 1995. Behavioral laws in histories of psychology: Psychological science, metascience, and the psychology of science. *Psychological Inquiry* 6: 89–114.

———. 1999. *Origins of genius.* New York: Oxford University Press.

———. 2000. Methodological and theoretical orientation and the long-term disciplinary impact of 54 eminent psychologists. *Review of General Psychology* 4: 13–21.

———. 2002. *Great psychologists and their times: Scientific insights into psychology's history.* Washington, D.C.: American Psychological Association.

Singer, B. F. 1971. Toward a psychology of science. *American Psychologist* 26: 1010–15.

Singer, T., Verhaeghen, P., Ghisletta, P., Lindenberger, U., and Baltes, P. B. 2003. The fate of cognition in very old age: Six-year longitudinal findings in the Berlin Aging Study (BASE). *Psychology and Aging* 18: 318–31.

Skinner, B. F. 1953. *Science and human behavior.* New York: Free Press.

Skoyles, J. R. 1999. Neural plasticity and exaptation. *American Psychologist* 54: 438–39.

Snow, C. P. 1959. *The two cultures and the scientific revolution.* Cambridge: Cambridge University Press.

Snyderman, M., and Rothman, S. 1987. Survey of expert opinion on intelligence and aptitude testing. *American Psychologist* 42: 137–44.

Soffer, O. 1985. Patterns of intensification as seen in the Upper Paleolithic of the central Russian Plain. In *Prehistoric hunter-gatherers: The emergence of cultural complexity,* ed. D. T. Price and J. A. Brown. Orlando: Academic Press.

Sokal, A. 1996a. Transgressing the boundaries: Towards a transformative hermeneutics of quantum gravity. *Social Text* 46–47: 217–42.

———. 1996b. A physicist experiments with cultural studies. *Lingua Franca* (May/June): 62–64.

Solms, M., and Turnbull, O. 2002. *The brain and the inner world: An introduction to the neuroscience of subjective experience.* New York: Other Press.

Sorokin, P. 1937. *Social and cultural dynamics.* New York: American Book.

Sparkman, R. 1994, January 17. "Nice scientists finish last." *Newsweek Focus* 4.

Spelke, E. 1990. Principles of object perception. *Cognitive Science* 14: 29–56.

Sperber, D. 1994. The modularity of thought and epidemiology of representations. In *Mapping the mind: Domain specificity in cognition and culture,* ed. L. A. Hirschfeld and S. A. Gelman. Cambridge: Cambridge University Press.

Staats, A. 1991. Unified positivism and unification psychology. *American Psychologist* 46: 899–912.

Stanford, C. B. 2001. The ape's gift: Meat-eating, meat-sharing, and human evolution. In *Tree of origin: What primate behavior can tell us about human social evolution,* ed. F. B. M. deWaal. Cambridge: Harvard University Press.

Stanley, J. C. 1988. Some characteristics of SMPY's "700–800 on SAT-M before age 13 group": Youths who reason *extremely* well mathematically. *Gifted Child Quarterly* 32: 205–9.

Stanley, J. C., Keating, D. P., and Fox, L. H., eds. 1974. *Mathematical talent: Discovery, description, and development.* Baltimore, Md.: Johns Hopkins University Press.

Stanovich, K. E. 1999. *Who is rational?* Mahwah, N.J.: Erlbaum.

Stanton, F., and Baker, K. H. 1942. Interviewer bias and the recall of incompletely learned materials. *Sociometry* 5: 123–34.

Starkey, P., and Cooper, R. G. 1995. The development of subitizing in young children. *British Journal of Developmental Psychology* 13: 399–420.

Steele, C. M. 1997. A threat in the air: How stereotypes shape intellectual identity and performance. *American Psychologist* 52: 613–29.

Steele, C. M., and Aronson, J. 1995. Stereotype threat and the intellectual test performance of African Americans. *Journal of Personality and Social Psychology* 69: 797–811.

Stenhouse, D. 1974. *The evolution of intelligence.* New York: Barnes and Noble.

Sternberg, R. J. 1985. *Beyond IQ: A triarchic theory of human intelligence.* New York: Cambridge University Press.

———. 1988a. A three-facet model of creativity. In *The nature of creativity,* ed. R. J. Sternberg. Cambridge: Cambridge University Press.

———. 1988b. The male/female difference is there: Should we care? *Behavioral and Brain Sciences* 11: 210–11.

———. 1988c. *The triarchic mind: A new theory of human intelligence.* New York: Viking Penguin.

Sternberg, R. J., and Gordeeva, T. 1996. The anatomy of impact: What makes an article influential. *Psychological Science* 7: 69–75.

Sternberg, R. J., Grigorenko, E. L., and Singer, J. L., eds. 2004. *Creativity: From potential to realization.* Washington, D.C.: American Psychological Association.

Sternberg, R. J., and O'Hara, L. A. 1999. Creativity and intelligence. In *The handbook of creativity,* ed. R. J. Sternberg. Cambridge: Cambridge University Press.

Sternberg, R. J., and Williams, W. M. 1997. Does the Graduate Record Examination predict meaningful success in the graduate training of psychologists? *American Psychologist* 52: 630–41.

Stevens, S. S. 1936. Psychology: The propaedeutic science. *Philosophy of Science* 3: 90–104.

———. 1939. Psychology and the science of science. *Psychological Bulletin* 36: 221–63.

Stiles, J. 2000. Neural plasticity and cognitive development. *Developmental Neuropsychology* 18: 237–72.

Stone, V. E., Baron-Cohen, S., and Knight, R. T. 1998. Frontal lobe contributions to theory of mind. *Journal of Cognitive Neuroscience* 10: 640–56.

Storr, A. 1988. *Solitude: A return to the self.* New York: Free Press.

Stuessy, C. L. 1988. Path analysis: A model for the development of scientific reasoning abilities in adolescents. *Journal of Research in Science Teaching* 26: 41–53.

Stumpf, H., and Stanley, J. C. 2002. Group data on high school grade point averages and scores on academic aptitude tests as predictors of institutional graduate rates. *Educational and Psychological Measurement* 62: 1042–52.

Stumpf, S. E. 1975. *Socrates to Sartre: A history of philosophy,* 2nd ed. New York: McGraw-Hill.

Stuss, D. T., Gallup, G. G., and Alexander, M. P. 2001. The frontal lobes are necessary for "theory of mind." *Brain* 124: 279–86.

Stuss, D. T., Picton, T. W., and Alexander, M. P. 2001. Consciousness and self-awareness, and the frontal lobes. In *The frontal lobes and neuropsychiatric illness,* ed. S. P. Salloway, P. F. Malloy, and J. D. Duffy. Washington, D.C.: American Psychiatric Publishing.

Subotnik, R. F., Duschl, R. A., and Selmon, E. H. 1993. Retention and attrition of science talent: A longitudinal study of Westinghouse science talent search winners. *International Journal of Science Education* 15: 61–72.

Subotnik, R. F., and Steiner, C. L. 1992. Adult manifestations of adolescent talent in science. *Roeper Review* 15: 164–69.

———. 1994. Adult manifestations of adolescent talent in science: A longitudinal study of 1983 Westinghouse Science Talent Search winners. In *Beyond Terman: Contemporary longitudinal studies of giftedness and talent,* ed. R. F. Subotnik and K. D. Arnold. Norwood, N.J.: Ablex Publishing.

Suedfeld, P. 1985. APA presidential addresses: The relation of integrative complexity to historical, professional and personal factors. *Journal of Personality and Social Psychology* 49: 1634–51.

Sulloway, F. J. 1996. *Born to rebel: Birth order, family dynamics, and creative lives.* New York: Pantheon.

———. In preparation. Introduction: The science of science. In *Testing theories of scientific change,* ed. F. J. Sulloway. Cambridge: Harvard University Press.

Tarr, M. J., and Cheng, Y. D. 2003. Learning to see faces and objects. *Trends in Cognitive Sciences* 7: 23–30.

Tarr, M. J., and Gauthier, I. 2000. FFA: A flexible fusiform area for subordinate-level visual processing automatized by expertise. *Nature Neuroscience* 3: 764–69.

Tattersall, I. 1997. *Becoming human: Evolution and human uniqueness.* San Diego, Calif.: Harcourt Brace.

Taubes, G. 1993. *Bad science: The short life and hard times of cold fusion.* New York: Random House.

Taylor, C. W., and Barron, F. 1963. *Scientific creativity: Its recognition and development.* New York: John Wiley and Sons.

Taylor, D. W. 1963. Variables related to creativity and productivity among men in two research laboratories. In *Scientific creativity: Its recognition and development,* ed. C. W. Taylor and F. Barron. New York: John Wiley and Sons.

Tellegen, A., Lykken, D. T., Bouchard, T. J., Wilcox, K. J., Segal, N. L., and Rich, S. 1988. Personality similarity in twins reared apart and together. *Journal of Personality and Social Psychology* 54: 1031–39.

Terman, L. M. 1925. *Genetic studies of genius,* vol. 1: *Mental and physical traits of a thousand gifted children.* Stanford: Stanford University Press.

———. 1954. Scientists and nonscientists in a group of 800 men. *Psychological Monographs* 68: Whole No. 378.

Tetlock, P. E., and Suedfeld, P. 1988. Integrative complexity coding of verbal behavior. In *Lay explanation,* ed. C. Antaki. Beverly Hills, Calif.: Sage.

Thagard, P. 1988. *Computational philosophy of science.* Cambridge: MIT Press.

Thagard, P., and Nowak, G. 1990. The conceptual structure of the geological revolution. In *Computational models of scientific discovery and theory formation,* ed. J. Shrager and P. Langley. San Mateo, Calif.: Morgan Kaufmann Publishers.

Thomas, A., Benne, M. R., Marr, M. J., Thomas, E. W., and Hume, R. M. 2000. The MBTI predicts attraction and attrition in an engineering program. *Journal of Psychological Type* 55: 35–42.

Thomas, A., and Chess, S. 1977. *Temperament and development.* New York: Brunner/Mazel.

Thompson, P. M., Cannon, T. D., Narr, K. L., van Erp, T., Poutanen, V., Huttunen, M., Lonnqvist, J., Standertskjold-Nordenstam, C., Kaprio, J., Khaledy, M., Dail, R., Zoumalan, C. I., and Toga, A. W. 2001. Genetic influences on brain structure. *Nature Neuroscience* 4: 1253–58.

Thorndike, E. L. 1920. Intelligence and its uses. *Harper's Magazine* 140: 227–35.

Thornhill, R. 1997. The concept of evolved adaptation. In *Characterizing human psychological adaptations,* ed. G. R. Bock and G. Cardew. New York: Wiley.

Tobias, P. V. 1987. The brain of *Homo habilis:* A new level of organization in cerebral evolution. *Journal of Human Behavior* 16: 741–61.

Tobin, K., Tippins, D. J., and Hook, K. S. 1995. Students' beliefs about epistemology, science, and classroom learning: A question of fit. In *Learning science in the schools: Research reforming practice,* ed. S. M. Glynn and D. Reinders. Hillsdale, N.J.: Erlbaum.

Tomasello, M. 2002. The emergence of grammar in early child language. In *The evolution of language out of pre-language,* ed. T. Givon and B. F. Maller. Amsterdam: John Benjamins Publishing.

Tooby, J., and Cosmides, L. 1992. The psychological foundations of culture. In *The adapted mind: Evolutionary psychology and the generation of culture,* ed. J. Barkow, L. Cosmides, and J. Tooby. Oxford: Oxford University Press.

Tramo, M. J. 2001. Enhanced: Music of the hemispheres. *Science* 291: 54–56.

Treffert, D. A., and Wallace, G. L. 2002. Islands of genius. *Scientific American* 286: 76–85.

Tremblay, J. M., Herron, W. G., and Schultz, C. L. 1986. Relation between therapeutic orientation and personality in psychotherapists. *Professional Psychology: Research and Practice* 17: 106–10.

Tweney, R. D. 1985. Faraday's discovery of induction: A cognitive approach. In *Faraday rediscovered: Essays on the life and work of Michael Faraday: 1791–1867,* ed. D. Gooding and F. James. New York: Stockton Press.

———. 1989. A framework for the cognitive psychology of science. In *Psychology of science: Contributions to metascience,* ed. B. Gholson, W. R. Shadish, Jr., R. A. Neimeyer, and A. C. Houts. Cambridge: Cambridge University Press.

———. 1991. Faraday's notebooks: The active organization of creative science. *Physics Education* 26: 301–6.

Tweney, R. D., Doherty, M. E., and Mynatt, C. R., eds. 1981. *On scientific thinking.* New York: Columbia University Press.

Tylecote, R. F. 1992. *A history of metallurgy,* 2nd ed. London: Institute of Materials.

Underwood, G. D. M., ed. 1996. *Implicit cognition.* Oxford: Oxford University Press.

Uylings, H. B. M., and Van Eden, C. G. 1990. Qualitative and quantitative comparison of the prefrontal cortex in rats and in primates, including humans. *Progress in Brain Research* 85: 31–62.

Valdes-Perez, R. E. 1994. Conjecturing hidden entities via simplicity and conservation laws: Machine discovery in chemistry. *Artificial Intelligence* 65: 247–80.

Valenza, E., Simion, F., Assia, V. M., and Umilta, C. 1996. Face preference at birth. *Journal of Experimental Psychology: Human Perception and Performance* 22: 892–903.

Vandenberg, S. G. 1988. Could these sex differences be due to genes? *Behavioral and Brain Sciences* 11: 212–14.

Van Zelst, R. H., and Kerr, W. A. 1954. Personality self-assessment of scientific and technical personnel. *Journal of Applied Psychology* 38: 145–47.

Velikovsky, I. 1950. *Worlds in collision.* New York: Macmillan.

Vernon, P. E. 1989. The nature-nurture problem in creativity. In *Handbook of creativity: Per-*

*spectives on individual differences,* ed. A. Glover, R. R. Ronning, and C. R. Reynolds. New York: Plenum.

Visalberghi, E., Fragaszy, D., and Savage-Rumbaugh, S. E. 1995. Performance in a tool-using task by common chimpanzees (*Pan troglodytes*), bonobos (*Pan paniscus*), and orangutan (*Pongo pygmeaus*), and capuchin monkeys (*Cebus apella*). *Journal of Comparative Psychology* 109: 52–60.

Vogeley, K., Bussfeld, P., Newen, A., Herrmann, S., Happe, F., Falkai, P., Maier, W., Shah, N. J., Fink, G. R., and Zilles, K. 2001. Mind reading: Neural mechanisms of theory of mind and self-perspective. *Neuroimage* 14: 170–81.

de Waal, F. B. M. 1998. *Chimpanzee politics: Power and sex among apes.* Baltimore, Md.: Johns Hopkins University Press.

Wahlsten, D. 1999. Single-gene influences on brain and behavior. *Annual Review of Psychology* 50: 599–624.

Walberg, H. J, Strykowski, B. F., Rovai, E., and Hung, S. S. 1984. Exceptional performance. *Review of Educational Research* 54: 87–112.

Wallach, M. A. 1970. Creativity. In *Manual of child psychology,* ed. P. H. Mussen. New York: Wiley and Sons.

Wallach, M. A., and Kogan, N. 1972. Creativity and intelligence in children. In *Human intelligence,* ed. J. McVicker Hunt. New Brunswick, N.J.: Transaction Books.

Waller, N. G., Bouchard, T. J., Lykken, D. T., Tellegen, A., and Blacker, D. M. 1993. Creativity, heritability, familiality: Which word does not belong? *Psychological Inquiry* 4: 235–37.

Waller, N. G., Lykken, D. T., and Tellegen, A. 1995. Occupational interests, leisure time interests, and personality: Three domains or one? Findings from the Minnesota Twin Registry. In *Assessing individual differences in human behavior: New concepts, methods, and findings,* ed. D. J. Lubinski and R. V. Dawis. Palo Alto, Calif.: Davies-Black Publishing.

Wallin, N. L., Merker, B., and Brown. S., eds. 2001. *The origins of music.* Cambridge: MIT Press.

Wason, P. C. 1960. On the failure to eliminate hypotheses in a conceptual task. *Quarterly Journal of Experimental Psychology* 12: 129–40.

———. 1966. Reasoning. In *New horizons in psychology,* ed. B. M. Foss. Harmondsworth, Eng.: Penguin Books.

Wason, P. C., and Green, D. W. 1984. Reasoning and mental representation. *Quarterly Journal of Experimental Psychology: Human Experimental Psychology* 36: 597–610.

Webb, R. M., Lubinski, D., and Benbow, C. 2002. Mathematically facile adolescents with math-science aspirations: New perspectives on their educational and vocational development. *Journal of Educational Psychology* 94: 785–94.

Weinsten, S., and Graves, R. E. 2001. Creativity, schizotypy, and laterality. *Cognitive Neuropsychiatry* 6: 131–46.

Weitzman, R. A. 1982. The prediction of college achievement by the Scholastic Aptitude Test and the high school record. *Journal of Educational Measurement* 19: 179–91.

Welsh, M. C., Pennington, B. F., and Groisser, D. B. 1991. A normative-developmental study of executive function: A window on prefrontal function in children. *Developmental Neuropsychology* 7: 131–49.

Wertheimer, M. 1959. *Productive thinking.* New York: Harper and Brothers.

Werts, C. E., and Watley, D. J. 1972. Paternal influence on talent development. *Journal of Counseling Psychology* 19: 367–73.

Wharton, C. M., Cheng, P. W., and Wickens, T. D. 1993. Hypothesis-testing strategies: Why two goals are better than one. *Quarterly Journal of Experimental Psychology* 464: 743–58.

Whewell, W. 1840. *The philosophy of the inductive sciences founded upon their history.* London: J. W. Parker.

———. 1856. *On the philosophy of discovery: Chapters historical and critical.* London: J. W. Parker.

White, B. Y., and Frederiksen, J. R. 1998. Inquiry, modeling, and metacognition: Making science accessible to all students. *Cognition and Instruction* 16: 3–118.

Wiener, N. 1953. *Ex-prodigy: My childhood and youth.* Cambridge: MIT Press.

Williams, G. C. 1966. *Adaptation and natural selection: A critique of some current evolutionary thought.* Princeton: Princeton University Press.

Willis, S. L., Jay, G. M., Diehl, M., and Marsiske, M. 1992. Longitudinal change and prediction of everyday task competence in the elderly. *Research on Aging* 14: 68–91.

Wilson, G. D., and Jackson, C. 1994. The personality of physicists. *Personality and Individual Differences* 16: 187–89.

Wise, L. L., Steel, L., and McDonald, C. 1979. *Origins and career consequences of sex differences in high school mathematics achievement.* Washington, D.C.: American Institute for Research.

Wiser, M., and Carey, S. 1983. When heat and temperature were one. In *Mental models,* ed. D. Gentner and A. L. Stevens. Hillsdale, N.J.: Erlbaum.

Wrangham, R. W. 1977. Feeding behavior of the chimpanzees in Gombe National Park, Tanzania. In *Primate ecology: Studies of feedings and ranging behaviour in lemurs, monkeys, and apes,* ed. T. H. Clutton-Brock. London: Academic Press.

———. 2001. Out of the *Pan,* into the fire: How our ancestors' evolution depended on what they ate. In *Tree of origin: What primate behavior can tell us about human social evolution,* ed. F. B. M. deWaal. Cambridge: Harvard University Press.

Wynn, K. 1992. Addition and subtraction by human infants. *Nature* 358: 749–50.

———. 1995. Origins of numerical knowledge. *Mathematical Cognition* 1: 35–60.

———. 1998. An evolved capacity for number. In *The evolution of mind,* ed. D. D. Cummins and C. Allen. New York: Oxford University Press.

Xie, Y., and Shauman, K. A. 1998. Sex differences in research productivity: New evidence about an old puzzle. *American Sociological Review* 63: 847–70.

Zachar, P., and Leong, F. T. L. 1992. A problem of personality: Scientist and practitioner differences in psychology. *Journal of Personality* 60: 667–77.

———. 1997. General versus specific predictors of specialty choice in psychology: Holland codes and theoretical orientations. *Journal of Career Assessment* 5: 333–41.

———. 2000. A 10-year longitudinal study of scientists and practitioner interests in psychology: Assessing the Boulder model. *Professional Psychology: Research and Practice* 31: 575–80.

Zeki, S. 1999. *Inner vision: An exploration of art and brain.* Oxford: Oxford University Press.

Zimmerman, C. 2000. The development of scientific reasoning skills. *Developmental Review* 20: 99–140.

Zuckerman, H. 1996. *Scientific elite.* 2nd ed. New York: Free Press.

Zuckerman, H., and Cole, J. R. 1975. Women in American science. *Minerva* 13: 82–102.

Zuckerman, H., and Merton, R. K. 1973. Age, aging, and age structure in science. In *The sociology of science,* ed. N. Storr. Chicago: University of Chicago Press.

# Index

Italicized letters *f* and *t* following page numbers indicate figures and tables, respectively